PNEUMATIC CONVEYING OF BULK MATERIALS

Milton N. Kraus was formerly with Colgate-Palmolive Co. where he was Bulk Material Handling Specialist—Worldwide in their Central Engineering Department and Supervisor of Project Engineering at their Jersey City Plant. He received a B.M.E. from Pratt Institute and is a licensed Professional Engineer in the states of New York and New Jersey.

PNEUMATIC CONVEYING OF BULK MATERIALS

Milton N. Kraus

Second Edition

CHEMICAL ENGINEERING

McGraw-Hill Publications Co., New York, N.Y.

Dedicated to the memory of Theodore R. Olive, Consulting Editor of Chemical Engineering, who conceived and encouraged the writing of this book.

Library of Congress Cataloging in Publication Data

Kraus, Milton N
Pneumatic conveying of bulk materials.

Includes index.
1. Pneumatic-tube transportation. 2. Bulk solids handling. I. Title.
[TJ1015.K7 1979] 621.5′4 79-18467

ISBN 07-606629-0 Chemical Engineering Magazine
ISBN 07-010724-6 Professional and Reference Books McGraw-Hill Book Co.

Preface to the Second Edition

The purpose of this volume is to present, in a single reference source, a fund of practical information that draws freely on the author's experience in the analysis, selection, installation, startup, and operation of almost every type of pneumatic conveying system.

This second edition reflects ten additional years of practical knowledge gained by the author, and thus embraces a total of twenty-seven years of experience in this field. The concise information herein has been extracted from case histories that could fill many volumes. Their lessons should be of value to all users of this book.

This book has been presented in three parts for ease of reference. The first part is essentially an introduction to the overall field of pneumatic conveying. Theoretical thermodynamics and fluid flow are discussed to the degree necessary for insight into the behavior of materials and into the limitations of theory in practical applications of pneumatic handling. The second part of the book is concerned chiefly with the hardware of conveying systems. Emphasis here is on the key phases of a system: transporting, conveying, storing, and controlling. The final part is devoted to the practical problems inherent in installing, inspecting, starting up, and operating a complete conveying system.

It is hoped that this book will fill a void in engineering literature and encourage prospective purchasers of this type of equipment to prepare proper specifications, without complete dependence on vendors. It is also hoped that vendors will find in this book new challenges to improve equipment for meeting the demands of more knowledgeable purchasers.

There is no single installation that can attain all of the desirable characteristics listed in this book for the various types of equipment. Compromises must be made to suit the properties of the materials handled and the limitations imposed by patent restrictions.

The omission of data on specific materials is intentional, so that there will be no temptation to apply such data to the handling of materials considered "equivalent." The author strongly recommends direct experimentation to ascertain the properties and characteristics of the materials to be handled, pilot plant tests to determine the flow behavior of the materials in a pipeline, and cooperation with a reputable vendor of pneumatic conveying equipment to obtain a successful installation.

Thanks are due to the various manufacturers who graciously furnished photographs and charts for the illustrations. Specific credits are given throughout the text. The author is also grateful to his former colleagues at Colgate-Palmolive Company's many plants for the information gleaned from their reports. Finally, he extends his most heartfelt thanks to his wife, Ingeborg, who typed the manuscript and persevered despite the obstacles presented by the mathematical symbolism.

MILTON N. KRAUS

Hillsdale, New Jersey
September, 1979

Contents

Notation

a area (in square inches)
acceleration (in feet per second per second)

A area (in square feet)

c_p specific heat at constant pressure

c_v specific heat at constant volume

C coefficients, as of discharge
clearance ratio in compressors
a constant
capacity (in tons per hour)

D diameter

f dimensionless friction factor

F force (in pounds)
filter rate in dust collectors

g gravitational acceleration

h hydraulic pressure head

H enthalpy (total heat) (in Btu per pound mass)
total hydraulic head

hp horsepower

bhp brake horsepower

ihp indicated horsepower

J Joule's equivalent (778 foot-pounds per Btu)

KE kinetic energy

k ratio of specific heats, c_p/c_v

L length of stroke of a piston (in feet)
length of pipeline (in feet)

m molecular weight

M mass (in pounds)

M' mass rate (in pounds per second)

n polytropic exponent

N number of working strokes per minute of a compressor
number of stages in a compressor

N_R Reynolds number

p absolute pressure (in pounds per square inch)

P absolute pressure (in pounds per square foot)

P_M total pressure of a mixture (in pounds per square foot)

Q energy transferred as heat (in Btu per pound mass)
volume of discharge (in cubic feet)
capacity (in tons per hour)

r radius

R gas constant
mixture ratio in a conveying stream

rpm revolutions per minute

s distance

S entropy per pound mass

SP static pressure

t ordinary temperature
time

T absolute temperature
tons per hour

v velocity (in feet per second)

v_c clearance volume in an air compressor (in cubic feet)

v_s piston displacement
swept volume in an air compressor (in cubic feet)

V specific volume (in cubic feet per pound mass)
volume (in cubic feet)

VP velocity pressure

w weight per unit volume (in pounds per cubic foot)

W work (in foot-pounds)
total weight (in pounds)

z potential hydraulic head
static head of a fluid (in feet)

α [*alpha*] expansion coefficient for gases
angle of inclination

μ [*mu*] absolute viscosity

ρ [*rho*] density

ϕ [*phi*] filter-rate correction factor

θ [*theta*] time (in minutes)

THEORY & DESIGN

1 Design Theory and Procedure

INTRODUCTION

Despite many years of research in and study of pneumatic conveying systems by manufacturers and users, this method of handling solids is still an art. Much that has been written on the theory of solids conveying is applicable to only a few selected materials of certain particle sizes transported in particular conveying systems. Further, there are no universally applicable empirical formulas for designing such systems.

The basic reason for the diversity of data on the pneumatic transportation of solids is the variety of the materials that can be conveyed. But, in turn, each material can be conveyed by a broad range of air velocities at various material-to-air loadings. The conveying air velocities are themselves a function of the particle-size spectrum and of the density, shape, and physical characteristics of the material, as well as of direction of flow—horizontal or vertical.

This dearth of practical information on pneumatic conveyors has left most users dependent on vendors for the design, selection, and installation of such equipment. Often, this has led to unfortunate results. The manufacturers and vendors of pneumatic conveying equipment possess most of the practical knowledge of these systems and obtain additional knowledge with each new system they install. Obviously, some of these installations become proving grounds for new types of equipment, with the purchaser bearing the cost—and, occasionally, the benefits.

The more progressive manufacturers will design a system to convey a particular material only after actual tests in a pilot plant, followed by scale-up and the judicious application of their knowledge from previous installations. The prospective purchaser should view each new installation as a cooperative effort. Best results are achieved when the purchaser can supply a complete specification covering his requirements, so that the vendor can determine his ability to fulfill them. Thus the purchaser must know what equipment he needs to accomplish a definite objective and must be able to interpret the results of the vendor's tests.

PNEUMATIC CONVEYORS

There are three basic categories of pneumatic conveyors.

Pipeline Conveyors. These transport particles of solid materials through vertical and horizontal pipelines by the conversion of kinetic energy in the airstream to dynamic pressure and aerodynamic lift. The majority of existing pneumatic conveyors for handling bulk materials are in this category. The pipeline and the equipment required to introduce air and material into it, and to receive and separate air and material at its terminus, comprise a pneumatic conveying system. The various types of pneumatic systems are described in Chapter 7.

Air-Activated Gravity Conveyors. These transport particles of solid fluidizable materials through near-horizontal chutes fitted with air-permeable bottoms. Air flows through the permeable bottom from a plenum chamber, aerates the material, and changes its normal angle of repose. The fluidized material flows through the conveyor seeking its new angle of repose and continues to flow since the conveyor is always set at an angle to the horizontal greater than the natural angle of repose of the fluidized material. This type of conveyor and its application are described in Chapter 14.

Tube Conveyors. These transport small packages or other items in a tubular container fitted with air-seal rings. The container is blown through the carrier tube by air pressure in the same manner as a piston. This type has a limited use in industry and is outside the scope of this work.

Advantages of Pipeline Conveyors. The principal reasons for installing bulk-handling pneumatic conveying systems are as follows:

1. Savings due to price differential between purchase of materials in bulk and purchase of materials in bags or small containers by elimination of packaging costs.
2. Savings due to reduced freight rate of bulk materials.
3. Savings due to reduction of labor costs for handling and storage of bags or small containers.
4. Reduction of losses due to torn sacks, spillage, residual materials in discarded bags or containers.
5. Delivery of dry materials via pipelines to remote plant areas which are economically inaccessible via mechanical conveyors.

6. Storage of bulk materials in large capacity bins and silos in plant areas otherwise unavailable for storage of packaged materials.
7. Increased plant safety due to elimination of man-handling bags or containers and elimination of dust when one is opening and dumping containers.
8. Reduced product contamination due to handling of materials in a closed system.
9. Elimination of fire protection systems usually required in areas where material in bags or bulk containers would be stored.

These objectives can be achieved only after a complete study and analysis to show the feasibility of this form of bulk handling.

To specify a pneumatic conveying system, one needs to know the following:

1. Material to be conveyed.
2. Method of transport from shipper to plant.
3. Operations necessary to meet plant production department requirements.
4. Type of equipment available for conveying and storing the material.
5. Degree of control required to meet operational requirements economically.
6. The vendor who will furnish the bulk of the conveying equipment.

EXPERIMENTAL BACKGROUND

The so-called "theory" of pneumatic conveying in pipeline conveyors is still in the development stage. It is the composite of data acquired by experimenters in determining the effect of the transfer of the kinetic energy of air to the flow of a variety of solid particles through horizontal and vertical pipelines. The formulas and numerical data obtained from these experiments cannot be extended or applied to the design of a commercial pneumatic conveying system for materials different from those tested. However, the observations made during these experiments reveal qualitatively how particles behave in vertical and horizontal airstreams under various loading conditions.

The following brief review of experimental work in the field of pneumatic conveying will provide background information on the state of the art. The first investigations in this field involved the determination of terminal velocities of various materials of particular particle sizes falling vertically through air. By equating the gravity force on the particles to the resistance to downward motion, the maximum free-fall velocity could be determined. From these experiments, it was concluded that, if the particles were immersed in a stream of air flowing upward at a higher velocity than the terminal velocity of the particles, they would move upward with a velocity equal to the difference between these velocities.

Similar investigations were made to determine the variation in resistance to motion of spheres of various sizes and densities immersed in various fluids. The fluids were liquid and gaseous, but the results of these tests were correlated by expressing the resistance to motion as a function of a dimensionless parameter, known as the Reynolds number, of the fluid density, and of the square of the velocity. Additional experiments extended the correlation for spheres to include particles of other geometric shapes by applying a sphericity correction factor. This factor was defined as the ratio of the surface area of a sphere, having the same volume as the geometric shape, to the surface area of the particle.

Further experimental work was devoted to the determination of the minimum flotation velocity of various materials. Flotation velocity was defined as that upward airstream velocity which would hold the material suspended without motion up or down. Formulas were developed for floating velocity in terms of material density and average particle diameter.

Experimental work on the transportation of solids in horizontal airstreams was undertaken, but was complicated by material falling out of suspension at floating velocities. It was found that higher velocities were required to transport a maximum proportion of the material through a horizontal line without settling of particles. Formulas were developed for minimum conveying velocities in terms of particle diameter, density, and specific gravity. Definite relationships were found between the relative velocity of spherical particles in a horizontal pipe, the terminal velocity of the same size particles in a vertical airstream, and the air velocity in a horizontal pipe.

All these investigations increased the knowledge of how materials behaved in an airstream, but the numerical data were of value only to investigators who were attempting to correlate the results of their experiments with data obtained by others.

The first extensive investigation of a complete pneumatic conveying system was made by Dr. J. Gasterstädt on a wheat conveying unit set up in the Machine Laboratory of the Technical High School in Dresden, Germany. The configuration of the test piping was arbitrary and the choice of pressure tap locations was unfortunate, since they did not permit isolation of the pressure drop in elbows, in vertical lines, and in the piping near the material inlet. However, the graphical representation of his data for pressure versus length of pipeline showed clearly the variations in pressure drop along the length of the line for various material loadings at constant air velocity (Figure 1–1). Most noticeable were (1) the large drop in pressure at the beginning of the conveying line due to acceleration of the material from entrance to conveying velocity, (2) the large drop in pressure at the elbows and in a vertical section of the line, and (3) the drop in pressure after the last

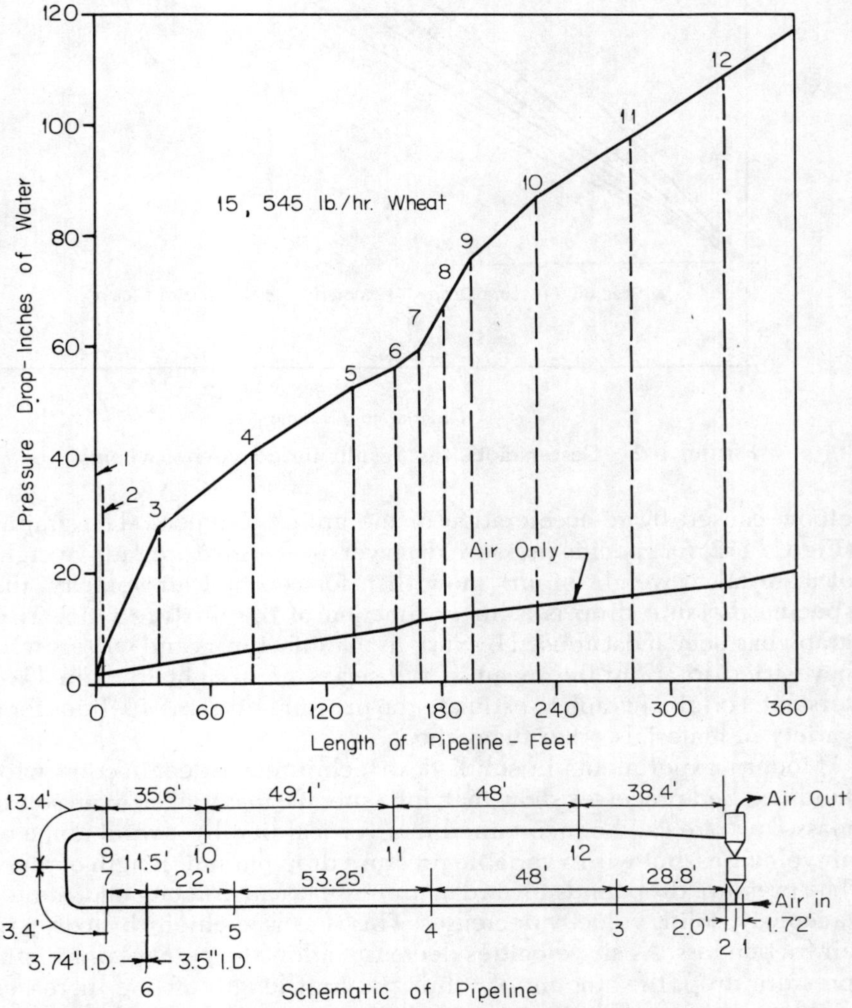

FIGURE 1–1. Schematic of Gasterstädt's piping arrangement for conveying wheat—typical pressure-drop data.

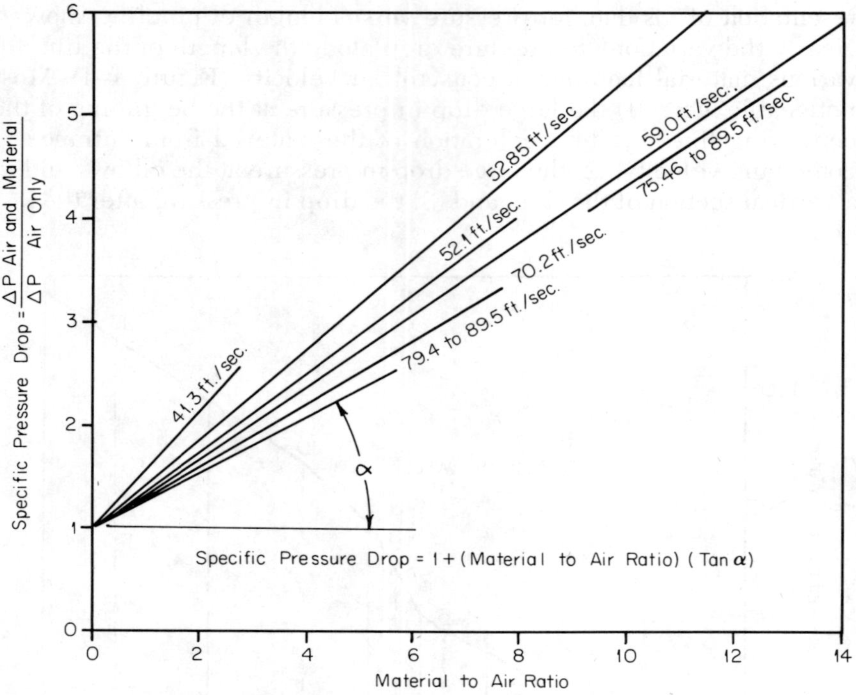

FIGURE 1–2. Gasterstädt's test results for conveying wheat.

elbow caused by re-acceleration of the grains of wheat. The graphs (Figure 1–2) for specific pressure drop versus the mixture ratio (weight of material to weight of air) show that, for constant air velocity, the specific pressure drop is a linear function of the mixture ratio. This graph has been most abused by various manufacturers and writers who have tried to apply the mean of the series of straight lines in Gasterstädt's original graph to estimate the pressure drop in pipelines for a variety of materials other than wheat.

Modern experiments in solids-flow techniques in connection with fluidized bed processes show that, for a specific material, a given solids mass-flow rate can be maintained in a vertical line by a wide range of air velocities, but with a variable pressure drop per unit length of pipe. The pressure drop tends toward a minimum as air velocity and, simultaneously, solids velocity decreases. This is caused chiefly by a drop in air friction loss. As air velocities decrease below the point of minimum pressure drop, the amount of solids per unit length of line increases and the pressure drop increases, but at a more rapid rate, due to friction of the material in the line, up to the point where slugging flow starts. The same phenomenon was observed when conveying a specific material in a horizontal line except that there was no well-defined minimum pressure drop. Instead, a discontinuous curve existed when air velocities decreased to the point where material settled out in the line, indicating an unstable condition. Each material tested, depending on its range of particle sizes, its shape, its density, and its physical characteristics, produced its own peculiar set of pressure-drop curves for a given mass rate of flow through a given range of velocities. The diversity in material characteristics shown by these experimental data indicates why it has been so difficult to correlate pressure-drop data obtained by various investigators for the flow of solids in an air stream. Figure 1–3 shows a typical set of curves obtained by one experimenter when conveying a particular material.

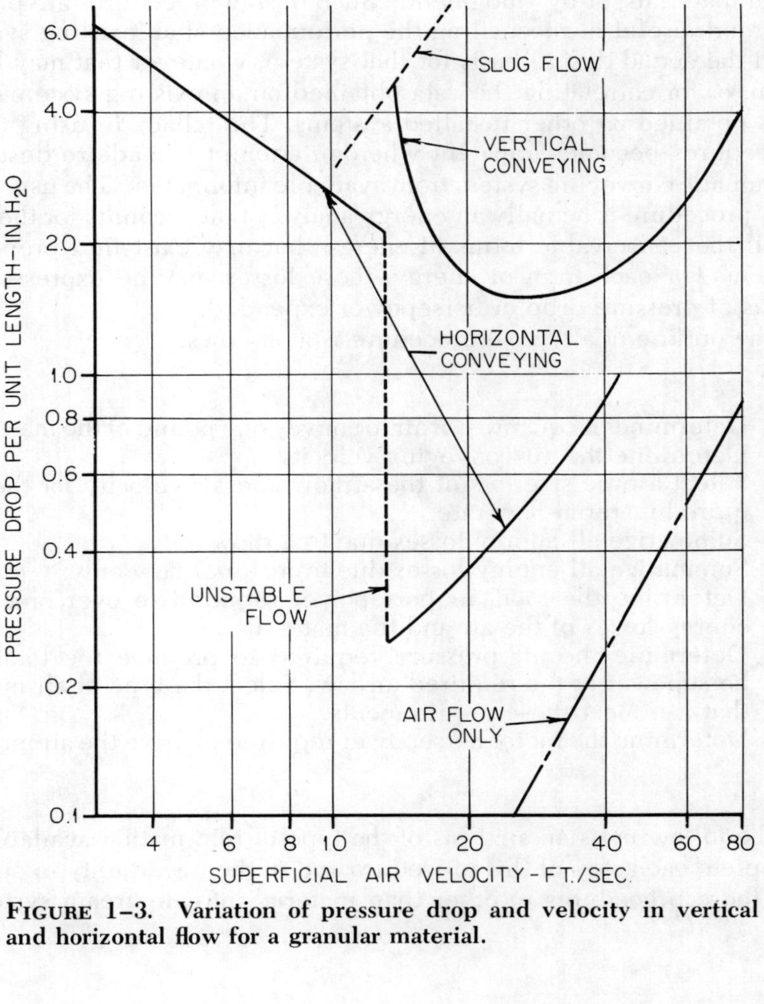

FIGURE 1–3. Variation of pressure drop and velocity in vertical and horizontal flow for a granular material.

DESIGN PROCEDURE

Published information on the design of pneumatic conveying systems is meager. Those articles that have appeared in the engineering literature are applicable only to material-into-air-stream systems. These articles present "design procedures" for such systems; the procedures seem to be straightforward and to lead logically to selection of a conveying line size and of the required capacity, pressure, and horsepower of an air mover. However, the information required to complete them must be obtained by actual test on a specific material, not by the selection of design factors from tabulated data or graphs relating to other materials, or by "judgment." Such design procedures are probably more useful in reconciling the performance of an existing system with the actual design basis for that system, whatever that may have been, or for correlating the data obtained on an existing system with data obtained on other installed systems. The fallacy in using these procedures becomes apparent when an attempt is made to design a pneumatic conveying system from available information. The usual design procedure is actually an energy analysis that accounts for the loss of all the conceivable forms of energy that may exist in a proposed system. For each form of energy, these losses may be expressed in terms of pressure drop or horsepower expended.

The outline of a typical procedure is as follows:

1. Determine the quantity of air to convey one pound of the material.
2. Determine the air conveying velocity.
3. Select a pipe size to suit the airflow and air velocity for the required material-flow rate.
4. Summarize all energy losses due to airflow only.
5. Summarize all energy losses due to material flow only.
6. Determine the total air horsepower required to overcome the energy losses of the air and the material.
7. Determine the air pressure required to produce the total air horsepower at the required airflow. Select the type of air mover that can meet these requirements.
8. Determine the motor horsepower required to drive the air mover.

The following is an analysis of the type of information available to complete each step of this procedure and of the uncertainty in applying these procedures to other than material-into-air-stream systems.

Air Required to Convey Material

There are only two sources of data in the literature that are intended to be used on systems where the material-to-air ratio is so low that several pounds of air are required to convey one pound of material. The more appropriate term "air-to-material ratio," in terms of cubic feet of air per pound of material, is used to express the concentration on such systems. The curves of Figure 1–4 show average air velocities

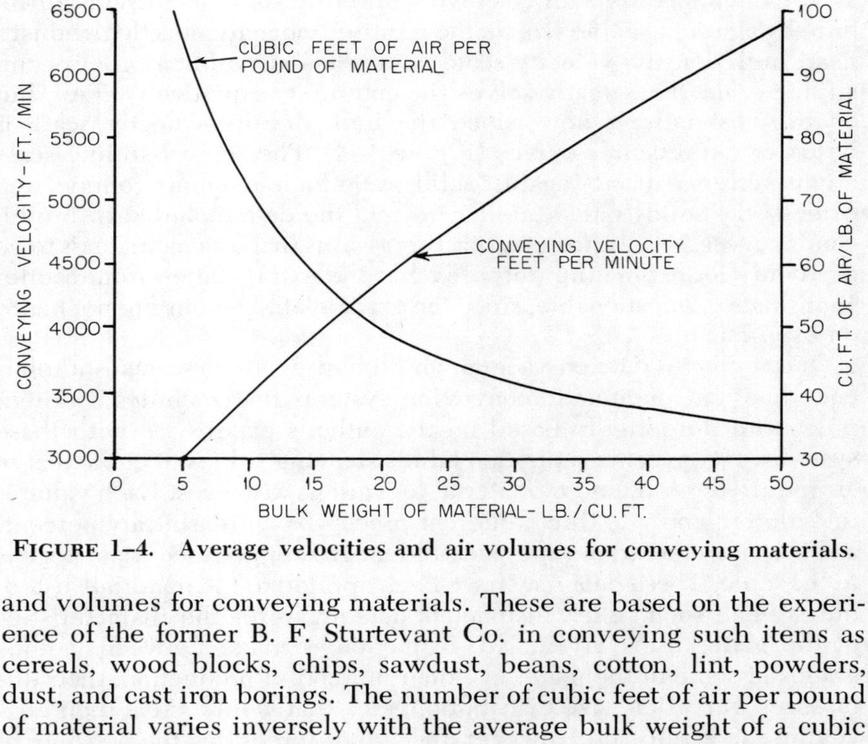

Figure 1–4. Average velocities and air volumes for conveying materials.

and volumes for conveying materials. These are based on the experience of the former B. F. Sturtevant Co. in conveying such items as cereals, wood blocks, chips, sawdust, beans, cotton, lint, powders, dust, and cast iron borings. The number of cubic feet of air per pound of material varies inversely with the average bulk weight of a cubic foot of material in the condition in which it is to be conveyed—not in terms of the true density of the material. These data, though somewhat obsolete due to their limited application, are still useful and safe for conveying some types of materials and are the basis for the design of many existing conveying systems.

The late W. G. Hudson, in several journal articles, had used a simple empirical formula, based on his experience, to approximate the airflow rate required to move material.

$$Q = wT$$

where Q is the cubic feet per minute of free air needed to move T tons per hour of material of specific weight w pounds per cubic foot. The fallacy here is that "specific weight" is defined as ". . . *the solid weight (not bulk density) of the material, increased more or less above the actual solid weight, as the experience of the engineer determines.*" This implies that w must be known from tests on existing installations handling similar materials; therefore this formula is useless to a designer without the background knowledge needed to determine the specific weight.

One manufacturer's air conveyor slide rule starts the design procedure by assuming a line size for the required capacity, sets this against a fixed bulk density–velocity scale, and reads air volume on a second sliding scale. This neatly solves the continuity equation $Q = Av$, but contributes nothing new, since the bulk density–velocity scale is based on Sturtevant's curves (Figure 1–4). The second sliding scale aligns system capacity against solids ratio for a given air volume, and extends the solids ratio scale far beyond the data included in Sturtevant's curves. Wisely, the manufacturer warns that his calculator is to be used only for estimating purposes, but the extrapolation from Sturtevant's data is questionable, since the extrapolated values are not based on experience.

One source of data on vacuum and low-pressure systems is a reference book on pneumatic conveying systems that includes a design procedure purportedly based on the author's experience with these systems. He presents saturation tables showing values of cubic feet of air required per pound of material for various distances. Each value is intended to apply to three different pipe sizes with arbitrary percentage increases for larger pipe sizes and for greater distances than appear in the table. These data may have been applied to the materials listed, but as in all such tables, insufficient data regarding the characteristics of the materials is given. No distinction is made between various grades of some of the materials, their method of production, their differences in particle size distribution, etc., that would affect their conveying characteristics. However, the author agrees that the behavior of a material under the influence of air "must be gleaned only from experience."

All commercial types of pneumatic conveyors involve dilute-phase conveying. Dilute-phase conveying is characterized by high-velocity flow of a stream of solid particles homogeneously dispersed in air. The concentration, or material-to-air ratio, and the velocity of the dispersed stream may vary over a wide range that is different for each specific material. As the air velocity in such a stream is reduced for a given concentration of solid particles, the solids velocity decreases until the entire dispersion of solids collapses and slug flow occurs. Systems in which several pounds of air are required to transport one pound of solids have sometimes been called "dilute-phase" systems in contrast to so-called "dense-phase" systems in which one pound of air may transport more than its weight of solids. The reality is that both types of systems may be dilute-phase systems.

Air Conveying Velocity

The most important factor in the transport of solid particles is the conveying velocity. The actual velocity of the particles of flowing material is unknown and difficult to determine, but the air velocity of the conveying stream is easily measured or calculated. In commercial applications, reference to the conveying velocity usually refers to the superficial air velocity—the velocity calculated from a measured airflow and the cross-sectional area of the conveying line. The minimum superficial air velocity that will convey at a given solids feed rate is called the "critical velocity." For vertical lift, this may be called "dropout velocity"; for horizontal conveying, this may be called "settling" or "saltation velocity." Since most systems have a combination of vertical and horizontal runs of piping, the only velocity of interest is the lowest one that will clean all materials out of the line.

The literature is replete with tables listing so-called "conveying velocities" for various materials, but their use is not recommended since there is no reference to the type of system in which the particles were conveyed nor is there a reference to the exact specification of the materials conveyed. The variables affecting the minimum conveying velocity are material-to-air loading (type of system), absolute density, particle size and shape, and physical characteristics such as tackiness, abrasiveness, moisture content, etc. Because of the unknown effect of these variables on the conveyability of a material, the best method of determining conveying velocity is to test the material-flow characteristics in a pilot plant using the type of equipment that is being considered for selection.

In some systems where the material-to-air-ratio is low and the weight of air predominates, it is safe to use conveying velocities based on a vendor's experience with a wide range of materials. Safety lies in the fact that most of the pressure drop in such a system is caused by airflow. The losses due to material flow are negligible insofar as power requirements are concerned. Furthermore, since these systems use fans, rather than positive-pressure blowers, for air supply, the selection of the fan may be based on an operating characteristic suitable for meeting variation in system resistance with minimum change in air delivery.

The existing data on conveying velocities, in addition to the usual tabulations of recommended velocities for various materials, consist of Sturtevant's curves (Figure 1–4) and of a tabulation of air-material ratio versus velocity, which appears in "Reference Data and Tables," issued by MikroPul Corporation. Sturtevant's curves show velocity and air-to-material ratio as a function of density, whereas the latter data relates conveying velocity to a range of air-to-material ratios, but with no reference to density. Sturtevant's curves are for application to a wide variety of materials; MikroPul Corporation data are intended only for company use in conveying finely powdered materials discharged from their pulverizers. A comparison of the data from these sources bears out a generally known phenomenon: namely, that coarse material having a small quantity of fine particles requires more air for conveying than material of the same bulk density containing a larger quantity of fine particles.

It is interesting to note that the bulk density–velocity scale on the aforementioned slide rule includes materials with bulk densities up to 120 lb/ft^3. This has been done by extrapolating Sturtevant's data, which applied only to materials with bulk densities up to 50 lb/ft^3, using the approximate formula

$$\text{Velocity} = 920\sqrt{\text{bulk density}}$$

Pipe Size

The cross-sectional area of the conveying line is calculated by application of the continuity equation to the selected air-to-material ratio and conveying velocity. The total air required is obtained by multiplying the quantity of material to be handled by the cubic feet of air required per pound. The total air requirement divided by the conveying velocity results in an area that must be compared to the actual cross-sectional areas of available piping or tubing for selection of the nearest line size—always favoring the area that would result in a higher velocity, if the required area lies between two pipe sizes. The selected line size may have to be adjusted later, if the calculated pressure drop is excessive or beyond the pressure range of available air movers.

Energy Losses Due to Airflow

The pressure drop in piping can be obtained from friction-loss charts that show pressure drop in inches of water per 100 ft of round piping versus airflow in ft^3/min for a range of pipe sizes and velocities. These charts are available in many handbooks and in the catalogs issued by manufacturers of air handling equipment. Figure 1–5 shows a typical chart for friction loss in round pipe. Since pressure drop varies inversely with the absolute pressure and directly with the flow to the 1.86 power (approximately), the values on the chart can be applied to pipelines under pressure with reasonable accuracy as long as the total pressure drop is less than 10 percent of the initial line pressure and airflow is entered as ft^3/min of free air. Thus, for a free airflow of 1,000 ft^3/min in a 6-in.-diameter duct, the chart shows a friction loss of 7 in. of water per 100 ft. If the line pressure at its entrance is 10 psig, the friction loss would be

$$\frac{14.7}{24.7} \times 7 = 4.16 \text{ in. of water (0.15 psi) per 100 ft}$$

Bend loss charts are also available in these sources of information, but, more often, an equivalent length of straight pipe for each bend is added to the total length of the straight pipe runs. The pressure drop must be determined for air supply piping as well as for the conveying line.

The pressure drop in the various types of air handling equipment is usually calculated by empirical formulas, developed by manufacturers of the equipment for their own use. These formulas always show pressure drop as a function of the velocity pressure and are applied to such equipment as silencers, cyclones, and dust filters.

The pressure loss at air entrance points such as at pickup nozzles and orifices is dependent upon the shape of the opening. The loss is estimated by reference to a chart showing hood entry losses for various configurations and expressing the loss as the velocity pressure multiplied by an appropriate factor.

The total pressure drop is calculated and expressed in inches of water, in pounds per square inch, or in any equivalent units that will be used to determine the air horsepower required by the system.

Energy Losses Due to Material Flow

The diverse characteristics of the materials handled in a pneumatic conveying line and the fact that the material-air mixture is not homogeneous or constant have resulted in much confusion regarding the calculation of pressure drop due to solids flow. Some investigators have tried to adapt equations derived for the flow of homogeneous fluids to the flow of airborne solids, even to the extent of developing friction factors. Many have tried to estimate the pressure drop by using a modified energy analysis that accounts for all work required to move the air and materials from its source to the point of discharge. Other

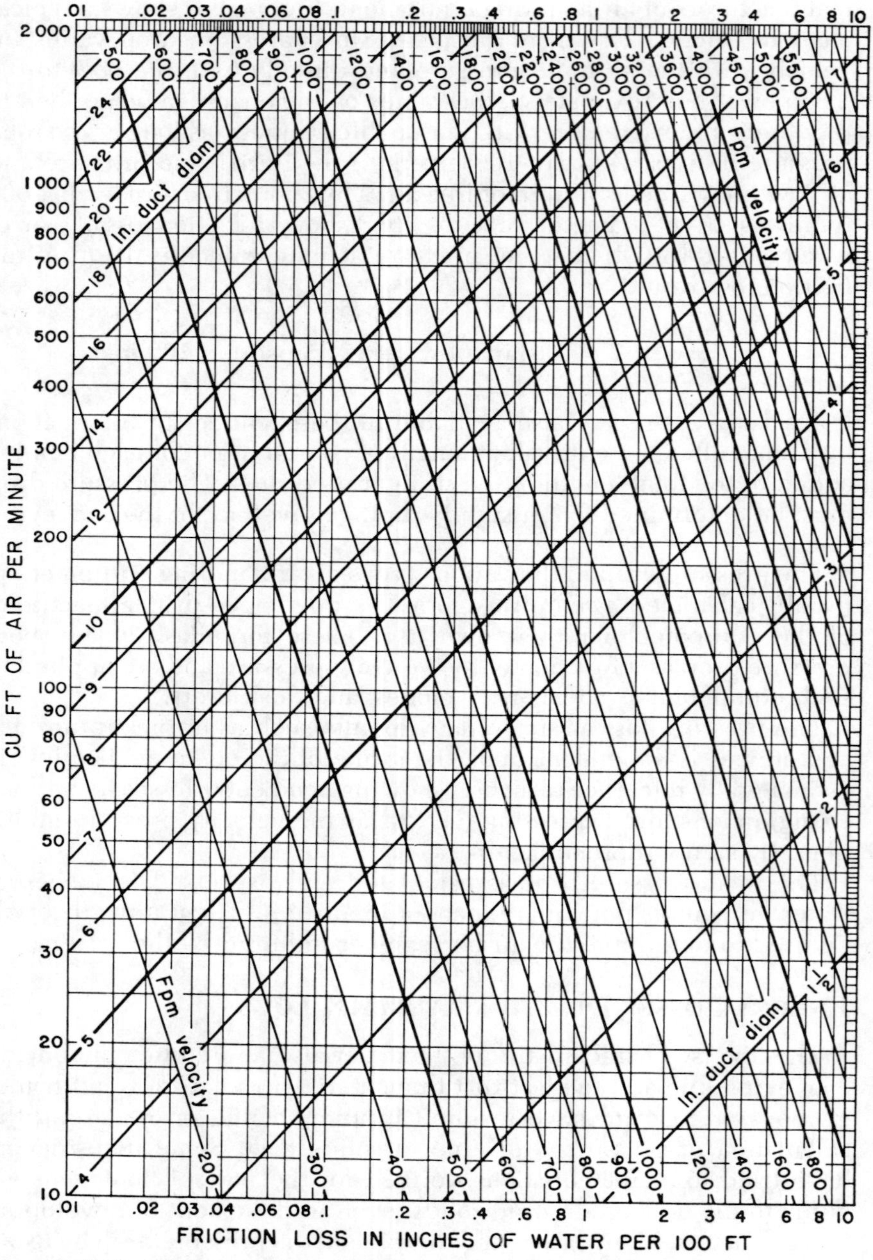

FIGURE 1–5. Friction loss in round pipes—straight ducts; this page, for volumes of 10 to 2,000 cfm; facing page, for volumes of 1,000 to 400,000 cfm. Based on standard air density of 0.075 lb/ft, flowing through average, clean galvanized metal ducts having approximately 40 joints per 100 ft. (Reprinted by permission from *ASHRAE Guide and Data Book 1967*.)

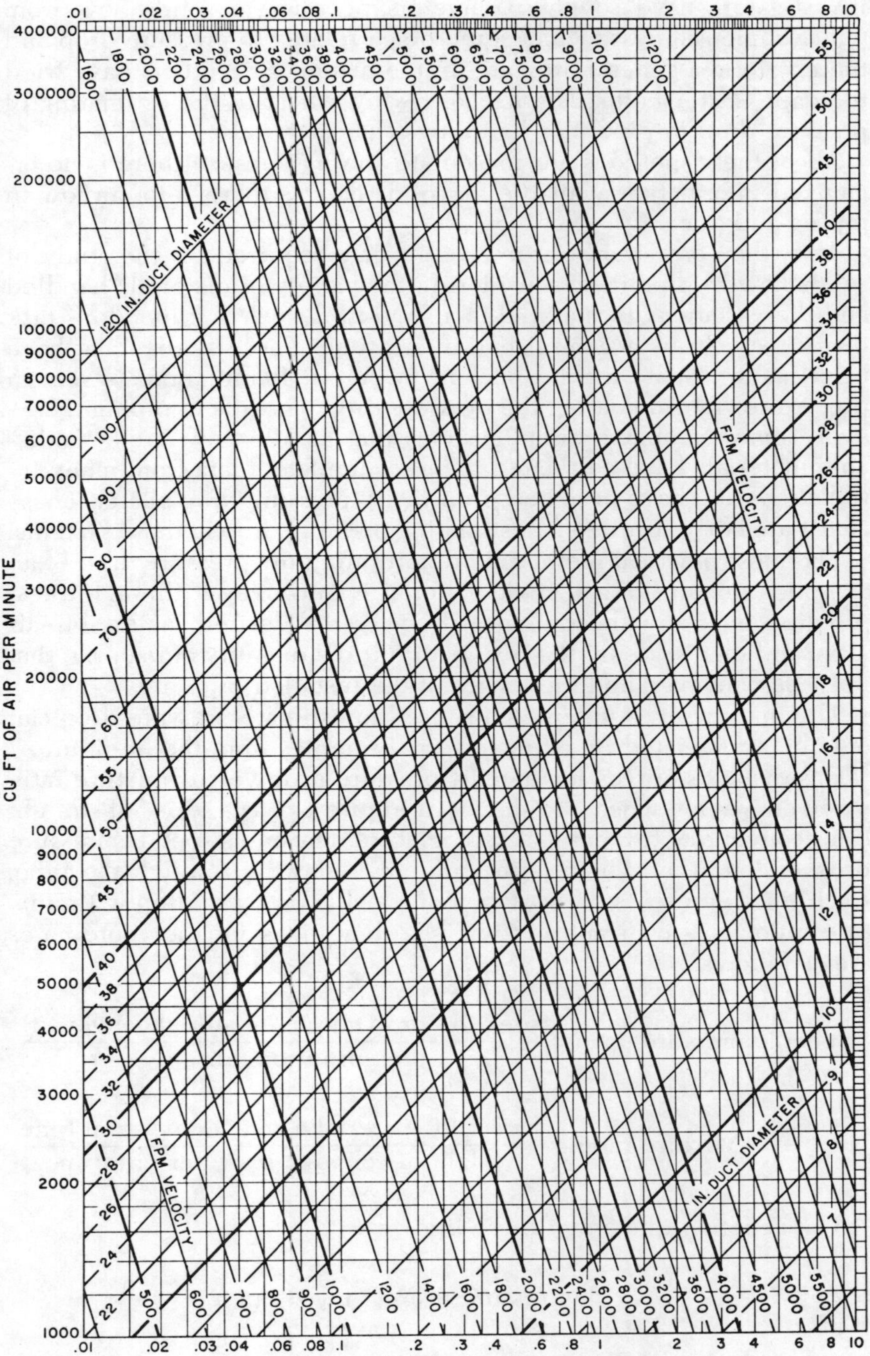
CU FT OF AIR PER MINUTE
FRICTION LOSS IN INCHES OF WATER PER 100 FT
IN. DUCT DIAMETER
120 IN. DUCT DIAMETER
FPM VELOCITY
FPM VELOCITY

investigators have sought parameters peculiar to solids flow from which dimensionless factors could be correlated to pressure drop and thereby form a general pressure-drop equation. Still others have tried to adapt Gasterstädt's data for conveying wheat to the estimating of pressure drop in pipelines handling other materials.

All of these methods are inaccurate simply because there is no organized information available for solids flow as there is for airflow in piping and ducts.

Also, the correlation of the type of data required for the study of solids flow in pipelines, considering the variety of materials handled and the variations in material-to-air ratios, is such a stupendous task that it may not be accomplished in our time. Those studies that have been made to date are insufficient to permit formulations of design procedures for calculating the pressure drop of solids in piping. However, some manufacturers of systems that handle powdered, fluidizable materials have correlated their accumulated data from many installations and have prepared proprietary pressure-drop charts. These charts are fairly accurate for estimating pressure drop in those systems.

The most practical method for determining the pressure drop of an air-solids mixture of a specific material is to convey it through a test pipeline, measure the pressure drop for the desired material-to-air ratio, and then scale up the results to suit the quantity of material that must be conveyed and the length of the installed pipeline.

The inaccuracy in the methods used to estimate pressure drop can only be appreciated by citing several examples from the literature.

Gasterstädt's curves for specific pressure drop versus mixture ratio when conveying wheat are shown in Figure 1–2 as a series of straight lines converging to a point at zero mixture ratio and a specific pressure drop of 1 unit, at which point only air is flowing through the pipe. Each line represents the variation of specific pressure drop with mixture ratio for a constant airflow. The definitions for the coordinates are as follows:

$$\text{Specific pressure drop} = \frac{\text{pressure drop (air and material conveyed)}}{\text{pressure drop (air only)}}$$

$$\text{Mixture ratio} = \frac{\text{weight of material conveyed per unit time}}{\text{weight of conveying air per unit time}}$$

The general equation of these lines was:

$$\text{Specific pressure drop} = 1 + R \tan \alpha$$

where α = angle of inclination of the lines on the chart
R = mixture ratio.

Tan α was demonstrated to be a function of the air velocity that increased at low velocities, but tended toward a common line at very high velocities. The pressure drop in these experiments ranged from ($R/0.92 + 1$) to ($R/2.03 + 1$). It is quite conceivable that a similar test using other materials would produce a different set of curves, but the incentive to conduct such tests is lacking. Therefore, curves based on Gasterstädt's data have appeared in the literature as a single sloping line from which the specific pressure drop has varied from the average of Gasterstädt's data, ($R/1.46 + 1$) in one manufacturer's catalog to ($R/3.43 + 1$) in another manufacturer's catalog.

A variation in the presentation of these data appears in one fan manufacturer's literature as a hyperbolic curve, with coordinate axes showing a friction multiplier versus cubic feet of air per minute per pound of material. A note on the curve requires that the friction for clean air be multiplied by the friction multiplier.

The air conveyor slide rule cited earlier contains a design factor based on Gasterstädt's data that is aligned with solids ratio to obtain the pressure drop in a pipeline. However, the error in using such a curve is compounded since the curves originate at a specific pressure drop of zero! This, by the definition of specific pressure drop, is impossible.

In the energy-analysis method, the designer tries to account for all work expended in moving the material. The energy required to accelerate material from rest and to move material a given distance can be computed from basic mechanical formulas for work and motion. However, the method breaks down when estimating pressure drop in piping and many assumptions are made that cannot be proven true. Hudson assumed hose and elbow pressure losses as being one-half of the loss due to acceleration. His piping losses were based on purely arbitrary factors of tonnage and height of lift and on only 20 percent of the length of the line! Fischer used a "finagle" factor to calculate the pressure drop in a horizontal line. This factor is defined as the tangent of the angle of slide between the material being conveyed and the material from which the duct is made. Any relationship between this factor and energy loss in a pipe where most of the material is airborne would be difficult to prove! He also uses this factor to estimate bend losses, assuming that all material slides against the wall.

A British handbook on pneumatic conveying states the following regarding pipeline friction: "*It is not easy . . . to deduce a rational formula which is applicable over the whole range of operations; such a formula would include many variables which are difficult to separate and to measure experimentally.*"

Most investigations on specific materials relate pressure drop to conveying velocity in a pipeline of given size so that a direct comparison can be made with the pressure drop when handling only air. The results are plotted on logarithmic charts so that the approximate pressure drop–velocity equation may be formed. In most cases, at high velocities, the plotted points form a straight line so that the slope or tangent of the angle made by the line will give the exponent or the power of the velocity that determines the pressure drop. The constant by which the variable is multiplied is the intercept of the extension of the sloping line at the abscissa corresponding to a velocity of one unit. The straight line on logarithmic paper represents the equation

$$p = b(\text{velocity})^n$$

where p and "velocity" are coordinates of any point on the straight line. The constant b may be determined mathematically, if it cannot be found graphically, by setting up the ratio of any two coordinates on the curve

$$\frac{Y}{Y_0} = \left[\frac{X}{X_0}\right]^n$$

then

$$b = \frac{Y_0}{X_0{}^n}$$

The Fanning Equation

Some writers have attempted to correlate pressure-drop data for flow of powdered materials in an airstream by applying the Fanning equation

$$P_1 - P_2 = f\frac{L}{D}\,\frac{v^2}{2g}\,\frac{\rho}{144}$$

where $P_1 - P_2$ = pressure drop in conveying line, lb/in.^{2}
f = a dimensionless friction factor
L = total length of line, ft
D = pipe bore, ft
v = velocity of mixture, ft/sec

ρ = density of the powder-air mixture, ft³/lb

$$\rho = \frac{(\text{weight of air}) + (\text{weight of powder})}{\begin{pmatrix}\text{volume of air at}\\ \text{average pipe conditions}\end{pmatrix} + (\text{volume of powder})}.$$

The original Fanning equation was developed for the flow of homogeneous fluids in turbulent flow, but the value of the friction factor for these fluids has been correlated and found to be a function of the Reynolds number:

$$N_R = \frac{\rho v D}{\mu}$$

There has been no such correlation for the flow of solids-air mixtures; the friction factors for some solids-air mixtures have been arbitrarily determined simply to equate pressure drop on test with the terms of the Fanning equation. The use of such friction factors without the background of test experience required for their determination is not recommended.

This equation has been modified by some manufacturers of systems to the simplified form

$$\frac{P_1 - P_2}{L} = \text{constant} \times \frac{W}{D}$$

where W = solids mass flow rate, lb/sec/ft².

This permits pilot-plant test data to be correlated so that charts of pressure drop per foot of pipe versus mass flow rate may be constructed for estimating the pressure drop in piping conveying dry, free-flowing fluidizable powdered or granular materials at a constant air flow rate. See Figure 1–6.

Since most pilot-plant conveying lines include bends, vertical risers, and horizontal lengths of piping, but only have provision for measuring pressure at the inlet and the outlet of the conveying line, a plot of pressure drop versus equivalent length of piping in a test run would be a straight line. From Figure 1–1, it is evident that if the broken line obtained by Gasterstädt were replaced by a straight line, the change in the average loss of pressure per foot of pipe would be relatively small. Thus, the pressure drop caused by acceleration at the entrance to the line, reacceleration after bends, lift of solids in vertical lines, and kinetic energy changes during flow are assumed to be part of the frictional loss in the pipe that varies linearly, with little error when used for estimating purposes. See Chapter 4.

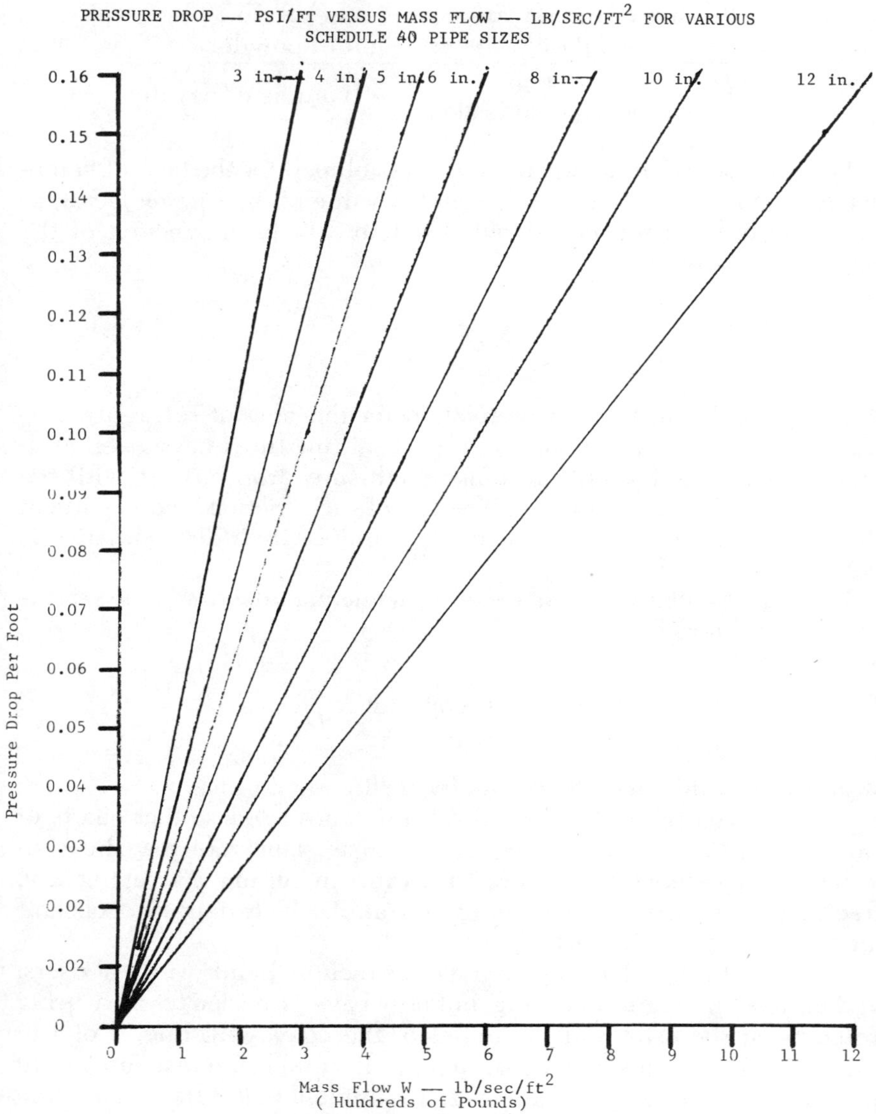

FIGURE 1–6. Pressure drop, psi/ft versus mass flow, lb/sec/ft² for various Schedule 40 pipe sizes.

Other assumptions made that permit use of the simplified Fanning equation are as follows:

1. There is negligible difference between the velocity of the air and the velocity of the airborne solids.

2. The velocity is constant since the airflow is constant. This is not

strictly true. The velocity increases as pressure decreases along the length of the piping due to expansion of the air. However, the increase in velocity counteracts the decrease caused by friction, impact, etc., so that this assumption may be considered valid.

3. The material-to-air ratio varies linearly with the density of the solids-air mixture. This is also not strictly true, but the variation in mixture density when handling solids weighing over 20 lb/ft^3 is relatively small.

The constant in the modified Fanning equation varies depending on the characteristics of the air supply—pressure and flow rate—supplied to a particular system. A value of 0.000137 for the constant seems to cover a wide range of systems, with material-to-air ratios between 10 and 40 to 1. The constant may be three-fourths of this in blow tank systems that use very high material-to-air ratios and very low conveying velocities. The constant may be 50 percent higher on systems where solids and air mix in a rotary feeder.

Pressure Drops in Pipe Fittings

Pressure drop losses in various types of fittings have been developed for homogeneous fluids, but the only fittings of interest in solids-air flow are 90-degree bends. There are no available data on the loss of pressure in bends having the bend-radius-to-pipe-diameter ratios recommended for use in pneumatic conveying lines—not even for a homogeneous fluid such as air. There are data available on the loss of water pressure in 90-degree bends with a relative radius r/D up to 20, but even such data by different investigators cannot be correlated; therefore it is difficult to recommend the use of such data for solids-air flow calculations. The data available for airflow may be applied to estimating losses in air piping, but they cannot be extended to include solids-air flow in conveying lines.

A comparison of the developed length of 90-degree bends to the equivalent resistance of the bends as compared to straight pipe, taken from the best available data on fluid flow, indicates that, for r/D ratios from 12 to 20, regardless of pipe size, a safe allowance for estimating purposes is to assume that the resistance is equal to 1.6 times the developed length of the bend, taken as straight pipe. Thus a 3-in. Schedule 40 conveying line having 100 feet of straight pipe, including bend tangents, and three 90-degree bends having an r/D ratio of 12 would have an equivalent length of

$$100 + \left[1.6 \times \left(\frac{2\pi[3 \times 12]}{12 \times 4} \right) \times 3 \right] = 100 + [1.6 \times 4.71 \times 3]$$
$$= 100 + 22.6 = 122.6 \text{ ft}$$

Determination of Air Pressure

The energy losses due to airflow are readily determined in terms of pressure drop in inches of water, since charts for the pressure drop in ducts and equations for the pressure drop through equipment express these losses in inches of water. The energy losses due to material flow may be given in terms of pressure drop in the piping, or as energy loss in foot-pound units due to acceleration, lift, and horizontal travel. The summation of the air and material losses is made after expressing all losses in terms of the same units—usually pressure drop in inches of water. This is done by utilizing the air horsepower equation for fans (see Chapter 2) as follows:

$$\text{hp} = \frac{QP}{33{,}000} = \frac{\text{cfm of air} \times \text{pressure (in psf)}}{33{,}000}$$

Since the numerator is the work done per minute and since one inch of water is equivalent to a pressure of 5.19 pounds per square foot, the air horsepower equation becomes

$$\text{hp} = \frac{\text{cfm of air} \times \text{pressure (in.} H_2O)}{6356}$$

This equation permits the conversion of energy losses from horsepower or foot-pound units to pressure drop in inches of water. The cfm is known at this stage of design from the selected pipe size and the conveying velocity. An alternate method used by some writers is to express all losses directly in terms of horsepower, and then, by use of the air horsepower equation, to convert the summation of these losses into pressure in pounds per square foot. The use of the air horsepower equation in this manner is limited to low-pressure systems using fans as air movers. Some writers have used this procedure, improperly, for systems using low-pressure positive-displacement blowers.

Determination of Drive Horsepower

The method of determining the horsepower for various gas processes is covered in Chapter 2. The drive motor must be sized to allow for additional losses in the air mover due to friction, turbulence, leakage, and mechanical friction, and for losses in the mechanical transmission due to slippage and friction. The brake horsepower of the fan must therefore be obtained from the equation

$$\text{bhp} = \frac{\text{air horsepower}}{\text{fan efficiency} \times \text{transmission efficiency}}$$

or, preferably, directly from selection charts in fan manufacturer's catalogs. The range of fan efficiencies is so broad that arbitrary selection of an efficiency is not recommended.

Transmission efficiency varies with the type of drive, but an overall efficiency of 98 percent may be used for V-belt drives—the usual type of transmission for fans when not directly connected to a motor.

2 Fundamentals of Fluid Flow

The purpose of this chapter is to present those fundamentals of fluid flow and thermodynamics that are necessary to a clear understanding of the phenomena taking place in a pipeline and in air handling equipment. Mathematics has been reduced to a minimum, and the derivations of several equations have been given to enable the reader to know their source and the limitations on their application.

FORCES AFFECTING FLOW

The forces affecting the flow of air and material in a pneumatic conveying pipeline are as follows:

Gravity. This is the external force caused by the gravitational attraction between the small mass of individual air and material particles and the huge mass of the earth. This force causes settlement of materials in horizontal pipelines and dust collectors, and must be overcome in upward vertical flow of air and material. It is proportional to the product of mass and the acceleration of gravity.

Pressure Differential. This is the force caused by a difference in pressure to initiate flow of air and material. Flow is always in the direction of the lower pressure. Flow from a low pressure area to a high pressure area can occur only in a machine that receives energy as shaft-work for the purpose of producing the required pressure increase.

Inertia. This is the force required to overcome the natural resistance of particles of air and material to change in speed and direction of flow. It is the product of the mass and acceleration of the fluid particles. This force must be exerted when air and material enter a system at low velocity and are accelerated to the velocity of flow, and when air and material change direction of flow as in pipe bends, and in cyclones, where material is separated from the air stream.

Shear. This is the force required to cause relative flow between adjacent particles of a homogeneous, viscous fluid. It is exerted due to differences in velocity between layers of fluid flowing through the center of a pipe and the outer layers that flow at lower velocity due to friction on the pipe wall.

Elasticity. This is the force exerted by the intrinsic tendency of a compressible gas to expand and flow from a high to a low pressure region. The density of the gas is reduced during this expansion. Expansion of air occurs in all pneumatic transport lines, but its effect is neglected at low pressure differentials.

All of these forces are considered when dealing with the flow of air or other conveying gas. Only the first three types of forces are considered when dealing with the flow of solid particles in an air or gas stream.

The relationship of these forces to the flow of air (or gas) can be determined mathematically with sufficient accuracy for all practical purposes and will be studied first, since air is the motivating medium for most pneumatic conveyors.

BASIC LAWS

THE LAW OF CONSERVATION OF MATTER

This law, applied to the steady flow of fluid through a pipe, states that the mass flow through any cross section of the pipe is constant. If A is the total cross section of the pipe in square feet, v is the average velocity in feet per second, and ρ is the density of the fluid in slugs per cubic foot, then Q_M the mass discharge in slugs per second at any cross section is

$$Q_M = \rho A v = \text{a constant}$$

Applying this law to successive piping cross sections results in the so-called "continuity equation," or equation for continuity of flow through the pipeline:

$$Q = \rho_1 A_1 v_1 = \rho_2 A_2 v_2 \cdots = \rho_n A_n v_n$$

If there is no variation in the density of the fluid between any two cross sections, the equation becomes the discharge equation:

$$Q = A_1 v_1 = A_2 v_2 \cdots = A_n v_n$$

since Q is the rate of flow in cubic feet per second. The density of air is considered constant in all flow problems when the pressure differential across a pipeline is less than 1 percent of the downstream absolute pressure. The error involved in this assumption is negligible in most problems.

The density of a fluid is the mass of a unit volume under specified temperature and pressure conditions and is related to specific weight by the equation

$$\rho = \frac{w}{g} \qquad \text{or} \qquad w = \rho g$$

where w is the specific weight in pounds per cubic foot, g is the acceleration of gravity in feet per sec^2 and ρ is measured in slugs per cubic foot. Since the acceleration of gravity is assumed to have a standard value of 32.2 feet per sec^2 and the unit of mass is defined as the mass in which a force of one pound will produce unit acceleration, then the unit of mass must be 32.2 times as large as the mass of one pound. This unit is called the "slug" in the English, or foot-pound-second, system of units.

The Law of Conservation of Energy

This law, applied to the steady flow of fluid in a pipe, states that the total energy in any mechanical system remains constant. Energy is defined as the capacity for doing work. In a conveying system, there are three forms of energy.

Kinetic Energy. This is the energy possessed by a mass due to its velocity and is given by the equation

$$\text{KE} = \frac{W}{2}\frac{v^2}{g}$$

where KE is kinetic energy measured in foot-pounds, W is the weight in pounds, v is the velocity of the mass in feet per second, and g is the acceleration of gravity as given above. This equation is often used to estimate the work required to accelerate the air and material entering a pipeline. It is derived from the equations for determining the work necessary to accelerate a particle from rest to a velocity v. Thus, if a constant force F acts on a mass M, the acceleration a will remain constant in accordance with the basic equation

$$F = Ma$$

During a time interval t, the mass will travel a distance s such that

$$s = \frac{1}{2} at^2 = \frac{1}{2}\left(\frac{v}{t}\right) t^2 = \frac{1}{2} vt$$

where v is the final velocity of the center of mass. The work done by this force through distance s is Fs; therefore

$$\mathrm{KE} = Fs = \frac{mv}{t} \cdot \frac{1}{2} vt = \frac{1}{2} mv^2$$

$$\mathrm{KE} = \frac{W}{2} \frac{v^2}{g}$$

The work required to accelerate 15,000 lb of material per hour to a velocity of 5,000 ft/min is

$$\mathrm{KE} = \frac{15{,}000}{2} \times \left(\frac{5{,}000}{60}\right)^2 \times \frac{1}{32.2} = 7{,}500 \times 6{,}940 \times \frac{1}{32.2}$$
$$= 1{,}618{,}000 \text{ ft–lb}$$

Since this energy is expended in one hour, the equivalent horsepower required for acceleration is

$$\frac{1{,}618{,}000}{33{,}000 \times 60} = 0.817 \text{ hp}$$

The acceleration energy is expended through a distance that is unknown and is difficult to determine unless some assumption is made of the force causing it. Thus, the frictional losses occurring in the acceleration section of a pneumatic conveying line would also be unknown. The distance required to obtain the conveying velocity can be determined by measuring the pressure drop along the line at closely spaced intervals and noting the point where the pressure gradient changes. Obviously, this can only be done on a long straight line uninterrupted by changes in direction that affect the pressure gradient. The losses caused by friction in the acceleration portion of a conveying line have been observed by investigators, but no numerical data is available on this subject.

Potential Energy. This is the energy possessed by a mass due to its position above a horizontal datum plane. A weight of w pounds falling from a distance z feet above the datum plane can perform wz foot-pounds of work. This is the same amount of work required to be expended in raising the weight to the height z.

Pressure Energy. This is the energy possessed by a mass of fluid due to its pressure above atmospheric pressure, or above the absolute zero of pressure, depending on the problem involved. The atmospheric pressure at sea level is assigned a value of 14.7 psi above absolute zero, although its true value fluctuates with the barometric pressure. Pressures may be expressed in various units on different scales, the choice of unit depending on the type of problem at hand. If a mass of incompressible fluid under pressure in a pipeline is connected to a vertical open-ended tube, it would flow up the tube until the pressure at the surface is equal to the pressure at the open end. The total height of fluid required to produce and balance the given pressure is

$$h = \frac{P}{w}$$

where P is in pounds per square foot, w is the weight per cubic foot of fluid, and h becomes feet of head of the fluid.

If a hole were placed in the side of the tube just below the free surface of the fluid, the fluid would flow continuously as long as pressure in the pipeline was maintained constant. In lifting a total weight W of fluid, the work done by the pressure is

$$\text{Pressure energy} = W\frac{P}{w} = Wh$$

Since the sum of these three forms of energy are constant, the total energy in a flowing fluid of weight W is expressed by the equation

$$\frac{Wv^2}{2g} + W\frac{P}{w} + Wz = \text{a constant}$$

Dividing through by W, the energy per pound of fluid becomes

$$\frac{v^2}{2g} + \frac{P}{w} + z = H$$

The terms in the left side of this equation have the dimension of length and are called velocity head, pressure head, and elevation head, respectively. The sum H is called the total head. The equation is known as Bernoulli's Theorem and may be used to analyze the transfer of energy in any fluid-flow problem by calculating the total energy at a point where each term is known and equating it to the sum of the terms of the point under consideration.

PERFECT GAS PROPERTIES

Air or any other gas used for pneumatic conveying is compressible and, by virtue of its expansive power, possesses intrinsic energy. The energy content of air (or gas) for a given state or condition is determined by its properties. The fundamental measurable properties are pressure, volume, and temperature. Other properties, such as density, internal energy, enthalpy, and entropy, may be derived from one or more of these. Equations which express the relations between these properties have been developed, based on the concept of a perfect or ideal gas. A perfect gas is defined as a gas that has two concurrent characteristics: (1) For a state change from one reproducible temperature to another, the ratio between initial and final pressure, when the state change is performed at constant volume, shall equal the ratio between the initial and final specific volumes for a constant-pressure state change. (2) The internal molecular energy of the gas shall depend only on its temperature. Expressed mathematically, the above characteristics are as follows:

$$\frac{T_1}{T_2} = \left[\frac{P_1}{P_2}\right]_{V\ \text{constant}} = \left[\frac{V_1}{V_2}\right]_{P\ \text{constant}}$$

Although real gases do not have these characteristics, the equations developed for perfect gases may be applied to them at conditions that are remote from the point at which they become liquefied.

PRESSURE

The pressure at any point in a fluid is defined as the limit approached by the ratio of force to area, and the engineer's unit is one pound per square inch (psi). In order to maintain consistency with the units used in the perfect gas equations, pressure should be converted to the pounds per square foot (psf) unit. Two scales of pressure are in use: (1) gage pressure, the pressure measured from the atmospheric pressure as the datum point; and (2) absolute pressure, the pressure measured from absolute zero. Thus the difference between absolute pressure and gage pressure is a constant, P_A, having a value of 2,116 psf or 14.7 psi at sea level under normal atmospheric conditions:

$$\text{Absolute pressure} = \text{gage pressure} + \text{atmospheric pressure}$$

A gage pressure of 20 psi would have an absolute pressure of

$$20 + 14.7 = 34.7 \text{ psi}$$

or

$$34.7 \times 144 = 5{,}000 \text{ psf}$$

The pressures used in the perfect gas equations, or laws, are in absolute pressure units; therefore gage pressures must be converted to these units before solving such equations. It is possible to have negative pressures—pressures less than zero—on the gage scale. The maximum possible negative pressure is P_A. The term "vacuum" is used to indicate negative pressures.

A gage pressure of minus 5 psi is equivalent to a vacuum of 5 psi or to an absolute pressure of

$$14.7 - 5 = 9.7 \text{ psi}$$

VOLUME

The volume used in the perfect gas laws is specific volume, or the volume occupied by a unit mass of a substance. Specific volume is expressed in cubic feet per pound. Although specific volume is considered a measurable property, it can only be determined in a laboratory. The inverse of specific volume is density. This is more easily measured; its usual definition is the mass of a unit volume of a substance. The units of measurement are the mass in pounds and the volume in cubic feet. The symbol for density is ρ.

TEMPERATURE

Temperature is a measure of the intensity of molecular energy in a fluid. The absolute zero of temperature is that temperature at which all molecular activity ceases. Although thermodynamically this point is conceivable as the receiver temperature for a heat engine having 100-percent thermal efficiency, it is unattainable in practice. However, the concept of an absolute zero of temperature permits formulation of a basis for a temperature scale depending upon energy relations in ideal, reversible heat-engine cycles, and independent of the characteristics of any particular substance. The perfect gas as defined above has the requisite characteristics of a standard thermometric substance for realizing the energy scale. Temperatures on that scale are known as absolute temperatures. The fact that several permanent gases conform to the perfect gas criterion at zero absolute pressure has enabled physicists to use the constant-volume hydrogen thermometer and helium thermometer as standard instruments for realizing the energy scale of temperature in practical thermometry.

By established custom, commercial thermometers are graduated with scales on which the fundamental interval is 100 degrees on the Centigrade (or Celsius) scale and represents the difference between the temperature of boiling water ($T_1 = 100°$) and the temperature of melting ice ($T_2 = 0°$) at normal atmospheric pressure. On a constant-volume thermometer, these points are in the proportion of 1.3662 to 1 regardless of the number of degrees assigned to the interval. Thus the zero of the Centigrade scale may be computed from

$$\frac{T_2 + 100}{T_2} = 1.3662 \qquad \text{and} \qquad T_2 = 273.1°\text{ C}$$

The ice point is thus 273.1° C above absolute zero, and any Centigrade temperature may be converted to absolute temperature by adding 273.1 to the reading. Similarly, since the difference between the steam point and ice point is 180 degrees on the Fahrenheit scale,

$$\frac{T_2 + 180}{T_2} = 1.3662 \qquad \text{and} \qquad T_2 = 491.6°\text{ F}$$

The ice point is thus 491.6° F above absolute zero and the location of zero on the Fahrenheit scale is (491.6 − 32) = 459.6° F above absolute zero. For most engineering computations, absolute zero is taken as 273° C or 460° F below the zero of the respective scales.

GAS LAWS

Boyle's Law

Robert Boyle's experiments with air showed that if the temperature of a quantity of gas is held constant, the volume of the gas varies inversely with the pressure during a change of state. This is expressed by the equation

$$\frac{P_1}{P} = \frac{V}{V_1} \qquad \text{or} \qquad P_1V_1 = PV = C \text{ (a constant)}$$

This relationship is known as Boyle's Law and is one of the defining characteristics of a perfect gas. Since the density of a gas varies inversely as its volume, it is proportional to the pressure, thus:

$$\frac{P}{\rho} = \frac{P_1}{\rho_1} \qquad \text{or} \qquad \frac{P}{w} = \frac{P_1}{w_1}$$

If a cylinder having a volume of 2 ft³ contains 15 lb of a gas at an absolute pressure of 100 psi and the gas is compressed to half its volume at constant temperature, its pressure will become

$$(100 \times 144)2 = P_2(\tfrac{1}{2} \times 2)$$

$$P_2 = \frac{28{,}800}{1} = 28{,}800 \text{ psf}$$

$$= \frac{28{,}800}{144} = 200 \text{ psia}$$

The density before compression was

$$\frac{15}{2} = 7.5 \text{ lb/ft}^3$$

The density after compression becomes

$$\frac{144 \times 100}{7.5} = \frac{144 \times 200}{w_2}$$

$$w_2 = \frac{200}{100} \times 7.5 = 15 \text{ lb/ft}^3$$

Gay-Lussac's or Charles' Law

Louis-Joseph Gay-Lussac verified experimentally Jacques Charles' theory that gases under constant pressure expand in proportion to an increase in temperature and that all gases have the same coefficient of expansion when heated through the same temperature range. The experiments also proved that the coefficient of expansion at constant temperature is equal to the coefficient of expansion at constant pressure. These coefficients are defined as the ratio of increase of pressure or volume to the pressure or volume at 0° C for a 1° change in temperature. These relationships form another defining characteristic of a perfect gas and are expressed by the following equations:

$$V_t = V_0(1 + \alpha_p t) \qquad \text{and} \qquad P_t = P_0(1 + \alpha_v t)$$

where V_t and P_t are volume and pressure, respectively, at temperature t

V_0 and P_0 are volume and pressure, respectively, at zero temperature

α_p and α_v are the coefficients of expansion at constant pressure and constant volume, respectively.

Combined Gas Law

The above relationships may be combined with Boyle's Law to form a more useful single relation between pressure, volume, and temperature that is known as the combined gas law, or equation of state for a perfect gas. The combined gas law is derived as follows. Consider a mass of gas at initial conditions of P_0, V_0, and $t_0 = 0°$, and that it is heated at constant volume V_0 until it arrives at the state P', V_0, t. Then from the above equations

$$P' = P_0(1 + \alpha t)$$

Next assume that the volume is changed, without changing the temperature, until the gas arrives at a third and final state, P, V, t. From Boyle's law

$$PV = P'V_0 \qquad \text{or} \qquad P' = \frac{PV}{V_0}$$

The intermediate pressure P' can be eliminated by equating its equivalent in each of the above equations, finally obtaining

$$PV = P_0V_0(1 + \alpha t)$$

This equation is rewritten without changing its value by multiplying both sides by $1/\alpha$, thus obtaining

$$PV = P_0V_0\,\frac{(1/\alpha + t)}{1/\alpha}$$

By putting $1/\alpha + t = T$, we introduce a temperature equivalent to t but referred to a zero lying $1/\alpha$ below the ice point. This is the absolute zero of temperature discussed above when defining temperature. Thus, from the definition of absolute temperature, the above equation becomes

$$\frac{PV}{T} = \frac{P_0V_0}{T_0} = R, \text{ a constant}$$

The constant R varies for each gas and with the pressure, volume, and temperature scales adopted as standard. In this work, specific volume is in cubic feet per pound, pressure is in pounds per square foot, and temperature is the absolute temperature in °F. The value of R for dry air is 53.3 ft per °F based on the specific volume at standard atmospheric pressure and 32° F.

The value of R for any other gas may be obtained from the universal gas constant. This constant is the number 1545 and represents the product of the molecular weight and the gas constant R for the permanent gases.

Although this equation is only true for an ideal gas, it is the common custom to employ these property relations in practical computations of the relations for a real gas. The equation is used to compute directly any one of the three properties, P, T, or V of a gas, given the other two and the gas constant of the gas in question. The above equation is for one pound of gas. For W pounds of gas, both sides of the equation must be multiplied by W.

GAS PROCESSES

The manner in which air (or gas) is expanded or compressed directly affects the amount of energy it can deliver or absorb. An ideal reciprocating air compressor with water-cooled cylinders is usually considered to compress air isothermally—that is, at constant temperature—with energy input from shaft work. A positive-displacement blower compresses air explosively without time for cooling and may be considered to compress air adiabatically—that is, without transfer of heat—or energy input from shaft work is retained in the air. In the expansion of air (or gas) through a pipeline or nozzle there is negligible heat transfer and the expansion may be considered adiabatic. The equations required to evaluate the energy change in these processes are derived from the perfect gas laws.

Isothermal Process

The work done by any gas exactly equals the heat imparted to it to maintain its temperature, or, conversely, the work done on the gas exactly equals the heat removed to maintain its temperature. From the perfect gas relation:

$$PV = C, \qquad P = \frac{C}{V}$$

$$\text{Work} = \int P\,dV = C\int_{V_1}^{V_2}\frac{dV}{V} = C\ln\frac{V_2}{V_1}$$

$$\text{Work} = P_1V_1\ln\frac{V_2}{V_1} \qquad \text{or} \qquad P_1V_1\ln\frac{P_1}{P_2}$$

in foot-pound units per pound of gas after expansion.

The factor V_2/V_1 is known as the expansion ratio r. If the expansion is reversed, r is known as the compression ratio.

For isothermal compression, the work required to compress the gas is identical with the work required to receive, compress and deliver the gas.

Since the term $\ln P_2/P_1$ equals $\int_{P_1}^{P_2}\frac{dP}{P}$ for a pressure range $P_2 - P_1$ which is relatively small, the logarithmic term may be replaced by the simpler term $(P_2 - P_1)/P_1$ for values of this ratio that do not exceed 1 percent. Thus for fans and low-pressure blowers, the flow work done by isothermal compression is

$$P_1V_1\left(\frac{P_2 - P_1}{P_1}\right) = (P_2 - P_1)V_1$$

Since the kinetic energy required to accelerate the air is of such relative magnitude in fans and low pressure blowers that it cannot be omitted, this term must be added to the flow work. The total work per pound of gas in foot-pound units therefore becomes

$$\begin{aligned}\text{Work} &= \frac{U_2^2 - U_1^2}{2g} + (P_2 - P_1)V_1\\ &= \text{velocity head} + \text{static head}\end{aligned}$$

Adiabatic Process

The work done by any gas exactly equals the internal energy it gives up since no external heat transfer occurs. Conversely, the work done on a gas exactly equals its increase in internal energy. The specific heat of the gas during the transfer of internal energy is assumed to remain constant.

From the combined gas law, PV = RT. If the pressure remains constant while the volume changes by dV, the temperature will change $P\,dV/R$ and the quantity of heat absorbed is the specific heat at constant pressure, c_p multiplied by the change in temperature $c_pP\,dV/R$. If the volume remains unchanged while the pressure varies by dP, the temperature will change by $V\,dP/R$, and the heat absorbed is the specific heat at constant volume c_v multiplied by the change in temperature $c_vV\,dP/R$.

When both volume and pressure change, the quantity of heat received by the gas is

$$dQ = \frac{1}{R}(c_pP\,dV + c_vV\,dP)$$

For an adiabatic process $dQ = 0$; therefore,

$$(c_pP\,dV + c_vV\,dP) = 0 \qquad \text{or} \qquad \frac{dP}{P} + \frac{c_p}{c_v}\frac{dV}{V} = 0$$

Integrating, $\ln P + c_p/c_v \ln V =$ a constant, but $c_p/c_v = k =$ ratio of specific heats; therefore

$$PV^k = \text{a constant} = P_1V_1{}^k$$

From

$$P_1V_1{}^k = P_2V_2{}^k$$

we obtain

$$\frac{P_1}{P_2} = \left(\frac{V_2}{V_1}\right)^k \qquad \text{or} \qquad \frac{V_2}{V_1} = \left(\frac{P_1}{P_2}\right)^{\frac{1}{k}}$$

Since $P_1V_1/P_2V_2 = T_1/T_2$, inversely, $P_2V_2/P_1V_1 = T_2/T_1$.

Substituting to eliminate V_2/V_1 nets the important equation for temperature rise during adiabatic compression of a gas:

$$\frac{T_2}{T_1} = \frac{P_2P_1{}^{\frac{1}{k}}}{P_1P_2{}^{\frac{1}{k}}} = \left(\frac{P_2}{P_1}\right)^{\frac{k-1}{k}}$$

The final temperature in °F can be obtained from the initial temperature in °F by

$$t_2 = \left(\frac{T_2}{T_1}\right)(t_1 + 460) - 460$$

The work done in expanding from V_1 to V_2 is

$$\int_{V_1}^{V_2} P\,dV = P_1 V_1{}^k \int_{V_1}^{V_2} \frac{dV}{V^k} = \frac{P_1V_1 - P_2V_2}{k-1}$$

in foot-pounds per pound of gas.

The ratio k of specific heats for air and inert gases is 1.4. The expression above can be reduced to either of the following convenient forms:

$$\text{Work} = \frac{P_1V_1}{k-1}\left[1 - \left(\frac{V_1}{V_2}\right)^{k-1}\right] \qquad \text{(used when } P_2 \text{ is unknown)}$$

or

$$\text{Work} = \frac{P_1V_1}{k-1}\left[1 - \left(\frac{P_2}{P_1}\right)^{\frac{k-1}{k}}\right] \qquad \text{(used when } V_2 \text{ is unknown)}$$

Should these expressions work out to be negative, it indicates that work is done on the gas and the process is a compression rather than an expansion.

These expressions represent the total work done in changing volume in a non-flow process within a cylinder. For steady-flow processes that take place in reciprocating compressors, reciprocating engines, or positive-displacement blowers, cognizance must be taken of the flow work involved in receiving and delivering the gas in addition to the work required to compress the gas. Thus, if the cylinder of a reciprocating engine receives a volume V_1 of gas at constant pressure P_1, the gas does work, by transmitted pressure, equal to P_1V_1. The gas supply is then cut off and the gas allowed to expand adiabatically to volume V_2 and pressure P_2, doing work equal to

$$\frac{P_1V_1 - P_2V_2}{k-1}$$

During the whole stroke the back pressure P_2, due to the expulsion of the gas from the other side of the cylinder, opposes the motion and requires work P_2V_2 to be expended. The total available work is

$$V_1P_1 + \frac{V_1P_1 - V_2P_2}{k-1} - V_2P_2$$

Bringing this expression to a least common denominator, it becomes

$$\frac{k}{k-1}\,[P_1V_1 - P_2V_2]$$

For a steady-flow compression process, this expression becomes negative.

HORSEPOWER REQUIRED TO COMPRESS GAS

Power is the rate of doing work; a unit of power is the horsepower. One horsepower is equal to a rate of 33,000 foot-pounds of work per minute (550 ft-lb-sec). Any of the foregoing expressions for work done during expansion (or compression) may be converted to horsepower by multiplying the work per pound of gas by the number of pounds of gas per unit of time and dividing the result by the foot-pound equivalent of one horsepower for the same unit of time.

The horsepower required for an isothermal compression is

$$\frac{MP_1V_1}{33{,}000} \ln \frac{P_2}{P_1}$$

where M = pounds of gas delivered per minute.

Since MV_1 is the volume of gas in cubic feet per minute at the compressor suction, the initial volume may be expressed as Q cubic feet per minute, and if p_1 and p_2 are initial and final absolute pressures, respectively, in pounds per square inch, the horsepower equation becomes

$$\text{hp} = \frac{144}{33{,}000} p_1 Q \ln\left(\frac{p_2}{p_1}\right)$$

If a compressor is rated at 500 cfm and compresses air from 15 psia to 100 psia, the horsepower required for isothermal compression is

$$\frac{144}{33{,}000} \times 15 \times 500 \times \ln \frac{100}{15}$$

$$= 32.5 \times \ln 6.67 = 32.5 \times 1.8976 = 62 \text{ hp}$$

For fans, the horsepower required for isothermal compression is

$$\frac{M[(P_2 - P_1)V_1 + (U_2^2 - U_1^2)/64.34]}{33{,}000}$$

where M = pounds of gas delivered per minute. This expression is sometimes simplified to the "air horsepower" expression

$$\text{Air horsepower} = \frac{CP}{33{,}000}$$

where C = cubic feet of air discharged per minute
P = total pressure (static + velocity) in pounds per square foot.

Since fan pressure differential is usually stated in inches of water (one inch of water is equivalent to 5.19 psf), this expression becomes

$$\text{Air hp} = \frac{\text{total inlet air, scfm} \times \text{pressure differential, in. } H_2O}{6{,}356}$$

The horsepower required to receive, compress adiabatically, and deliver a quantity of gas occupying a volume V_1 before compression is

$$\frac{\frac{k}{k-1} MP_1V_1 \left[1 - \left(\frac{P_2}{P_1}\right)^{\frac{k-1}{k}}\right]}{33{,}000}$$

Again, since MV_1 may be expressed as Q cubic feet per minute and p_1 and p_2 are initial and final absolute pressures, respectively, in pounds per square inch, the horsepower equation becomes

$$\text{hp} = \frac{144}{33{,}000} \frac{k}{k-1} p_1 Q \left[1 - \left(\frac{p_2}{p_1}\right)^{\frac{k-1}{k}}\right]$$

since $k = 1.4$ for air this may be simplified to

$$\text{hp} = 0.0153 p_1 Q \left[1 - \left(\frac{p_2}{p_1}\right)^{0.286}\right]$$

For multistage air compressors having the same inlet temperature at each stage, this equation becomes

$$\text{hp} = 0.0153 N p_1 Q \left[1 - \left(\frac{p_2}{p_1}\right)^{\frac{0.286}{N}}\right]$$

where N = number of stages.

If a positive-displacement blower is rated to compress 500 cfm of free air to 15 psig, the horsepower required for adiabatic compression is

$$\frac{144}{33{,}000}\left(\frac{1.4}{1.4-1}\right) \times 14.7 \times 500 \left[1 - \left(\frac{29.7}{14.7}\right)^{\frac{1.4-1}{1.4}}\right]$$

$$= 112[1 - 2.02^{0.286}]$$

$$= -22.9 \text{ hp}$$

The negative sign indicates that a compression process was involved.

3 Moisture in Conveying Air

The effect of moisture in the conveying air is neglected in most pneumatic conveying installations because the time of contact with the material is very short. In some cases, the quantity of material handled relative to the quantity of conveying air used is so large that the amount of moisture pickup by the material is negligible. However, there are applications where moisture must be removed to prevent degradation of the product, deliquescence, or condensation within pipelines and equipment.

ATMOSPHERIC AIR

Atmospheric air contains moisture in an amount determined by the atmospheric pressure and temperature. It is a mechanical mixture of a volume of dry air, considered as a perfect gas, and the same volume of water vapor jointly and independently occupying the same space. All the properties of this mixture may be determined from two equations, a table of saturated vapor properties, and two physical constants. The two equations are (1) the perfect gas law derived in Chapter 2 and (2) Dalton's law of partial pressures, which states that the total pressure of the mixture is equal to the sum of the partial pressure of the gas (dry air) plus the partial pressure of the water vapor. The table of saturated vapor properties is a standard steam table for saturated steam at various temperatures. The two physical constants are (1) the gas constant R, defined in Chapter 2 and (2) the specific heat of a gas at constant pressure c_p. These data are used to construct a psychrometric chart for standard atmospheric pressure.

PSYCHROMETRIC CHARTS

The properties shown on the psychrometric chart (see Figures 3–1 and 3–2) are in terms of an arbitrarily defined unit of one pound of dry air plus the weight of water vapor associated with it. A psychrometric

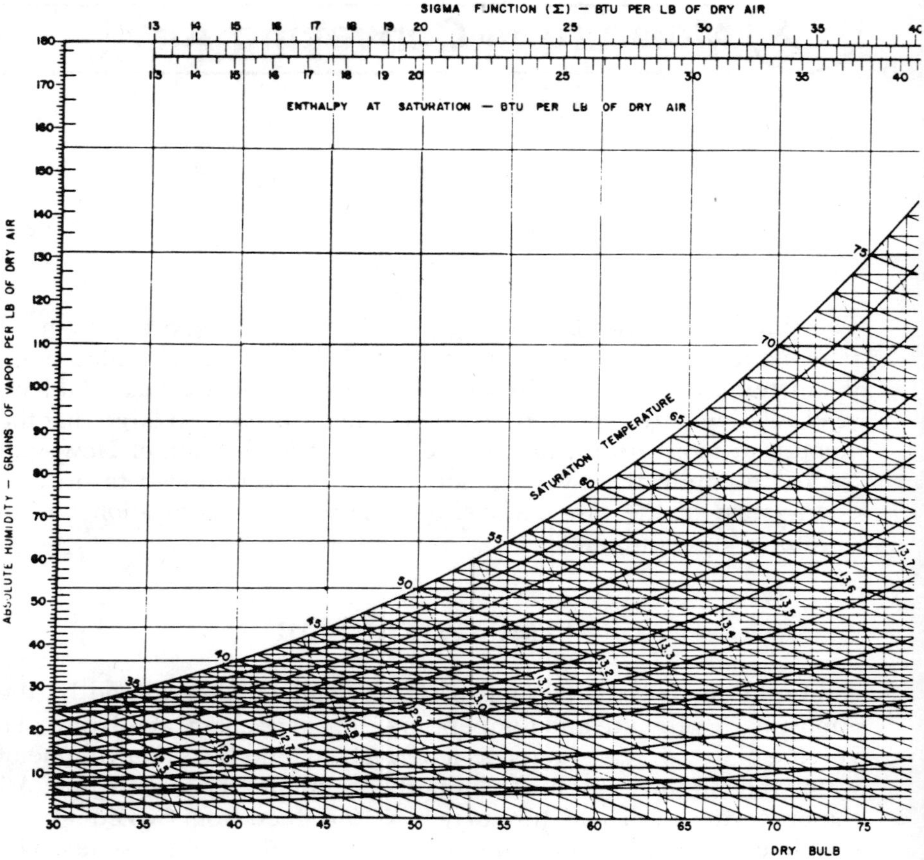

FIGURE 3–1. Standard psychrometric

chart reduces to a minimum the effort needed to calculate atmospheric air properties since a knowledge of the numerical value of two properties permits fixing the state of the vapor and the determination of all other properties at that state. The two properties most often used to determine the moisture content are the dry- and wet-bulb temperatures of the air. The dry-bulb temperature is simply the temperature reading of an ordinary thermometer exposed to the air. The wet-bulb temperature is the temperature reading of a thermometer with a wetted, wick-covered bulb that is exposed to a stream of air. The two thermometers may be combined to form a sling psychrometer that is whirled rapidly in the air; readings of each thermometer are taken immediately after stopping the motion. The two readings are used

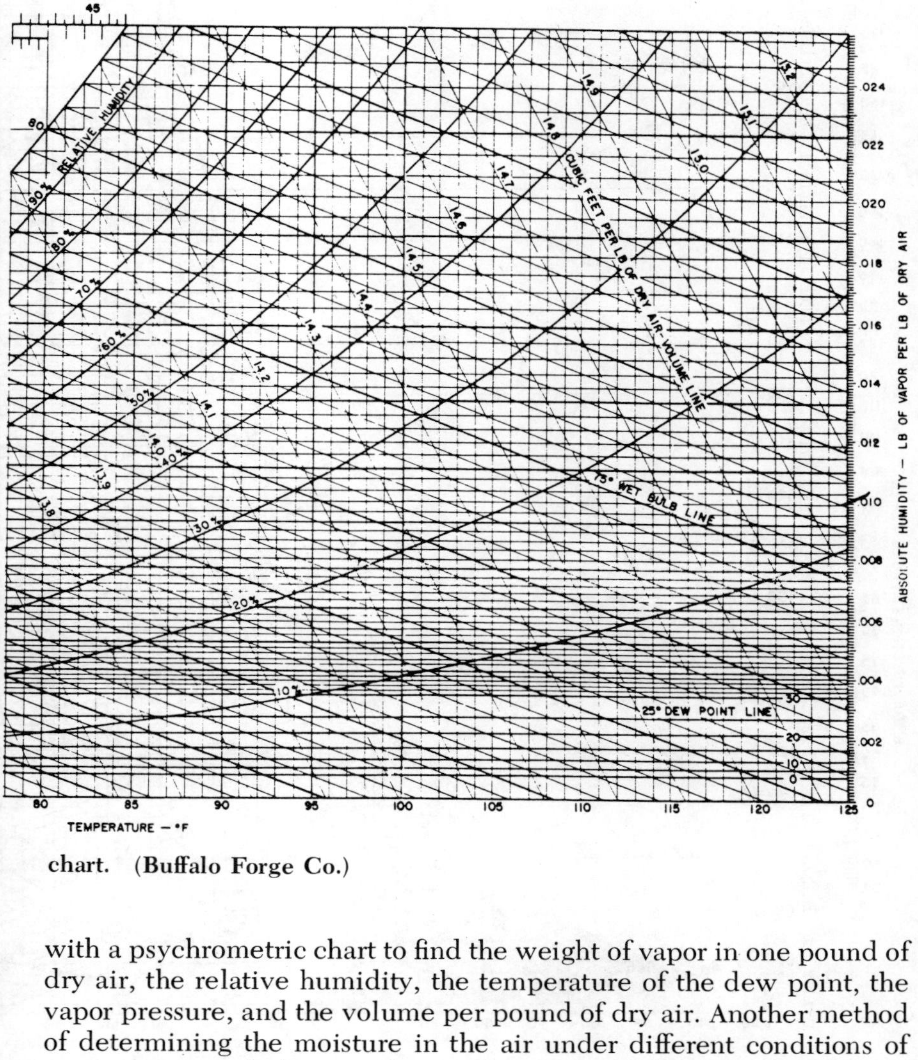

chart. (Buffalo Forge Co.)

with a psychrometric chart to find the weight of vapor in one pound of dry air, the relative humidity, the temperature of the dew point, the vapor pressure, and the volume per pound of dry air. Another method of determining the moisture in the air under different conditions of temperature and degree of saturation is to use the same two psychrometer readings and refer to *Psychrometric Tables,* issued by the Weather Bureau (but available from the Superintendent of Documents, Washington, D.C.) to obtain the dew point, vapor pressure, relative humidity, and the weight of a cubic foot of vapor. These tables are prepared for use at several barometric pressures, ranging from 30 inches to 23 inches of mercury, and are therefore useful at high elevations.

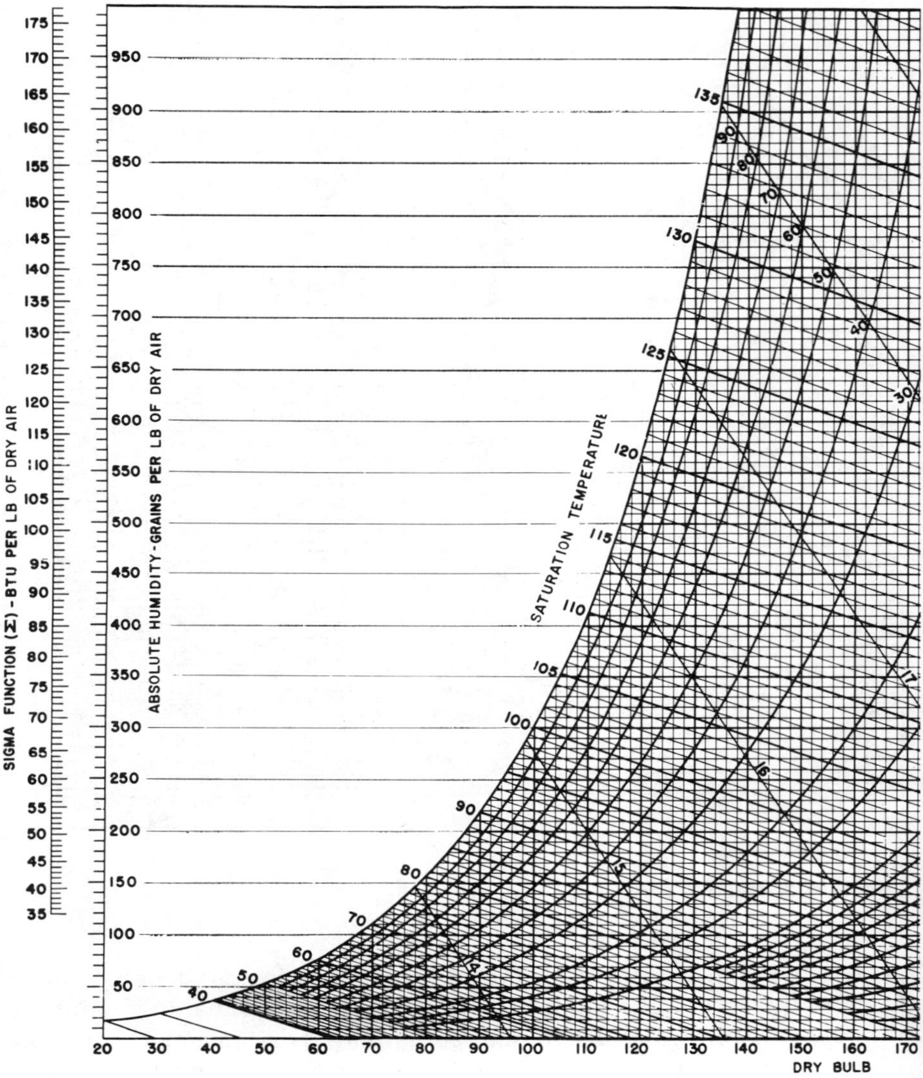

FIGURE 3–2. High-temperature

DEW POINT

The maximum amount of water vapor that can occupy the same space as a given volume of dry air is entirely a function of its temperature. The temperature fixes its vapor pressure. The amount of vapor is given in the saturated steam tables. The air-vapor mixture at this con-

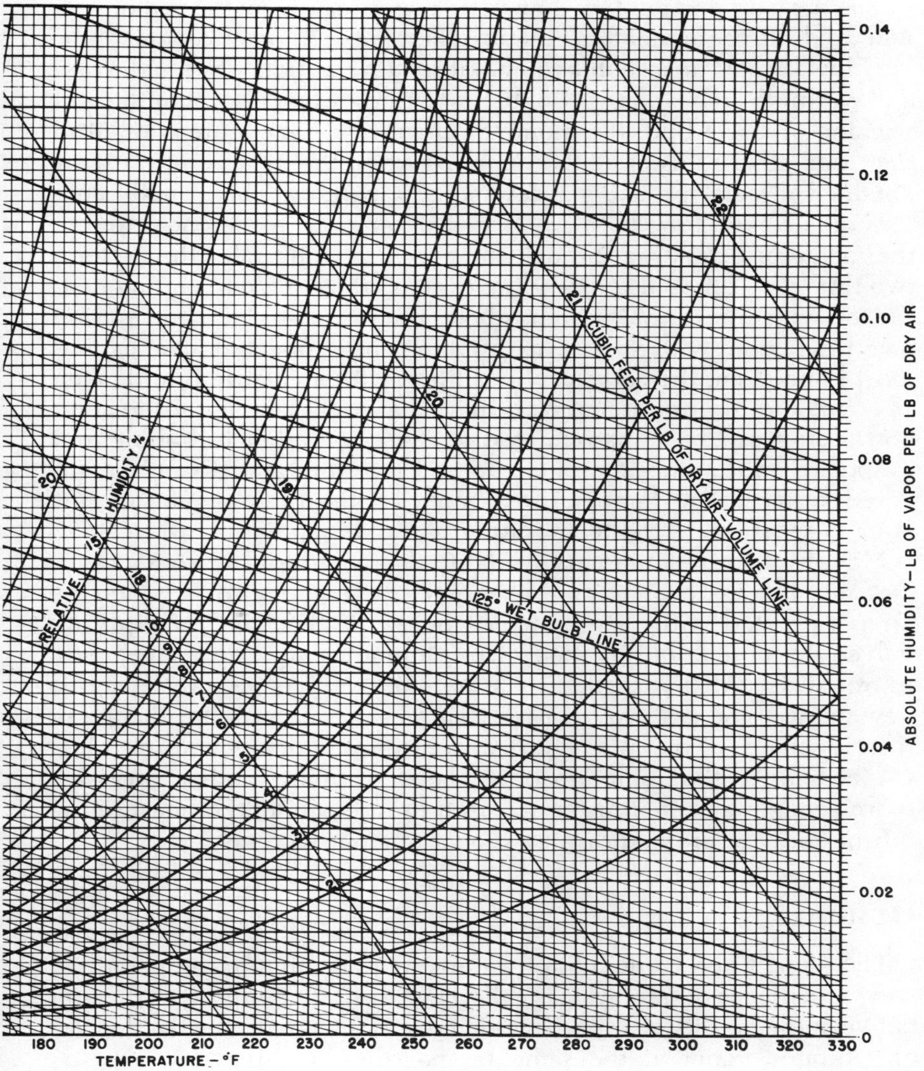

psychrometric chart. (Buffalo Forge Co.)

dition is saturated with heat and, incidentally, moisture. Any removal of heat at a given vapor pressure will cause condensation of saturated vapor. The corresponding temperature is known as the dew point. Any addition of heat at the given vapor pressure will increase the temperature of the mixture and cause it to expand, thus increasing the temperature of the dew point and superheating the vapor.

The measurement of low dew points on dried air by the use of wet- and dry-bulb thermometers is almost impossible in practice, due to the very low, almost unreadable temperature differential between the dry- and wet-bulb indications. Also, the moisture evaporated from the wet wicking of a wet-bulb thermometer changes the moisture content of the dried air. In this case, the temperature of the dew point must be obtained directly by using a dew-point apparatus. This usually consists of: a corked tube with a polished silver surface cemented to it; a thermometer extending through the cork to a point near the bottom; two thin glass tubes also extending through the cork, one ending near the top with an external rubber tube to remove gases, and one ending near the bottom with an external rubber tube connected to an aspirator. In operation, the corked tube is filled with a volatile liquid, such as ether, and inserted into a well-lighted air duct near an observation port. The aspirator is manipulated to cause the ether to bubble and evaporate until a fine film of condensed moisture appears on the silvered surface. The temperature shown by the thermometer at this point is regarded as the temperature of the dew point.

Modern instrumentation is available that can draw air from a dryer and measure its dew point or frost point (the temperature at which the air is saturated with respect to ice below 32° F). The principle involves the optical observation of the formation of a dew layer on a mirror surface that is chilled by a thermoelectric cooler to the actual dew point of the gas. An optical sensing bridge detects the change in the reflectance of the mirror surface and proportionally controls electric current to the cooler to track the dew point. Both a precision thermocouple and thermistor are used to independently measure the mirror temperature and display it on a readout meter.

HUMIDITY

The weight of a cubic foot of water vapor at different temperatures and percentages of saturation is called the absolute humidity. The ratio between the density of superheated vapor in the mixture to the density of saturated vapor at the same temperature is called the relative humidity (R.H.). This ratio is also the ratio of the grains of moisture contained in a cubic foot of mixture at a given temperature to the grains of moisture that could be contained in a cubic foot of mixture if it were saturated.

Example. What is the temperature of the dew point, vapor pressure, specific volume, and absolute humidity of atmospheric air at 75° F and 50% R.H.?

Solution. From steam tables, the vapor pressure of saturated water vapor at 75° F is 0.4298 psia and its specific volume is 740 ft³/lb.

From the definition of relative humidity, the specific volume of the vapor at 50% R.H. is 2 × 740 = 1,480 ft³.

Inspection of steam tables indicates that this volume corresponds to that of saturated steam at 54° F. Therefore 54° F is the temperature of the dew point and the pressure of the superheated vapor is 0.2064 psia.

The specific volume of the mixture is

$$V = \frac{RT}{P} = \frac{53.3(75 + 460)}{(14.696 - 0.2064)144} = 13.7 \text{ ft}^3/\text{lb}$$

The absolute humidity of the mixture is

$$\frac{13.7}{1{,}480} = 0.00925 \text{ lb vapor/lb dry air}$$

$$= 0.00925 \times 7{,}000 = 64.8 \text{ grains vapor/lb dry air}$$

AIR DRYING

From the information above, it is obvious that moisture in the air must be removed by cooling the air below the temperature of the dew point, or by lowering the vapor pressure of the air. Simple cooling does not change the specific humidity and is ineffective for moisture removal unless the temperature is reduced below the dew point. The three methods used for moisture removal are as follows:

1. Surface Dehumidification. The air is passed over a pipe coil with a surface temperature reduced to the desired dew point by the flow of cold water, refrigerated water, or a refrigerant. The condensed moisture drops into a pan and overflows to a drain. Any mist picked up by airflow is removed by impingement on a series of separator baffles.

2. Adsorption. The air is passed through a bed of silica gel or activated alumina where vapor condensed on the surface of the adsorbent is drawn into capillary openings, thereby reducing the vapor pressure at the surface. The pressure gradient causes a mass transfer of vapor to the working surface. As the capillaries fill with water, the rate of dehumidification decreases. The latent heat of vaporization of the condensed vapor raises the temperature of the adsorbent and the process of adsorption evolves heat so that there is a net increase in temperature of the discharged dry air. When dehumidification ceases, the moisture must be extracted from the adsorbent bed by direct heating of the bed, or by passing heated air through it, thus regenerating it. For continuous drying of air, two drying beds are used, arranged to operate alternately on an automatically timed cycle, one bed is drying air while the other bed is being regenerated (Figure 3–3). A continuously rotating drum dehumidifier is available that rotates slowly, passing the adsorbent through sealed chambers, so that part of the bed is actively drying air while other parts are being regenerated and cooled.

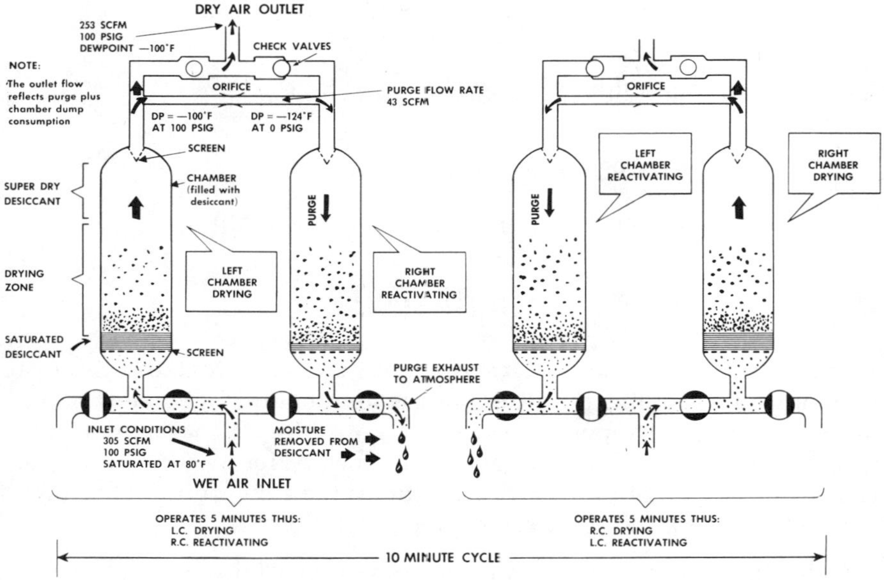

FIGURE 3–3. Operation of a heatless air dryer. (Pall Trinity Micro Corporation, a subsidiary of Pall Corporation.)

3. ***Absorption.*** The air passes through a spray of a lithium chloride or calcium chloride salt solution that reduces the vapor pressure below the saturation pressure of water vapor so that moisture is absorbed by the solution. As dehumidification occurs, the concentration of the salt solution decreases and the vapor pressure increases. The salt solution must be reconcentrated by evaporation in external equipment, or by overflowing a portion of the solution at a constant rate, equal to the moisture extraction rate, and by adding salt. The heat of solution tends to raise the temperature of the discharged air. The dried air must pass through moisture impingement baffles before entering a duct system.

MOISTURE IN HIGH-PRESSURE AIR

Conveying systems using compressed air above 20 psig must be equipped with moisture-removal equipment beyond the compressor discharge connection in order to prevent direct injection of water into the conveying line. The mass ratio of an air-vapor mixture is unchanged during compression, and saturation occurs at a relatively low pressure. Continued compression to a higher pressure condenses more moisture, which is discharged into the receiver and downstream equipment. In addition, the heat of compression must be removed by the use of air- or water-cooled inter- and after-coolers. These heat exchangers compound the problem, since the moisture will condense within them and must be removed. The air discharged from each of these heat exchangers should be passed through a moisture separator equipped with an air trap in the drain line that can discharge the accumulated moisture. Since all ordinary moisture separators are not 100-percent effective, an air dryer should be located in the air discharge line from an air receiver. The air dryer may be simply a closed chamber filled with an adsorbent that is regenerated periodically. The outside of the chamber is wrapped with electrical heating cable. During regeneration, the cover of the chamber is removed and heat is applied to drive off the adsorbed moisture. After regeneration, the heat is turned off and the cover is replaced. On large compressed air systems, a continuously operated air dryer—one of the types described above—would be required to dry the air.

DETERMINING THE AMOUNT OF CONDENSED MOISTURE

The usual procedure for determining the amount of moisture condensed during the compression of air to a high pressure is to refer to a curve showing moisture content of saturated air at various temperatures and pressures. See Figure 3–4. Since this curve presumes that the entering air and discharged air are at the same temperature and saturated with vapor, the calculation used to prepare the curve is relatively simple. The procedure used is as follows:

1. Obtain the pounds of moisture per cubic foot of air at atmospheric pressure and at the given air temperature by referring to the saturated steam table for the specific volume of the saturated vapor at the air temperature; the reciprocal of this volume is the weight of moisture in a cubic foot of air.
2. Note the absolute vapor pressure listed in the steam table at the

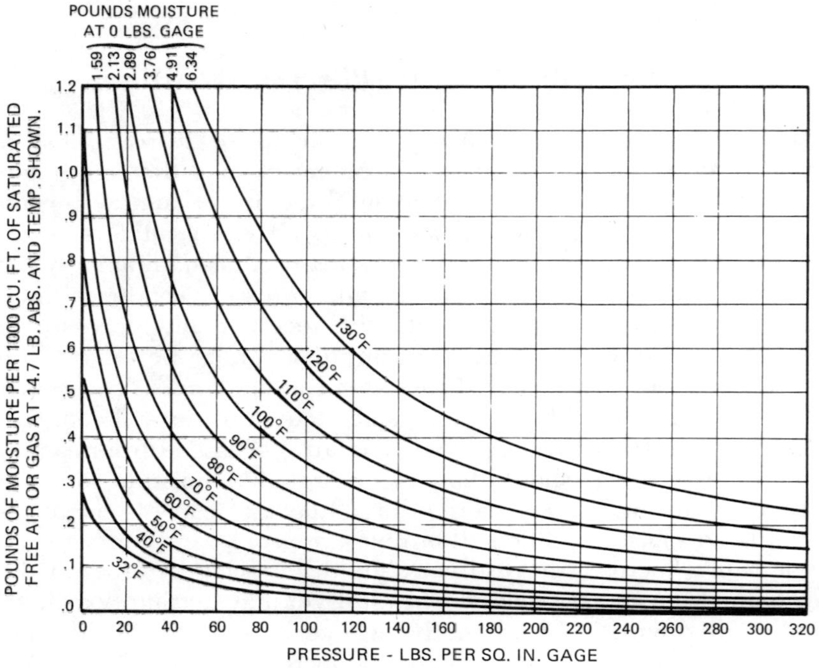

FIGURE 3–4. Moisture content of saturated air at various temperatures and pressures. (Compressed Air and Gas Data 1971.)

inlet air temperature and subtract this from the absolute inlet and discharge pressures of the compressor to obtain the partial pressure of the air at these points.

3. Multiply the weight of moisture in a cubic foot of air at the inlet by the ratio of the partial air pressure at the inlet to the partial air pressure at the discharge to obtain the pounds of moisture in a cubic foot of air at discharge conditions.

4. Subtract the moisture at discharge conditions from the moisture at inlet conditions to obtain the pounds of moisture condensed per cubic foot of dry air.

This procedure is an approximate calculation since the moisture content is not proportional to the partial air pressures at inlet and outlet and also because the inlet air is not always saturated. However, the procedure is simple and the results obtained are on the safe side for sizing moisture removal equipment.

Example. Saturated atmospheric air at a temperature of 90° F enters a compressed air system and is discharged at a pressure of 100 psig and 90° F. How many pounds of moisture is removed per 1,000 cu ft of free air?

Solution. Reference to a steam table shows the following:

Saturation: Temperatures

	Abs. Pressure		Specific Volume		
Temp. Fahrenheit	Lb/ Sq in.	In. Hg	Sat. Liquid	Evap.	Sat. Vapor
90°	0.6982	1.4215	0.01610	468.0	468.0

Lb moisture/ft³ of air at inlet = 1/468 = 0.00214

Partial pressure of air at inlet = 14.7 − 0.6982 = 14.0018 psia

Partial pressure of air at outlet = (100 + 14.7) − 0.6982 = 114.0018

Lb moisture/ft³ of air at discharge

= 14.0018/114.0018 × 0.00214 = 0.0002625

Moisture condensed per 1,000 ft³ of air

= 1,000(0.002140 − 0.000262) = 1.878 lb

MOISTURE IN INERT GAS

Inert gas is formed by the perfect combustion of a simple hydrocarbon fuel such as natural gas or propane. These gases are of the form C_xH_y and the composite chemical reaction equation may be formed from the following:

$$C_xH_y + \left(X + \frac{Y}{4}\right) O_2 + 3.78\left(X + \frac{Y}{4}\right) N_2$$

$$= XCO_2 + \frac{Y}{2} H_2O + 3.78\left(X + \frac{Y}{4}\right) N_2$$

As in all chemical equations, the relative volume of each gas is proportional to the coefficient of its chemical symbol. The relative weight of each component is proportional to the product of its coefficient and the molecular weight of the component. Thus, for natural gas, which has the molecular form CH_4, the equation becomes

$$CH_4 + 2O_2 + 7.56N_2 = CO_2 + 2H_2O + 7.56N_2$$

This indicates that one volume of natural gas requires 2 volumes of oxygen from the air to form one volume of CO_2 and two volumes of water vapor. The nitrogen is inert and appears on both sides of the equation unaffected by the combustion process. The equation also shows that, gravimetrically, the following amounts of each molecular component are involved in the process:

$$(12 + 4) + (2 \times [16 \times 2]) + (7.56 \times [14 \times 2]) = (12 + [16 \times 2]) + (2 \times [2 + 16]) + (7.56 \times [14 \times 2])$$

thus

$$16 \text{ units } CH_4 + 64 \text{ units } O_2 + 211 \text{ units } N_2 = 44 \text{ units } CO_2 + 36 \text{ units } H_2O + 211 \text{ units } N_2$$

The units may be expressed in terms of any weight scale, but the same scale must be used consistently throughout the equation. The molecular weights are formed by combining the atomic weights of each component. The atomic weights commonly used in gas combustion calculations are as follows:

Carbon	12
Hydrogen	1
Oxygen	16
Nitrogen	14

In the above equation, the inert gas consists of nitrogen and carbon dioxide. The water vapor is condensed in a heat exchanger and removed from the gas by separators.

Similarly for propane gas, which has the molecular form C_3H_8, the composite chemical reaction equation becomes

$$C_3H_8 + 5O_2 + 18.9N_2 = 3CO_2 + 4H_2O + 18.9N_2$$

It is important to note that these gases form large amounts of water vapor, which exists as superheated steam at combustion temperatures. A large portion of this may be condensed and removed when cooled to ambient temperatures, but the portion that is not cooled will have a high dew point and may condense in a conveying line unless removed by cooling, by refrigeration, or by a dessicant prior to entrance into the line.

From these equations, it may be observed that one volume of natural gas produces 8.56 volumes of inert gas, whereas one volume of propane gas produces 21.9 volumes of inert gas—more than 2½ times as much as natural gas. Also, although a unit of propane gas produces twice as much water vapor by volume as the same amount of natural gas, the percentage by weight of water vapor is less. The water vapor produced by each gas is: natural gas—12.38 percent; propane—9.82 percent.

The amount of moisture remaining in the inert gas at standard pressure and temperature may be estimated as follows using the products of combustion of natural gas as an example:

$$CO_2 \text{ produced} = \frac{44}{16} = 2.75 \text{ lb/lb fuel}$$

$$H_2O \text{ produced} = \frac{36}{16} = 2.25 \text{ lb/lb fuel}$$

$$N_2 \text{ produced} = \frac{211}{16} = 13.2 \text{ lb/lb fuel}$$

$$\text{Total products} = 18.20 \text{ lb/lb fuel}$$

Gaseous constituents 18.20 − 2.25 = 15.95 lb/lb fuel. Partial pressure of vapor in products if at or above dew point is

$$P_M \times \frac{M_x/m_x}{M_x/m_x + M_y/m_y + M_z/m_z}$$

where P_M = total pressure of mixture
M_x, M_y, M_z = mass of each constituent
m_x, m_y, m_z = molecular weight of each constituent.

Therefore

$$p_x = 14.7 \times \frac{2.25/18}{2.25/18 + 2.75/44 + 13.2/28}$$

$$= 14.7 \times \frac{0.125}{0.6585} = 2.79 \text{ psi}$$

The following data is obtained from steam tables: the saturation temperature of water vapor (dew point of products mixture) at 2.79 psia = 138.65° F; the partial pressure of vapor and condensate at

70° F = 0.36 psia and the volume of vapor = 867.9 ft³/lb. Therefore, the partial pressure of residual gases at 70° F = 14.7 − 0.36 = 14.34 psia and the volume of residual gases at 70° F and 14.34 psia = MR_mT/P. Equivalent R_m for the mixture is

$$\frac{1{,}545(2.75/44 + 13.2/28)}{15.95} = \frac{1{,}545(0.5335)}{15.95} = 51.6$$

(1,545 is the universal gas constant). Therefore

$$V = \frac{15.95 \times 51.6 \times 530}{14.34 \times 144} = 211.5 \text{ ft}^3$$

This volume is also the volume of the water vapor in the products at 70° F; thus the specific volume of the vapor must be

$$\frac{211.5}{2.25} = 94 \text{ ft}^3/\text{lb}$$

and the quality of the vapor becomes

$$100 \times \frac{94}{867.9} = 10.82 \text{ percent}$$

4 Pilot Plant and Field Testing

TEST SETUP

A typical pilot plant for pneumatic conveying tests comprises an assembly of equipment as follows:

1. Several loops of piping of convenient size to handle relatively small samples of material. The piping is usually arranged with flexible hoses and quick-couplings for rapid changes in conveying line length and for switching to different discharge points. The length of line could be varied from 50 feet to as much as 400 feet in increments of 50 to 100 feet, as desired. Valved hose connections for plant air are located at strategic locations in the piping to permit blowing out a plugged line. Occasionally, a duplicate piping arrangement, one or two sizes larger, may be installed for testing samples for large capacity systems.

2. A feed hopper discharging into a rotary airlock feeder, or into a blow tank. A discharge valve is connected to the feeder or blow tank to prevent blow-back of high-pressure air when unplugging the conveying line. The hopper is vented to a dust filter and may be equipped with a sight port to observe the feeding action.

3. A motor-driven air-supply blower or compressor with a variable speed transmission. The discharge piping includes a check valve to prevent back flow of material from the feeder, a relief valve to protect the blower, and pressure and temperature indicators to determine air properties. The blower inlet piping is equipped with an air filter and an airflow meter. The flow meter is usually a head meter—a venturi tube, a flow nozzle, or an orifice plate—equipped with a manometer for measuring the pressure differential.

4. A receiver. This may be a weigh hopper vented to a dust filter, or simply a dust filter, when a scale-mounted blow tank is used to feed the conveying line.

5. Switching connections for recycling material from the dust filter outlet to the feed hopper in order to establish steady-flow conditions, or to discharge material to the weigh hopper. See Figure 4–1.

The test setup for a vacuum system differs from that described

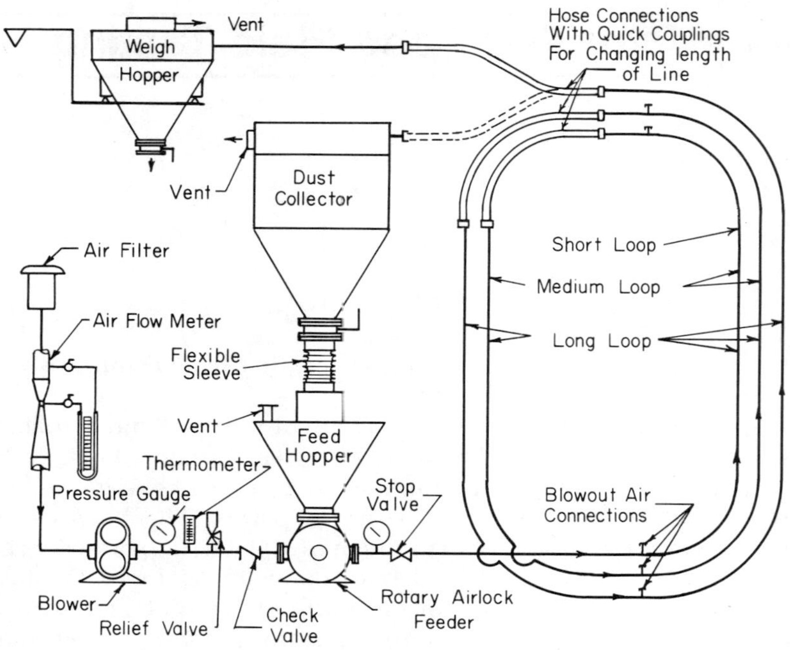

FIGURE 4–1. Typical pilot plant arrangement for a pressure system.

above. Material may be drawn into the line by suction, and dropped out of a separator into a drum for weighing, a damper in the suction line to the exhauster may be used to regulate airflow, or the exhauster speed may be changed. The discharged air from the exhauster is metered to obtain the airflow. Figure 7–15 shows test data from a controlled-feed blow-tank that was mounted on a pneumatic weight transmitter, or so-called load cell.

TEST PROCEDURES

The test procedure varies with the vendor and the scope of the test. It may consist of an operator dumping a five gallon pail of the material into a feed hopper while the observer watches it disappear and the vendor's representative shouts "There it goes!" The purpose of this test is to show that the material can be conveyed. After the observer is gone, detailed data will be obtained by the vendor so that he can prepare his bid. These data will become the vendor's property.

A blow-tank test may be completely automated with airflow, air pressure, and material-discharge rate recorded on a chart and presented to the customer together with the recommended line size and the airflow and pressure requirements for a full-size system.

A pressure system may be tested by the "brute force" method. Here, material will be dumped into a feed hopper and elaborate data taken for a complete range of flow rates, feeder speeds, blower pressures, and line lengths. A graphical plot of derived data will show a wide disparity in material-to-air ratios and superficial air velocities. The plotted points will be scattered and will show no orderly or systematic relationships between the variables. Selection of the optimum material-to-air ratio and air velocity is thus made difficult.

When the proposed layout of a system is known, the length of the test circuit should be as close to the proposed length as possible and include as many bends as may be required. The ratio of vertical piping to horizontal piping should be similar if possible, but this is not necessary, since the effect of vertical lift on pressure drop is easily estimated.

When no layout is available, the test must be performed using the various lengths that can be assembled from the available test loops. Tests may be run on circuits with equivalent lengths of 50, 100, 200, 300, and 400 feet.

When it is proposed to increase the capacity of an existing system or to design a similar new system to convey the same material, data from the existing plant may be used in preference to obtaining pilot-plant test data.

When it is proposed to utilize an existing system to convey a different material than that for which it had been designed, tests for capacity may be devised and the system used for testing. However, each component in that system should be analyzed and equipment for the proposed new system selected with due regard to the difference in the physical properties and flow characteristics between the two materials. See Chapter 5.

When a friable material must be conveyed with a minimum of degradation, additional bends may be installed into the test loop, or the material may be recycled several times through the loop in order to obtain test samples for particle-size analysis. A sample from each cycle through the loop would be checked for percentage of fines below a specified screen size.

PERFORMING THE TEST

The test procedure should be performed systematically when possible. Usually all that is required is sufficient data for scale-up when conveying at the maximum possible material-to-air ratio and the lowest air velocity that will ensure complete clean-out of the pipeline without settling of material within the horizontal sections of the conveying line, or drop-out in the vertical sections of the line. The material-to-air ratio during conveying is not always equal to the ratio of material feed to the airflow as shown on the flow meter. Cognizance must be taken of air leakage, particularly when the system is equipped with any form of airlock feeder that is operated at a high pressure differential.

The objective of the test is to obtain the maximum material flow and the minimum airflow that can be passed through the test loops without plugging the system or leaving residual material within the system, particularly at the bends. The associated pressure drop through each test loop would be measured up to the limit imposed by the available air mover. These data would then be scaled-up in order to size the conveying line and the air mover to meet the capacity requirements of a proposed system. The pressure drop for airflow only must be determined before feeding material into the system, since this is the datum above which conveying conditions are determined. This pressure drop is maximum when there is no material feed into the line. Any rise in pressure after starting material flow has negligible effect on the airflow-only pressure drop, since pressure drop varies inversely as the absolute pressure in the pipeline.

Air supply from a positive-displacement blower is recommended in order to maintain a relatively constant superficial air velocity at the pickup point for all rates of material flow. The data obtained can then be plotted in order to obtain the constant for the modified Fanning equation described in Chapter 1.

When centrifugal fans or blowers are used to supply air during a test, it is almost impossible to maintain constant velocity at the pickup point because the airflow follows the characteristic curve of the air mover as the resistance in the pipeline changes. (See Chapter 13.) When a fan is used, the pressure drop in the test loop must be obtained for several rates of airflow, with and without material flow. The ratio of specific pressure drop to material-to-air ratio for each rate of airflow would then be used to scale up the test data.

Conveying tests furnish data only for scaling up the size of the conveying pipeline. Other factors required for determining the air-mover pressure and airflow requirements must be estimated from pressure-drop and air-leakage data on the actual components that will be installed in the proposed system. These items include the following: air-mover inlet and discharge filters, silencers, flow nozzles, dehydrators, rotary or double-gate airlock feeders, and all piping and pipeline components that handle air only and no material. The layout of the air-supply, vent, and conveying lines is required in order to obtain complete data.

Conveying Pressure and Air Leakage

All airlock feeders leak air. The amount of leakage can only be estimated by treating the clearance spaces as an equivalent orifice area with a low discharge coefficient. A test feeder may leak more air than a new feeder of the same design due to the effects of previous testing with other materials. A check of feeder clearances should be made if excessive dusting and backflow of air is observed at the material inlet or feed hopper. In addition, the pressure at the entrance to the conveying line should be obtained and used to calculate actual conveying air volume rather than using the air pressure at the discharge connection of the air-supply blower. The pressure within the blow tank of an air-into-material system is essentially the line entrance pressure. The discharge pressure at the end of the conveying line is normally assumed to be atmospheric, but, when the material enters a dust filter under negative pressure, a correction should be made in the calculation for the total pressure drop in the conveying line. Each bend in a pipeline is assumed to be equivalent to a number of feet of straight piping, the allowance will vary with the type of system under test, the vendor's experience, and the size of the line. No attempt is made to segregate the pressure drop in vertical and horizontal portions of a test line. Despite theoretical considerations that indicate a higher pressure requirement for vertical conveying than for horizontal conveying at a given rate of material flow, experience with some materials may indicate otherwise, since velocities must be maintained high enough to prevent settling out of materials in a long horizontal line.

Material Control

The quantity of material dumped into the feed hopper should be weighed and all material accounted for, after it is blown through the line. This accounting is most important on materials that may smear and build up in the pipe bends. The dust filters should be brought to equilibrium conditions during the recycle phase of the test. They should be cleaned in the normal manner prior to conducting the data-collection phase of the test. The dust collected during the test should be weighed after each conveying run. The interior of the dust filter should be inspected for hang-up of material and appearance of the filter fabric.

During the recycle phase of the test, the rates may be varied in order to determine the range of conditions required to obtain smooth flow—flow without surges, sluggishness or plugging of the line. When friable, tacky, abrasive or any other problem material is tested, it may be desirable to make a prolonged conveying run in order to determine if the feeder performance and mechanical clearances are adversely affected by action of the material, or if buildup occurs in the conveying line. A larger than normal sample must then be provided for the test since recycling may destroy the properties of the material that may be the cause of these adverse effects. It should be remembered that the residence time of material in a conveying line is very short, and that recycling may therefore produce conditions that would never occur in an actual system: friable materials may become completely pulverized; abrasive materials may become degraded; sharp corners of pulverized particles may become rounded and the particles less abrasive; tacky materials may become surface conditioned by the air and lose their tackiness; etc.

Blow-Tank Procedures

Some of the testing techniques described above are applicable to blow-tank systems, except those that apply to recycling the material. The operation of a blow tank is essentially intermittent and no airlock feeder is used. The data for a blow-tank test must be obtained quickly since a 25-cubic-foot test tank may be completely discharged in several minutes and a 2-cubic-foot test tank within one minute. A complete record of the events occurring within the tank is required and may be obtained by elaborate test equipment using high speed recording charts; by a series of simultaneous observations using several technicians to take data at short intervals; or by a combination of observed and recorded data. The essential test data required is the variation of airflow into the tank starting from the air injection period up to the air cut-off point; air pressure variation during the complete discharge cycle; the variation in weight of material in the tank during the filling and discharge periods; and the total elapsed time of each of these events. A sight port is installed on some test tanks for observing the manner in which material enters the conveying line and to detect lumps blocking the entrance to the line. Some blow tanks require auxiliary air, which is admitted through nozzles as a transport jet in the conveying line, or as fluidizing jets located in the side of the tank. The pressure required to operate the jets must be determined and the amount of air used by them must be accounted for and added to the total air requirement for a blow-tank system.

SCALE-UP OF TEST RESULTS

The method of scaling up test results for sizing a conveying line is essentially the same regardless of the type of system tested. The known factors are the size of the test line, its length, the rate of flow (it should be the same, if possible) of free air entering the line—with and without material—the rate of flow of material through the line, and the pressure at the entrance to the line or the pressure drop across the line with and without material. These data may meet the exact requirements for a relatively small system and, in this case, no scale-up is necessary. However, a larger-capacity system will require scale-up of the test data.

The first step is to determine the superficial velocity of the airflow at the material pickup point from the given data: airflow, in cubic feet of free air, corrected for the density increase caused by blower pressure and temperature increase, divided by the internal area of the line, in square feet. The total air volume must be reduced by the known or estimated leakage through the airlock feeder, if one is used to feed material into the piping.

Determine next the new pipeline size as follows: From the weight rate of material flow and the area of the test pipe, calculate the mass rate of flow in pounds per second per square foot of pipe area. The ratio of the required rate of material flow to the mass rate of flow is the area of the scaled-up pipeline in square feet. Compare this area with the areas of commercial sizes of piping and tubing and select the size that corresponds closely with the required area.

Determine the actual pressure drop for conveying material by subtracting the airflow-only pressure drop from the total pressure obtained at maximum material load. Use this pressure and the known equivalent length of the test loop to calculate the pressure drop per foot of pipe. Since pressure drop varies inversely as the ratio of internal pipe diameters, multiply the pressure drop per foot of test pipe by the ratio of test pipe diameter to scaled-up pipe diameter in order to obtain the scaled-up pressure drop per foot of pipe.

If the selected size of pipe differs from that calculated from the mass rate of flow, the actual conveying rate will be increased or decreased depending on whether the selected size is larger or smaller, respectively, than the calculated size. If it is desired to maintain the design conveying rate, the scaled-up pressure drop per foot must be recalculated, using the ratio of the calculated area of the pipe to the selected area of the pipe, since mass flow is inversely proportional to pipe area.

The adjusted pressure drop may then be multiplied by the total equivalent length of the proposed conveying line in order to obtain the total pressure required to convey the material through the line. An estimate for airflow-only pressure drop through the scaled-up line must be made and added to this in order to determine the scaled-up airflow requirements.

The scaled-up airflow requirement at the material pickup point is determined from the calculated velocity and the internal area of the selected pipe. This flow may be converted to airflow in terms of cubic feet of standard air so that the estimate for airflow-only pressure drop may be made using the method shown on page 15; or by reference to air-pressure drop tables in any compressed air data book.

For sizing the air mover required for the proposed system, the scaled-up airflow and pressure requirements calculated above must be increased as described previously in order to take cognizance of all components that constitute the conveying system.

Example. Given the following data from a test in which there was no settling of material:

Test line—300-ft-long, 3-in. Schedule 40 pipe including the equivalent resistance of four 90° bends having a radius of 12 times the nominal diameter.

Airflow—230 cfm at standard conditions (14.7 psi absolute pressure, 70° F temperature) as read from a venturi meter at inlet to the blower.

Material flow—300 lb/min of a semiabrasive material used in a sanitary product.

Blower data—Discharge pressure at inlet to the conveying line: air only, 1 psi; with material at maximum rate, 13.25 psi; air temperature 130° F.

Required: To size a line to handle 15 tons/hour of the same material over a straight-line distance of 350 ft. The line will have five bends with a radius of 12 times the nominal diameter. Determine the scaled-up airflow and pressure requirements.

Solution. *For conveying velocity:* The required conveying velocity is computed from the actual volume of air at the entrance to the test conveying line. Since no airlock feeder loss was given, assume that it is 10 percent of the measured airflow. The cross-sectional area of a 3-in. Schedule 40 line is 0.0153 ft^2.

$$\text{Actual volume of air at inlet} = 230 - 23 = 207 \text{ cfm standard air}$$

Volume at pipe conditions is

$$207 \times \frac{144 \times 14.7}{144\,(14.7 + 13.25)} \times \frac{460 + 130}{460 + 70} = 207 \times 0.526 \times 1.113$$

$$= 121.2 \text{ cfm}, \ \frac{121.2}{0.0513} = 2{,}362 \text{ fpm}$$

For pipe size: From the cross-sectional area of the test pipe and the material flow rate, the mass flow is

$$\frac{300}{60 \times 0.0513} = 97.5 \text{ lb/sec/ft}^2$$

The cross-sectional area required to handle 15 tons/hour is

$$\frac{15 \times 2{,}000}{60 \times 60 \times 97.5} = 0.085 \text{ ft}^2$$

The nearest commercial sizes of stainless steel pipe and tubing are

4-in. Schedule 40 pipe—0.0884 ft²
5-in. Schedule 40 pipe—0.1389 ft²
4-in. Schedule 10 tubing—0.099 ft²
5-in. Schedule 10 tubing—0.153 ft²

Select 4-in. Schedule 10 tubing as closest to the required size and because stainless steel is best for sanitation and abrasion resistance.

For pressure drop per foot in 4-in. tubing: Pressure drop in test line, internal diameter—3.068 inches, is

$$\frac{13.25 - 1.00}{300} = 0.0408 \text{ psi/ft}$$

Pressure drop in 4-in. tubing, internal diameter—4.26 inches, is

$$0.0408 \times \frac{3.068}{4.26} = 0.0294 \text{ psi/ft}$$

Adjust for mass flow required to maintain conveying rate

$$0.0294 \times \frac{0.085}{0.099} = 0.0252 \text{ psi/ft}$$

For air pressure: The air pressure required at the entrance to the 4-in. Schedule 10 line when handling 15 tons/hour is obtained as follows:

Total equivalent length of 4-in. tubing is

$$350 + 5 \times 1.6\left(\frac{2\pi\ (4 \times 12)}{12 \times 4}\right) = 350 + 50.26 = 400.26 \text{ ft}$$

Pressure drop when conveying material is

$$400.26 \times 0.0252 = 10.08 \text{ psi}$$

Assume pressure drop for airflow only is 0.3 psi until the actual airflow is determined; the tentative pressure at the pickup point is therefore

$$10.08 + 0.3 = 10.38 \text{ psi}$$

For airflow in 4-in. tubing: The airflow at the pickup point is calculated from the required conveying velocity and the area of the tubing

$$2{,}362 \times 0.099 = 233.84 \text{ cfm}$$

The pressure drop for airflow only may be obtained by converting this flow to standard air and referring to Figure 1–5 as follows:

$$233.84 \times \frac{14.7 + 10.38}{14.7} \times \frac{460 + 70}{460 + 130}$$
$$= 233.84 \times 1.706 \times 0.898 = 358.3 \text{ scfm}$$

From Figure 1–5, pressure drop per 100 ft piping at this flow is 5.8 in. of water at 14.7 psia. At 10.38 psi (25.08 psia) the total pressure drop is

$$5.8 \times \frac{14.7}{25.08} \times 4 \times 0.036 = 0.49 \text{ psi}$$

(0.036 is the psi equivalent of 1 in. of water.)
The final pressure at the pickup point is

$$10.08 + 0.49 = 10.57 \text{ psi}$$

The previous calculations need not be corrected, since the difference between the assumed and calculated pressure drop for airflow only will have little effect on the scaled-up airflow.

ESTIMATING AIRLOCK LEAKAGE

The conveying line air pressure and flow requirements are thus established, but the size of the air mover must include an allowance for leakage through the airlock feeder. This may be estimated as airflow through the clearance space of the full size unit treated as an equivalent orifice. The air pumped out of the system as the feeder rotates must be added to this. This is estimated by multiplying the volume of air displaced per revolution, corrected for pressure and temperature conditions within the system, by the number of revolutions of the feeder per unit time.

The amount of air displaced by the feeder per revolution depends upon the type of feeder used. A double-entry feeder would have its pockets filled with material for almost three-quarters of a revolution with some air trapped because of leakage past the vanes. After the material is discharged, the vanes are filled with air at some pressure between the full discharge pressure and the feed hopper pressure. Thus a one cubic foot per revolution, 12 vane feeder, rotating at 30 rpm and discharging material into a line at 15 psig would have an estimated pocket volume air leakage of

$$0.25\ \text{ft}^3 \times 30\ \text{rpm} \times \frac{29.7}{14.7} \times \frac{530}{590} = 13.6\ \text{scfm}$$

Drop-through and blow-through feeders would have greater pocket volume leakage because less material is trapped between the material inlet and outlet nozzles. Also, the pockets between these points are only partially full, the amount depending on the judgment of the system designer. (See Chapter 9.)

Additional leakage is caused by the higher-pressure air flowing directly through the radial and axial clearance spaces between the rotor and its housing. The leakage is assumed to occur through the clearance between the housing and the projected area of the rotor. No cognizance is taken of the fact that the tips of the rotor vanes act as labyrinth seals so that leakage decreases progressively at each pocket between the feeder-outlet and the material-inlet connection. Some rotors are equipped with shrouds that close off the ends of the pockets. These shrouds may be free or may have partial or full peripheral seals bearing on them in order to reduce peripheral leakage of air.

If the feeder described above had an 11.25-in. × 18-in. rotor without shrouds and vane-end and vane-tip clearances of 0.002 and 0.003 in., respectively, the leakage area would be

$$2 \times 0.002\ \text{in.} \times 11.25\text{-in. rotor diameter} = 0.045\ \text{in.}^2$$

$$2 \times 0.003\ \text{in.} \times 18.00\text{-in. vane length} = 0.108\ \text{in.}^2$$

$$\text{Total clearance area} = 0.045 + 0.108 = 0.153\ \text{in.}^2$$

The amount of air leakage is estimated by treating this clearance area as a very inefficient orifice and substituting values in some simplified equation for the flow of a compressible gas through a circular orifice. For the range of airlock-pressure differentials used in pneumatic conveying, the following equation, though intended for maximum flow at critical pressure, is the simplest form that gives results closely agreeing with many other equations intended for less-than-critical flow

$$W = 0.323\, a\, C_d \sqrt{\frac{p_1}{V_1}}$$

where a = clearance area, in.^{2}
p_1 = high pressure, psia
V_1 = specific volume at the higher pressure, ft^3/lb
C_d = coefficient of discharge

The coefficient of discharge for a circular orifice is usually taken as 0.6 because of the uncertainty of the location of the vena contracta. For estimating purposes, this coefficient may be used to reflect the reduction of flow caused by the devious airflow path through the airlock.

The estimated amount of leakage for the above feeder due to clearance is determined as follows:

$$V_1 = 13.33 \times \frac{14.7}{29.7} \times \frac{590}{530} = 7.344 \text{ ft}^3/\text{lb}$$

(13.33 ft^3 is the specific volume of one pound of standard air)

$$W = 0.323 \times 0.153 \times 0.6 \sqrt{\frac{14.7 + 15.0}{7.344}}$$

$$W = 0.06 \text{ lb/sec} \qquad \text{or} \qquad 13.33 \times 0.06 \times 60 = 48 \text{ ft}^3 \text{ standard air}$$

$$\text{Total leakage} = 13.6 + 48 = 61.6 \text{ scfm}$$

In terms of the loss in volume at the conveying-line entrance pressure, this becomes

$$61.6 \times \frac{14.7}{29.7} \times \frac{590}{530} = 34 \text{ cfm}$$

Total estimated maximum leakage is 61.6 cfm. This amount is much greater than the actual leakage since the tips of the rotor blades act as labyrinth seals and the leakage pressure decreases progressively in each pocket from the feeder outlet to the material inlet connection.

When a gate lock is used, the leakage is estimated by multiplying the volume between the upper and lower gates, corrected for pressure and temperature conditions within this space, by the number of discharge cycles per unit time and adding this to the estimated leakage through the clearance spaces between the gates and their housing.

SCALEUP OF BLOW-TANK TESTS

The maximum capacity of a blow-tank system is fixed by its volume and the time available for its discharge in accordance with the formula

$$C = \frac{3\rho V_2}{100t}$$

where C = capacity, tons/hr, ρ = loose density, lb/ft^3, V_2 = blow-tank volume, ft^3, and t = total cycle time, min. The total cycle time, in this case, is also the total discharge time. The pipeline size, working volume and total cycle time are so interrelated that a trial and error approach is used to obtain a balanced design that would allow sufficient time in the cycle for filling and, where required, for air pressure recovery. A standard-size blow tank may be selected based on space requirements, cost considerations or previous experience. The discharge line is sized for the same entrance velocity and mass flow as the test line as well as for a discharge period compatible with the capacity requirements and available air pressure as in the following example.

Example. A 10-ft^3 controlled-feed blow tank discharged 400 lb of a particular material through 150 ft of 1½-in. piping in 2 min. The test curve showed an average pressure of 8.2 psig during the blow period and 1.5 psig during clean-out of the line (air only). Air temperature averaged 130° F. A flowmeter recorded an airflow of 68 cfm of free air during this period. Size a blow tank and piping to discharge 30,000 lb/hr of the material through a line 320 ft long. Loose density of the material is 75 lb/ft^3.

Solution. Try a 25-ft^3 tank:

$$\text{Total cycle time} = \frac{3 \times 75 \times 25}{100 \times 15} = 3.25 \text{ min}$$

This may not allow enough time for filling or for air pressure recovery. Try a 35-ft^3 tank.

$$\text{Total cycle time} \times \frac{3 \times 75 \times 35}{100 \times 15} = 5.25 \text{ min}$$

The discharge time must be less than this in order to allow time for filling. The entrance air usage and velocity during the test averaged

$$\text{Usage} = 68 \times \frac{14.7}{22.9} = 43.6 \text{ cfm at } 8.2 \text{ psig}$$

$$\text{Velocity} = \frac{43.6}{0.01414} = 3.082 \text{ fpm}$$

Material-to-air ratio during the test was

$$\frac{400 \times 13.35}{68 \times 2} = 39.3 \text{ lb material/lb air}$$

The mass rate of flow on test was

$$\frac{400}{2 \times 60 \times 0.0141} = 236.4 \text{ lb/sec/ft}^2$$

Allowing 1.25 min for filling and pressure recovery, the discharge time would then become 4 min. If the weight capacity of the tank is 2,625 lb, the line size for the same mass flow and the required discharge rate becomes

$$\frac{2625}{4 \times 60 \times 236.4} = 0.0463 \text{ ft}^2$$

The nearest standard sizes are

2½-in. Schedule 40 pipe—0.0332 ft^2
3-in. Schedule 40 pipe—0.0513 ft^2

The 2½-in. size would increase the discharge time as follows:

$$0.0332 \times 236.4 \times 60 = 471 \text{ lb/min}$$

$$\frac{2{,}625}{471} = 5.57 \text{ min}$$

This exceeds the total cycle time. The 3-in. size would decrease the discharge time as follows:

$$0.0513 \times 236.4 \times 60 = 727 \text{ lb/min}$$

$$\frac{2{,}625}{727} = 3.61 \text{ min}$$

Use 3-in. Schedule 40 pipe.

For Pressure Drop Per Foot in 3-in. Pipe. Pressure drop per foot in 1½-in. test line, 1.61-in. inside diameter (airflow-only pressure drop is 1.5 psig):

$$\frac{8.2 - 1.5}{150} = 0.0447 \text{ psi/ft}$$

Pressure drop per foot in 3-in. pipe, 3.068-in. inside diameter:

$$0.0447 \times \frac{1.61}{3.068} = 0.0234 \text{ psi/ft}$$

Adjust for mass flow required to obtain the conveying rate

$$0.0234 \times \frac{0.0463}{0.0513} = 0.0212 \text{ psi/ft}$$

For Air Pressure at Entrance to Line. The air pressure required in the 3-in. Schedule 40 line when handling 30,000 lb/hr is obtained as follows:

Pressure drop when conveying material is

$$320 \times 0.0212 = 6.784 \text{ psi}$$

Assume pressure drop for airflow only as 0.7 psi until the actual airflow is determined. Tentative pressure at the pickup point becomes

$$6.784 + 0.7 = 7.484 \text{ psi}$$

For Airflow in 3-in. Schedule 40 Line. The airflow at the pickup point is calculated from the conveying velocity and the pipe area:

$$3{,}082 \times 0.0513 = 158.1 \text{ cfm}$$

The pressure drop for airflow only may be obtained by converting this to standard air and referring to Figure 1–5 as follows:

$$158.1 \times \frac{14.7 + 7.484}{14.7} \times \frac{460 + 70}{460 + 130}$$

$$= 158.1 \times 1.509 \times 0.898 = 214.25 \text{ scfm}$$

From Figure 1–5, pressure drop at this flow is off the chart. At 200 scfm, the pressure drop per 100 ft piping at this flow is 10 in. of water; at 214.25 scfm, the pressure drop is

$$\left(\frac{214.25}{200}\right)^{1.86} \times 10 = 11.37 \text{ in. of water}$$

At 7.484 psig (22.184 psia), the total pressure drop is

$$11.37 \times \frac{14.7}{22.184} \times 3.2 \times 0.036 = 0.867 \text{ psi}$$

(0.036 is psi equivalent of 1 in. of water.)

The final pressure at the pickup point is

$$6.784 + 0.867 = 7.65 \text{ psig}$$

Final selection would be a 35-ft^3 tank discharging into a 3-in. Schedule 40 pipeline, requiring an average flow of 215 scfm of air and operating at an average pressure of 7.65 psig within the tank.

SCALE-UP OF AIR-LIFT TEST

Air lift is a variant of the negative-pressure vacuum system used to elevate friable materials that would be degraded by impact in bends. The test setup would consist of a vertical pipe having a bell-mouth entrance at the bottom for drawing in air and a side entrance just above it for a material conveyor. The top of the pipe discharges into a disengaging chamber designed to reduce the velocity of the material and release the air with a minimum capture of fine particles of material. The released air is drawn through a fabric dust filter, a control damper, and a fan that has a steep operating characteristic curve. (See Chapter 13.) The disengaged material is dropped through an airlock discharge gate into a bulk container or into a conveyor that delivers the material to a process. The airflow may be measured on the discharge side of the fan by a Pitot tube traverse or by a venturi meter.

Pressure drop is measured at two points on the lift pipe: at the top, close to the disengaging chamber inlet; and at the bottom, about 2 ft above the material entrance opening. Pressure readings may be taken across the control damper and the fan for record, but these do not enter into the scale-up calculations. The material conveyor is usually a natural-frequency vibrating conveyor with an adjustable vertical gate built into it at the entrance opening. The conveyor is calibrated by weighing out the discharged material for several positions of the gate before starting up the fan. The airflow measurements in the fan discharge duct are used to establish the superficial velocities within the airlift pipe.

Example. Given the following data from a test in which velocities were adjusted until incipient fallout of material was observed:

Test pipe—4-in. inside diameter; 51 ft between top and bottom pressure taps; bottom tap, 2 ft above material entrance opening.
Airflow—Determined by setting superficial velocities for four values.
Material flow—Determined from four conveyor gate settings

Superficial Air Velocity, ft/sec	Solids Feed Rate, lb/min	Pressure Tap Readings, in. of water			
		P_1 Air Only	P_1 Mixture	P_2 Mixture	$P_1 - P_2$ Mixture
44.5	22.3	0.4	1.6	4.5	2.9
50.5	31.0	0.7	2.3	6.0	3.7
57.0	50.0	1.0	3.4	8.6	5.2
67.3	71.2	2.0	5.8	13.0	7.2

Required: To size an airlift pipe to elevate 16,000 lb/hr of a friable material to a height of 120 ft.

Solution. *Plot data on graph paper.* A plot of solids feed rate versus superficial velocity shows a linear relationship between these coordinates. A plot of $P_2 - P_1$ versus superficial velocity also shows a linear relationship between the coordinates. A plot of P_1 versus superficial velocity for air only and for the mixture is a straight line only when plotted on logarithmic paper; therefore the scale-up for entrance loss must be determined separately from the pipe loss.

A separate particle-size analysis of material collected from each test showed high attrition above 57 ft/sec velocity; therefore scale-up will be based on this value as a limit.

For pipe size: Mass flow rate at 57 ft/sec is

$$\frac{50}{60 \times 0.0872} = 9.556 \text{ lb/sec/ft}^2$$

The cross-sectional area required to lift 16,000 lb/hr is

$$\frac{16{,}000}{3{,}600 \times 9.556} = 0.465 \text{ ft}^2 \qquad \text{(about 9.25-in. diameter)}$$

The nearest commercial duct size is 10-in. diameter (0.545 ft²). Use 10-in.-diameter stainless steel duct in order to prevent rust contamination.

For airflow in 10-in. duct: From the test velocity, the airflow is

$$57 \times 60 \times 0.545 = 1{,}864 \text{ cfm}$$

For pressure drop per foot in 10-in. duct: Pressure drop in test line, internal diameter—0.333 ft, for airflow only is obtained as follows:

$$\text{Velocity} = 57 \times 60 = 3{,}420 \text{ fpm}$$
$$\text{Airflow} = 3{,}420 \times 0.0872 = 298 \text{ cfm}$$
$$\text{Pressure drop per 100 ft of pipe, from Figure 1–5} = 5 \text{ in. of water}$$
$$\text{Pressure drop in 51 ft of pipe} = 5 \times 0.51 = 2.55 \text{ in. of water}$$

Pressure drop per foot in 4-in. pipe is

$$\frac{5.20 - 2.55}{51} = 0.052 \text{ in. of water/ft}$$

Pressure drop per foot in-10 in. duct is

$$0.052 \times \frac{0.333}{0.833} = 0.021 \text{ in. of water/ft}$$

Adjust for mass flow to maintain the conveying rate

$$0.021 \times \frac{0.465}{0.545} = 0.018 \text{ in. of water/ft}$$

For air pressure (vacuum): The negative pressure required in the 10-in. duct when lifting 16,000 lb/hr of material is obtained as follows:

Duct loss, airflow only, from Figure 1–5 is 1.6 in per 100 ft at 2,000 cfm and 3,420 fpm. At 1,864 cfm, the loss per 100 ft is

$$\left(\frac{1{,}864}{2{,}000}\right)^{1.86} \times 1.6 = 0.877 \times 1.6 = 1.4 \text{ in. of water}$$

$$\text{Loss in 117 ft of duct} = 1.4 \times 1.17 = 1.64 \text{ in. of water}$$

$$\text{Duct loss for the mixture} = 0.018 \times 117 = 2.11 \text{ in. of water}$$

$$\text{Total loss in straight duct} = 2.11 + 1.64 = 3.75 \text{ in. of water}$$

Entrance loss in the 10-in. duct would be the same as in the 4-in. duct for the same entrance velocity and for the same geometrical configuration of bell-mouth entrance nozzle and side opening for material entry. However, correction must be made for the reduced material-to-air ratio required to maintain the conveying rate in the 10-in. duct.

The material-to-air ratio on test was

$$\frac{50}{298/13.33} = 2.24$$

The material-to-air ratio in the 10-in. duct is

$$\frac{16{,}000/60}{1{,}864/13.33} = 1.91$$

Specific pressure drop at entrance, from test data, is

$$\frac{3.4}{1.0} = 3.4$$

Since specific pressure drop is directly proportional to material-to-air ratio, the scaled-up specific pressure drop is

$$\frac{1.91}{2.24} \times 3.4 = 2.90 \text{ in. of water}$$

$$\text{Duct loss} + \text{entrance loss} = 3.75 + 2.90 = 6.65 \text{ in. of water}$$

Fan selection: Since the airflow must be maintained constant at 1,864 cfm, the pressure drop within the airlift would vary between 2.64 in. of water for airflow only and 6.65 in. of water at full material load. However, additional pressure drop must be estimated in order to determine the static pressure required for a fan to serve this system. The following items must be considered: (1) pressure drop in the disengaging chamber, (2) pressure drop in the fabric dust filter, (3) pressure drop in the ductwork between the air-separating equipment and the fan, (4) pressure drop across the control damper that will be required to obtain the system operating point on the characteristic curve of the fan, and (5) pressure drop across the fan discharge duct and venturi tube required to control the damper. The total pressure drop of these items may be two to three times the air-lift pressure drop.

When the total fan static pressure has been estimated, a study must be made of the characteristic curves for a series of wheel sizes of fans that have a steep operating characteristic—small change in flow with large change in pressure. Considerable judgment must be exercised in reconciling operating conditions with the available fan curves. (See Chapter 13.)

Where the conveying velocity is not limited because of attrition of the material, the selection of a fan is simplified, since the damper may be eliminated. However, the no-load airflow will determine the horsepower required to drive the fan.

AIRFLOW METERING

One of the most important factors in the testing of pneumatic conveying systems is the measurement of airflow into the system. Airflow in pilot plant work is measured by so-called head meters in which the flow through a reduced cross section of a pipe is accelerated, producing a large reduction in static pressure at that section. The static pressure differential between the inlet to the reduced section and at the reduced section becomes a measure of the loss in pressure energy and can be equated to the gain in velocity energy in accordance with Bernoulli's Theorem, assuming constant density:

$$\frac{P_1}{w} + \frac{v_1^2}{2g} = \frac{P_2}{w} + \frac{v_2^2}{2g}$$

$$\frac{P_1}{w} - \frac{P_2}{w} = \frac{v_2^2 - v_1^2}{2g}$$

From the equation of continuity, the quantity of fluid flowing through each cross section is the same:

$$A_1 v_1 = A_2 v_2$$

from which

$$v_1 = \frac{A_2}{A_1} v_2 = n v_2$$

where $n = A_2/A_1$. Substituting this value of n in the energy equation above and combining terms, the velocity at the reduced section is found to be

$$v_2 = \sqrt{2g\left(\frac{P_1}{w} - \frac{P_2}{w}\right)\left(\frac{1}{1-n^2}\right)}$$

The theoretical discharge, or quantity of fluid flowing through the meter, is

$$Q_2 = A_2v_2 = A_2\sqrt{2g\left(\frac{P_1}{w} - \frac{P_2}{w}\right)\left(\frac{1}{1-n^2}\right)}$$

where Q_2 is measured in cubic feet per second when the areas are given in square feet, pressures in pounds per square foot, and density in pounds per cubic foot, at the pressure, temperature, and humidity at the primary element.

When $P_1 - P_2$ is given in inches of water, this expression may be simplified to

$$Q_2 = 1096.4cA_2\sqrt{\left(\frac{\Delta P}{w}\right)\left(\frac{1}{1-n^2}\right)}$$

where c is an experimental discharge coefficient.

Example. If a venturi tube having a 3-in. inside line diameter and a 2-in. throat diameter shows 8 in. on a differential-pressure gage when air is entering at 7 psig and 150° F, what is the airflow?

Solution.

A_1 = area of 3-in. inlet = 0.0491 ft²

A_2 = area of 2-in. throat = 0.0218 ft²

$$n = \frac{A_2}{A_1} = \frac{0.0218}{0.0491} = 0.444$$

$n^2 = 0.197$

$c = 0.98$ for a venturi meter

$$w = 0.075 \times \frac{530}{610} \times \frac{21.7}{14.7} = 0.0962 \text{ lb/ft}^3$$

$$Q = 1096.4 \times 0.98 \times 0.0218\sqrt{\left(\frac{8}{0.0962}\right)\left(\frac{1}{1-0.197}\right)}$$

$$= 23.4\sqrt{83.2 \times 1.247}$$
$$= 23.4\sqrt{103.7} = 238 \text{ cfm}$$

Primary Elements

For actual flow, the theoretical discharge must be modified by an experimental coefficient that accounts for friction and imperfect energy conversion. This coefficient of discharge reduces the theoretical flow. Its value varies with the type of primary element used to reduce the cross section. The three types of primary elements are, in order of decreasing efficiency, the venturi tube, the bell-mouthed flow nozzle, and the thin-plate orifice.

The venturi tube is a single unit in which the constricted section, or throat, is located between a short converging inlet and a long diverging outlet. Its efficiency of energy conversion is very high and, although its discharge coefficient varies with Reynolds number, it may be taken as 0.98 for the usual range of application. It is always installed directly in a pipeline with a long length of straight section preceding it.

The flow nozzle is a short cylinder with a flared inlet leading tangentially into it. In effect, it is equivalent to a venturi tube without the diverging outlet. Its coefficient of discharge is between 0.95 and 0.98, but varies with the type—long-radius A.S.M.E. nozzle or an I.S.A. short nozzle.

The thin-plate or sharp-edged orifice is a circular hole in a thin flat diaphragm that can be inserted between pipe flanges. There is no guided entrance that can form the fluid to a definite reduced cross section, therefore the fluid shapes its own cross section, or vena contracta, beyond the orifice plate. The difficulty in locating the smallest cross section of the stream for a static pressure tap accounts for the large variation in the coefficients of discharge used with this element. Three types of tap locations are used, each having its own coefficient of discharge that is also a function of Reynolds number and the ratio of orifice diameter to pipe diameter. Flange taps are located one inch from the upstream face of the orifice and one inch from the downstream face. The coefficient of discharge varies from about 0.80 to 0.60. Vena contracta taps are located about one pipe diameter from the upstream face of the orifice and the downstream tap is located at the position of minimum pressure as estimated from a curve showing location as a function of diameter ratio (see A.S.M.E. "Fluid Meters Report"). The coefficient of discharge varies from about 0.82 to 0.60.

Pipe taps are located about 2½ pipe diameters from the upstream face of the orifice plate and 8 pipe diameters downstream from the same face. The coefficient of discharge ranges from over 1.30 to 0.60.

The flow nozzle and the orifice plate may be located at the end of a pipe or in the side of a large receiver when measuring airflow into the atmosphere.

ADIABATIC FLOW

When the pressure differential is small, as it usually is when inserting a primary metering element in a pipeline, the assumption that the air density remains constant is valid. However, where the pressure differential is high and the change in air density will affect the accuracy of the flow calculation, the adiabatic flow formula for compressible gases must be used to obtain the exact flow.

The derivation of the adiabatic flow formula is rather lengthy and involves the same basic procedure used to derive the flow formula for air at constant density, namely, equating the change in kinetic energy to the change in internal energy and flow work, assuming that the air neither receives nor imparts energy from or to surrounding bodies.

$$v_2 = \left[\frac{2gP_1V_1 \left(\frac{k}{k-1} \right) \left(1 - r^{\frac{k-1}{k}}\right)}{1 - \left(\frac{A_2}{A_1} \right)^2 r^{\frac{2}{k}}} \right]^{1/2} \text{ ft/sec}$$

where r is the pressure ratio P_2/P_1.

CRITICAL PRESSURE

The rate of flow of air or gas through an orifice or nozzle is a maximum when the upstream pressure is more than twice the downstream pressure. The point of minimum pressure to which an expansive fluid will expand in passage through an orifice or nozzle is called the critical pressure. The critical pressure is $P_{cr} = .53\ P_1$, and, regardless of how much lower the downstream pressure may be, the weight rate of flow will remain constant. If the critical pressure ratio is introduced into the flow formula, the following equation results for rate of flow of air through a circular orifice:

$$W = \frac{0.532ap}{\sqrt{T}}$$

where W = rate of flow, lb/sec
a = area of orifice, in.^{2}
p = upstream pressure, psia
T = upstream absolute temperature, °F.

PRESSURE MEASUREMENT IN AIRFLOW METERING

The primary metering devices depend on the correct measurement of static pressure for accuracy. For airflow through a pipe or duct, the pressure at the wall is measured by a U-tube manometer in which air pressure is balanced by the height of the column of liquid that it can support. The tube connection is made flush with the inside of the wall and in a section of pipe where flow is known to be parallel to the wall. This eliminates errors in readings due to eddy currents caused by protrusions or components of flow velocity at changes in direction. The specific gravity of the liquid within the manometer can be selected to suit the height of available instruments and the pressure range to be measured. Manometer fluids range from mercury, with a specific gravity of 13.6, to organic compounds varying from ethyl alcohol with a specific gravity of 0.789, to acetylene tetrabromide with a specific gravity of 2.96. The zero of a manometer scale is set at the lowest liquid level and the vertical height of column read in inches of fluid up to the free surface. This is then converted to pressure in any convenient units by using the required conversion factors. The density of air on the pressure side of the manometer is so low compared to the density of the manometer fluid that its effect is neglected. The standard manometer always indicates the pressure difference between the atmosphere and the pressure within a pipe or duct.

The differential pressure between two points in a pipeline, or across a primary metering device, may be obtained directly by using a differential manometer instead of measuring the static pressure at each section. In this device, each leg of the U-tube is connected to the pressure tap on each side of the primary element, and the pressure differential measured by the vertical height between the free surfaces of the manometer fluid. One leg of the U-tube may be inclined at an angle to magnify the differential height, but it is the vertical height that determines the head, or inclined scale reading. The use of a differential reading for head on flowmeters is considered to be a more accurate procedure than measuring the difference between separate static readings at each pressure tap.

AIRFLOW BY PITOT TUBE MEASUREMENT

The measurement of airflow in large pipelines or ducts by the standard metering devices may not be feasible due to the configuration of the piping system or the shape of the ducts. In this case, a Pitot-static tube is used to measure velocity at a particular section indirectly and the flow calculated from the known cross sectional area.

The Pitot tube consists of two concentric tubes with the outer tube tapered and sealed against the inner tube at one end. The tapered end is bent at right angles for insertion through the wall of a duct so that the open end of the inner tube can be faced upstream, parallel to the wall, to measure impact pressure. The outer tube has several small side holes through the wall, back of the tapered section, for measuring static pressure. The protruding end of the assembly has the total pressure connection located in line with the tube and the static pressure connection located at right angles to the tube.

The tubes may be connected individually to a manometer for static pressure readings in which case the center tube will measure

$$\text{Total pressure} = \text{static pressure} + \text{velocity pressure}$$

and the outer tube will measure only static pressure. A differential manometer connected to these tubes will read velocity pressure directly.

Pitot Tube Traverse

Since the velocity across a section of pipe or duct varies considerably, due to the effect of deflection at bends in the line and to the type of flow—laminar or turbulent—a complete traverse of a cross section must be made and the average velocity obtained. The usual method of making a Pitot tube traverse is to divide a circular cross section into five or ten annular rings of equal area, then divide each of these rings into two others of equal area. The circles that divide each ring into two equal areas become the stations or measuring points for taking readings across two diameters of a duct located at right angles to each other. The velocity pressure reading at each station is converted to velocity by using a table of velocity pressures or the formula

$$v = 1096\sqrt{\frac{\text{VP}}{\text{Weight of a cubic foot of air}}}$$

where VP = velocity pressure, inches of water
v = velocity, fpm.

The calculated station velocities are then averaged and the average velocities multiplied by the area at the cross section to obtain the quantity of air flowing.

The Pitot tube traverse in a square or rectangular duct is made by dividing the area into 16 parts or multiples of 16 so that the centers of each of these areas are not more than six inches apart. However, trying to locate the Pitot tube at these stations is very difficult and some other method of obtaining flow may have to be devised.

ESTIMATING AIRFLOW INTO DUCTS

Occasionally it is necessary to check sources of airflow or leakage into exhaust or negative pressure systems. A rough method of estimating such flow is based on the principle of the orifice, namely, the measurement of the static suction near the suction hood or air entrance nozzle. One end of a thick-walled soft rubber hose is connected to a U-tube manometer and its other end placed against small drilled holes located around the periphery of the duct at a section one diameter downstream from the throat of tapered entrances, and three diameters downstream for abrupt entrances. The volume rate of flow is then determined by the formula

$$Q = 4005 C_e A \sqrt{\text{SP}}$$

where Q = airflow, cfm
C_e = coefficient of entry, from hood entry loss charts
A = area of duct, ft^2
SP = static pressure reading, inches of water.

When an elbow intervenes between the entrance and the pressure tap, an estimate of the pressure loss caused by the elbow should be made from standard ductwork elbow loss data and the loss subtracted from the manometer readings.

5 Flow Characteristics of Materials

EFFECTS OF PROPERTIES AND CHARACTERISTICS ON CONVEYING

The properties and characteristics of materials will determine whether or not they can be successfully conveyed pneumatically. They also will determine the detailed design of the conveying system equipment to an extent not fully appreciated by the average user. The properties that affect gravity flow do not necessarily affect the flow of materials in a pneumatic conveying system. The variety of materials that are air conveyed easily leads one to the conclusion that almost any material that can be introduced into a pipeline can be conveyed pneumatically. This is true to a certain extent, but there are pitfalls in handling some materials. The materials must be able to pass through piping and equipment without clogging and without injuring the material. They must be readily separated from the conveying air.

Even when a material is successfully transported through a pilot plant size conveying system, the running time of the test is so short that troubles may appear later on a full-size system. This usually occurs simply because some unusual property or characteristic of the material was not considered. Not only must the characteristics of the conveyed material be studied, but also the characteristics of the fines collected in the dust filter.

Conveying Problems Due to Material Characteristics

The following examples of the problems that can occur should emphasize the necessity for knowing the materials handled. Some materials will insidiously build up on the outer radius of pipe bends and, after several days of operation, will block the piping. Some mixtures of material will classify during transport and most of the fine particles will accumulate in the dust filter. Some mixtures will become demixed because one ingredient absorbs moisture from the air and sticks to the pipe wall, while the remainder of the mixture is conveyed through the line. A dry powdery mixture that pours readily out of a container may become a pasty mass in the conveying line due to the conveying air absorbing the heat of compression from the air blower. Perfumed materials may lose their odor during conveying, whereas other materials may absorb odors carried into the air inlet of the conveying system. An abrasive material that is successfully conveyed without erosion suddenly causes excessive erosion because the source of the material, and the method of pulverizing it, has changed.

In some cases, friable materials become degraded when air-conveyed through a pipeline that has several bends. Such a material may be successfully air-conveyed vertically, disengaged from the air,

TABLE 5–1
Effect of Material Characteristics on Selection of Equipment

Property or Characteristic	Used to Determine
Specific gravity	Power and air requirements
Settled specific weight	Power and air requirements
Aerated specific weight	Bin and hopper volumes; feeder capacities
Sieve analysis—complete spectrum down to +325 mesh and average particle size in microns through 325 mesh	Dust collection requirements; feeder clearances; type of seals at points of air pressure differential; minimum conveying velocity.
Abrasiveness relative to a commonly known abrasive material	Type of conveying system; materials of construction; type of feeder; type of seals at points of air pressure differential; type of bearing protection required; power requirements.
Moisture content and rate of deliquescence	Type of conveying system; dust collection requirements; storage turnover requirements; air drying and bin venting requirements; type of feeder; necessity for flow inducers in bins and hoppers.
pH or corrosiveness relative to some well-known acid	Materials of construction; air drying and venting requirements
Tackiness relative to some well-known sticky material	Type of conveying system; air drying and venting requirements; type of dust collection system.
Aeration and deaeration characteristics	Type of conveying system; type of flow inducers in bins and hoppers; type of level indicator; design of chutes from bins and hoppers; requirements for deaeration.
Angle of repose	Design of bins and hoppers; type of flow inducers in bins and hoppers.
Toxicity	Type of dust collection; type of venting system in bins and hoppers.
Temperature limitations	Requirements for cooling conveying air; insulation requirements for bins, dust collectors and ducts.
Crystal structure or form of particles	Type of conveying system; type of feeder; form of piping system.
Absorption of odors	Location and type of conveying air filters.

and delivered to its final destination by gravity or by an auxiliary mechanical conveying system.

Many dry, finely pulverized materials will become fluidized in the conveying system and will flow uncontrollably from a bin and may even flush through screw conveyors. Fine dust collected in filter hoppers also may become fluidized and flush through downstream equipment.

Plastics at certain stages during processing will require an inert gas for conveying to prevent oxidation, whereas, at another stage in processing, clean, atmospheric air may be used for conveying.

The properties and flow characteristics of the material to be conveyed should be investigated thoroughly and the information tabulated under a heading of "Material Specification." If the material has already been handled within the plant in drums, bags, or other containers, most of its flow properties will be known from experience. If the material is new to the company, consult the manufacturer or shipper about his methods of in-plant handling and loading of bulk transport equipment. Ask him also for the physical properties of the material. Many materials that are handled mechanically by the manufacturer or shipper may be handled pneumatically by the user. In this event, he may even prevail upon his supplier to ship the material in transport equipment especially designed for pneumatic unloading.

When all possible physical data pertaining to the material has been assembled, the purchaser should determine a few things for himself by testing.

Table 5–1 lists the material properties or characteristics that the vendor wants to know and how he will use them for designing a pneumatic conveying system.

TESTS FOR PROPERTIES AND CHARACTERISTICS

Some tests for determining the properties of materials are standardized; other tests may have to be devised to determine properties and characteristics that would enable evaluation of the handling and storage problems that may occur in a pneumatic conveying system. The most common properties of materials and methods for their determination follow.

SPECIFIC GRAVITY

The specific gravity of a material is the ratio of its density to the density of water. This measurement gives a true comparison of the relative density of materials with respect to each other. This test consists of determining the weight of the volume of water displaced by the material. Several methods are covered in standard physics textbooks for materials lighter and heavier than water.

The specific gravity of powders that are soluble can be estimated by vibrating them in a container until they form a densely packed mass. The volume and weight of this mass is measured and the density compared with the known density of an equal volume of water. This ratio may be considered the specific gravity of the material. The specific gravity of hollow beads may be determined by sink or float methods. Here the material is dropped into various liquids of known specific gravity that will not dissolve the beads. The specific gravity of the liquid that can support the beads without their floating to the surface or sinking to the bottom is considered to be the specific gravity of the bead.

Aerated or Loose Specific Weight

This is the weight per unit volume of powdered or granular material after it is poured gently into a container of known volume. The filled container is weighed, an allowance made for tare and the density computed. The manner of pouring may be modified by screening the material as it is poured. This density may be used to determine (1) the approximate weight of material discharged by a feeder with a known volumetric capacity, and (2) the approximate weight of material in a bin of known volume immediately after filling it pneumatically, or (3) the volume of a bin or hopper required to receive a specified weight of material from a pneumatic conveying system.

A more accurate method of obtaining the aerated density for fluidizable materials is as follows: Construct a small deep test bin having an aeration pad built into the bottom and a plenum chamber for introducing air beneath the pad. Connect a manometer at the bottom and top of the bin above the aeration pad. Fill the bin with a known weight of the material to be tested. Apply air pressure to the plenum and increase the pressure gradually while observing the manometer readings. Bring the powder to the state of incipient fluidity—the point at which there is no increase in pressure drop across the bed of material. The known weight divided by the measured volume is the fluidized density.

Settled Specific Weight

This is the weight per unit volume of material in a container of known dimensions after the material has been vibrated for a predetermined time. The period of time and the manner of vibrating the sample are usually standardized by a plant laboratory to suit the types of material used in processing operations for a particular industry. This density may be used to determine the work required to deliver material to its destination at a specified rate—and ultimately, the total air requirements for the conveying system.

Particle Size

The distribution of particle size of granular and powdered materials is determined roughly by dividing a sample into many fractions, closely limited in size, by passing it through a shaking nest of graded wire mesh sieves. This procedure gives size distribution based on the weights of the fractions retained or passed through the screens. The results are expressed as percent by weight between sieve intervals and percent finer than the finest sieve used.

The sieves are made to standards of the American Society of Testing Materials or of the United States Standards Institute and sizes may be expressed as U.S. Sieve No. or Tyler Sieve No. Sizes range from No. 3⅓ to No. 325 (0.2210-in. opening to 0.0017-in. opening). The number and sizes of screens used to obtain the size spectrum varies and depends on the standards established for a particular process by a quality control group. Some industries may specify that all material be more, or less, than a specified screen size, or must be within a range of sizes. In some cases, the fines may be ignored, insofar as the ultimate use is concerned, and may not be specified. However, for pneumatic conveying work, the percentage of fines must be known. The amount of fines may determine the design of dust filters, equipment seals and clearances, and conveying velocities.

The smallest screen size is No. 325, but many materials may have a high percentage of particles passing through this screen and will require a special test to determine the size distribution in the sub-sieve range. All sub-sieve size tests must be performed by a trained laboratory technician.

The most direct method of sub-sieve size analysis is by using the optical microscope. The sample is placed on a slide or a special graticule and the diameter of a large number of particles measured in order to obtain a representative sample. The particles must be spherical or nearly spherical because the dimension perpendicular to the slide cannot be measured.

Another method is to use an elutriator composed of a series of vertical tubes of varying diameters through which dispersed particles in a stream of fluid are moved upward at a known velocity. Particles with a terminal velocity greater than the fluid velocity fall to the bottom of each tube while the remaining particles move upward. The size ranges in each tube are calculated from Stokes Law and the collected powder weighed. The final discharged material is caught on filter paper and weighed in order to account for all the material introduced into the tubes. The size range in each cut are fairly wide and the method has other defects, but is suitable for comparative tests under controlled conditions. Another sub-sieve sizing method is the sedimentation method using liquid or a gas. In gas sedimentation, dust is dispersed by gas at the top of a long column and the particles fall downward onto the pan of a recording balance. The size distribution is calculated from a weight-time curve by direct application of Stokes Law. In liquid sedimentation, the powder is uniformly dispersed in a liquid that will not dissolve the material and the sedimentation followed in a given layer of liquid by (1) measuring the change of concentration with time by extracting a sample periodically, (2) measuring the change of concentration with time by photometric methods, and (3) measuring the weight of material accumulating on a balance pan at the bottom of the column. (See Figure 5–1.)

Several types of analytical instruments are available for speeding up the determination of sub-sieve size ranges of powdered materials.

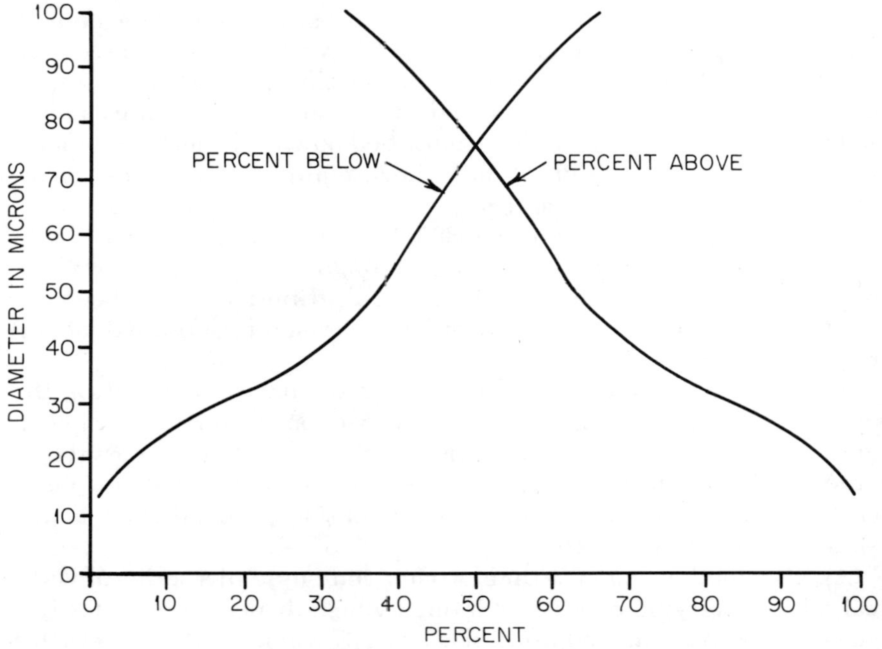

FIGURE 5–1. Typical sedimentation-test results.

Abrasiveness

The abrasiveness of a material is determined by its effect in wearing away the surfaces of the metal with which it comes in contact by impingement and deflection during conveying. The abrasive characteristics of certain cereal grains is known and provision can be made for resisting erosion. The abrasiveness of pulverized minerals depends on the shape of a particle after its crystals are fractured, the size spectrum of the material, the source of the material and the velocity of impingement. In general, coarse crystals are more abrasive than fine crystals. Crystals conveyed at high velocity are more abrasive than crystals conveyed at a lower velocity. A test of relative abrasiveness is to discharge a sample of the material, through a standard sandblasting gun, at a weighed target set a known distance away from the gun. The loss in weight of the target after a fixed period of time at a set blasting pressure would be a measure of the abrasiveness of the material, or a measure of the target's resistance to abrasion.

An abrasive material would require handling at low conveying velocities to reduce impingement effects; handling in a dense mass to avoid deflection against the walls by air currents; and air purging or tight sealing of equipment bearings and seals to avoid enlargement of clearance spaces by erosion or rubbing contact.

The equipment may have to be designed for regular replacement of parts that are eroded. The cone bottoms of bins would have wear resisting liners; rotary feeders would have the entire interior hard chrome plated; diverter gates would be made of meehanite castings; piping would have wearback elbows; etc.

Moisture Content

The moisture content of many materials increases by exposure to moist air and changes their flow properties. Deliquescent materials may form a hydrate on the particle surfaces that may cause them to bind together and prevent them from flowing. Grains may ferment and become tacky. Fine dusts may adhere to bin walls and lose their free-flow and fluidization properties.

Each material handled in a particular industry will have its own range of moisture tolerance. Controls to preserve its flow characteristics will have to be determined to suit the process requirements. Some materials may be conveyed and stored under a dry air blanket to eliminate moisture effects. Outdoor equipment may be insulated, or maintained at a high temperature by recycling hot air, in order to prevent condensation from affecting the materials handled. Some materials may be turned over—that is, transferred from one bin to another—in order to prevent fermentation or setting up solidly. Some deliquescent materials may be mixed with an additive to prevent hydration. The storage period for some materials may require close control in order to prevent lumping or hardening in bins.

In general, hygroscopic and deliquescent material requires dry storage, continuous movement, and limited storage volume gauged to short term process requirements. In extreme cases, the storage bins may have to be constructed of wood or bric-tile and residue of material in spouting, gates and feeders may have to be removed regularly.

Moisture content is commonly determined by the loss in weight of a sample size of the material after exposure to a given temperature for a preset time.

Corrosiveness

The corrosion of metals shortens the life of conveying equipment. The materials handled in a pneumatic conveying system are considered dry, but, when certain materials contact moisture from the atmosphere, because of their deliquescent properties or because of condensation, the material dissociates and forms ions—atoms that are electrically charged.

The positive-charged atoms are called hydrogen ions and the negative-charged atoms are called hydroxyl ions. The relative concentration of these ions in a solution determines its acidity, or aggressiveness in reacting with metals. The acidity of a material is a measure of its hydrogen-ion concentration, since it is the hydrogen ion that can displace the ions in metal to cause corrosion. The hydrogen potential or hydrogen-ion index is expressed in terms of pH value. pH is the common logarithm of the number of liters of the given solution that are needed to furnish one gram-ion hydrogen ion. Pure distilled water contains 1 gram H^+ in 10,000,000 (or 10^7) liters of water. Therefore, the pH of pure water is 7. When the material contains more hydrogen ions than distilled water, the water is acid—the lower the pH value, the more acid it is. If the water is alkaline (less acid), the pH value is above 7—the higher the pH, the greater the alkalinity.

The simplest measurement of pH of a solution may be made by adding a few drops of an indicator to it. An indicator is a chemical that changes color at a definite pH value or through a limited working range of pH values.

Commercial instruments to measure pH values are electrical instruments that measure the electromotive force, or potential, set up by a hydrogen electrode dipped into a test solution. In general, these instruments use a null method to balance the electromotive force of the electrode against the electromotive force of a standard cell. The balance point is reached when a galvanometer shows no deflection; then the voltage of the electrode is read on a voltmeter.

The fabrics used in dust filters are also attacked by the solutions formed when moisture is absorbed by the filter cake. The choice of fabric thus depends on the pH value of the filter cake material.

Tackiness

This is the tendency of a material to stick to surfaces on which it is deposited and to cling to adjacent particles. A rough test for tackiness is to try to form a ball of the material by rolling it between the palms of the hands. If the ball falls apart, it is not tacky. If it sticks together, it may not be suitable for pneumatic conveying. If it can be smeared onto a metal surface and built up so that it cannot slide off, it may not be suitable for pneumatic conveying in a regular pipeline with horizontal and vertical runs, but may be conveyed in a vertical lift designed to limit the smearing action. A tacky material issuing from a process may sometimes be conditioned by circulation of air so that the surfaces of the particles become dry and the particles become free flowing. Mixtures containing tacky ingredients may be made free-flowing by proper proportioning of the dry and tacky ingredients. Some tacky materials, though conveyable in a pneumatic system, may blind the fabrics in a dust filter and may require the use of a loop system of conveying that will eliminate the fabric filter, or the use of a multi-stage cyclone collector when a loop system is impractical.

Aeration and De-Aeration Characteristics

The ability of the material to be fluidized by aeration determines how it will be handled in a conveying system. Fluidized material is ideal for handling in systems that use high material-to-air ratios. It can be transported in air-activated gravity conveyors. However, its tendency to flush uncontrollably may require special flow controls or the use of deaerating devices at its final destination. Rotating-paddle-wheel types of level indicators may be unaffected by the presence of the material. The material may tend to flow through equipment doors, covers, joints, etc. It may reduce the weight capacity of bins and feeders or cause these to vary considerably.

Nonfluidizable materials may require impact-type flow inducers to start the discharge of material from hoppers or bins. The material may settle in horizontal conveying lines and defy all methods of restarting flow, short of dismantling the piping. It may lodge in equipment that has internal velocities that are lower than conveying velocities.

The tests for aeration and de-aeration characteristics are qualitative and are intended to predict the nature of the problems that may appear when handling the material, so that provisions may be made to solve them—if they occur. A few simple tests are as follows: (1) Shake the material in a partially filled container and note if it swells in volume owing to aeration. (2) Stand the container of aerated material on a flat surface and note the time it takes to deaerate of its own accord. (3) Rapidly tap a container of aerated material at a constant rate and note the time it takes to de-aerate. (4) Shake the material in a partially filled closed cardboard tube having a taped hole in the side near the bottom, stand the tube vertically on the edge of a table, remove the tape; note the trajectory of the stream of material. A fluidized material will issue like water; a dead material may just ooze out and plug the orifice.

Angle of Repose

This is the angle between the horizontal and the natural slope formed by a pile of material dropped from some elevation. This angle determines the maximum amount of material that can fill a bin from a given inlet spout. When a pile has been built up to the spout no more material can enter the bin without some means of leveling the pile. When material is discharged from a bin, the flowing material slides, forming a sloping surface downward to the outlet. This angle of slide may be less than the angle of repose because the momentum and aeration of a sliding mass of material will carry additional material along with it. The proper slope sheet angle for assuring flow from a bin or hopper is one that is greater than the angle of repose. If the material tends to form cliffs, its angle of slide is nearly 90 degrees and the material will not flow without the use of flow inducers. Therefore, knowledge of the angle of repose is useful for designing bins and for selecting flow inducers that will assure free discharge of material. It is an indicator of the flowability of the material.

Measurement of the angle of repose is comparative since no standards exist. A measured sample of material is slowly dropped from a known elevation onto a plate to form a pile and the angle to the horizontal formed by its slope is measured with a protractor. The angle of slide is produced by dropping a standard weight onto the plate a predetermined number of times thus jarring the pile and causing the surface material to flow. The new angle to the horizontal is then measured. These tests are made to show relative flowability of materials. The more free-flowing a material is, the lower these angles will be. A fluidizable material will have much lower angles of repose and slide than a de-aerated or "dead" material. Two rough tests that reveal the action of the material under flow conditions are as follows: (1) Dip a hand into a container of the material and note the resistance to pressure and the manner in which the material trickles through the fingers. (2) Shovel the material from a pile and throw it against a wall and against another pile. Note its angle of repose, how it breaks away from the pile, how it lands when it hits the wall, and how it lands on another pile of the same material.

Toxicity

The poisonous nature of any material should be established so that proper handling and venting systems can be designed to protect personnel. The necessity for servicing bins and dust filters will require that men be equipped with protective clothing and oxygen masks before entering these units. In some cases, a dust filter may have to be hosed clean with water before men can enter it to replace bags. Most manufacturers of hazardous or toxic materials warn purchasers of the danger involved when using their product. In addition, many of these manufacturers issue bulletins describing safe handling methods, and procedures for treating injuries that may be caused by contact with these materials. All information of this type should be procured in advance of designing a pneumatic conveying system so that safety provisions could be designed into the system.

Temperature Limitations

These should be determined for each material in order to ensure that it can be conveyed without danger of plugging up a pipeline. Temperature is related to tackiness for some materials and the comments relating to tackiness apply here. In a few cases, the warm material from a process may be cooled by pneumatically conveying it to the next stage of the process. The quantity of conveying air will then be determined by the cooling requirements for the material rather than by its transport requirements. Bins, dust collectors and ducts will then require insulation in order to prevent condensation within the system.

Crystal Structure, Form of Particles

These may be determined by examining the material under a magnifying glass or microscope and noting the structure of the particles. The structure may appear fragile and indicate that the material may disintegrate during conveying. The materials may be deliquescent and particles will appear bound together due to bonding of the surface hydrate. A flat, needle-like structure will indicate that materials may lock together in bins and require the use of flow inducers to get them out. The sharp edges of hard crystalline materials will indicate the possibility of erosion of piping and equipment. A microscopic examination is so simple that it should be one of the first steps in analyzing a material. The forms revealed may indicate the necessity for more detailed study of some of the other properties and characteristics of the material.

Absorption of Odors

This property may be determined by blowing odorous air through a deep layer of material and noting whether the material retains the odor when it is removed from the source of perfume. The source of the odor may be from an adjacent process. The odor may be objectionable and may affect the use of the material, if it is picked up by the conveying air. Test equipment is easily improvised from pipe nipples, a pair of screwed companion pipe flanges and a pipe cap. A small sheet of heavy canvas belting or other air permeable material is bolted between the flanges. A nipple and cap are assembled on one side of the flanges to form a plenum into which a cloth dipped in the odor producing material is placed. The cap is tapped to receive a connection from an air supply hose. A long nipple or pipe is assembled to the opposite side of the flanges and the assembly supported vertically so that the long pipe may be filled with the material to be tested. The air is turned on and allowed to percolate through the material for a desired period of time. The material is then removed and checked for retention of odor.

The inverse test for loss of odor may be made by using the same equipment, but with the plenum chamber empty. This test may be desirable when it is intended to air convey perfumed materials.

6 Methods of Transport

FACTORS THAT DETERMINE SELECTION

The type of transport used for bulk material that will be conveyed pneumatically is based primarily on economics. Transports range from primitive, covered-hopper railroad cars, or trucks with gravity discharge, to the modern self-unloading, covered hopper car or trailer truck. An economic study must start with an estimate of the total amount of material required to be transported annually and its lading density. The lading density is assumed to be the average of the loose, aerated bulk density and the packed bulk density of the material. The number of transports required is then calculated using the maximum allowable load limits in pounds for the types of units under investigation. The load limits should be obtained from capacity curves supplied by the manufacturer for the specific type transport under study. These curves generally show the loading capacity versus material density up to the load limit, also height of lading below ceiling or eaves versus material density up to the load limit. The maximum volume of the transport should not be used for this calculation since, as the curves indicate, the height of lading below the ceiling decreases as the material density increases. The annual investment in transportation equipment may then be computed from the leasing rate or purchase price for each specific type of unit. This study should also include consideration of a number of other physical and economic factors.

Distance from the Shipping Point

The load limits imposed by state laws economically confine bulk-trailer operations to short hauls (under 500 miles). On long hauls, the bulk trailer cannot compete with a railroad car that has three times its load-carrying capacity. The relatively smaller loads that can be hauled in a trailer would require provision to handle more transport units within a plant for a given annual tonnage. Since the trailer ladings are limited by the state laws, the lowest load requirements must prevail where several state lines must be crossed. Most trailer manufacturers issue a compendium of size and weight limits by regions. These should be checked to determine the capacity of the trailer selected for study.

Access into the Plant

In some plants, material can be delivered by railroad car or trailer and, in many cases, both types of transport may be used to insure continuity of material supply. In older plants, the track radii and clearances may be too small to accommodate large hopper cars. Where no railroad siding exists, or where no vacant car spot is available, a team-track operation may be arranged by routing a railroad car to the nearest available siding, or to an integrated rail-truck distribution terminal, and transferring its contents into bulk trailers that can be unloaded pneumatically at the plant. Railroad sidings must be laid out permanently in parallel lines with large radii curves, switches, bumpers, etc. in order to spot the cars for unloading. Bulk trailers, being more maneuverable, may be spotted at docks laid out to suit available space conditions. They may be laid out in parallel or diagonal rows or in continuous lines of unloading spots. The trailers may enter or leave on public streets, but consideration must be given to local traffic conditions and neighbors' access to their property. Strikes may preclude the entrance of trailers, but railroad cars may be able to enter the property and vice-versa.

Capital Investment

Tables 6–1 and 6–2 list features of the various types of transports that may influence their selection. Figures 6–1 to 6–14 illustrate the principles of operation of typical railroad cars and truck trailers. It will be noted that self-unloading trailers require the least investment in plant equipment, but high motor-freight costs may negate any savings on this account and may even warrant shipment by railroad car, even though this may require a large investment in a plant conveying system for unloading the railroad car. Where usage is low, a plant may be forced to use truck trailers since it would have no facilities to handle railroad carloads of material. However, where built-in storage facilities are limited, railroad cars may be lined up at several car spots for temporary outside storage within the demurrage limitations of the type of car in use. For some classes of material, a railroad car or bulk trailer may serve in lieu of a day bin for a process requiring large amounts of material daily. In this manner, the material has to be trans-

TABLE 6–1
Types of Rail Cars for Bulk Handling of Solid Materials

Type of Transport	Volume (ft^3)	Means of Discharging Cargo	Equipment Required To Discharge Cargo	Equipment Required in Plant for Conveying
R.R. box car	3,900 to 5,800	Manually guided vacuum pickup nozzle	Vacuum unloading hose and pickup nozzle	Vacuum or V-P conveying system
Container car	1,400 to 2,400	High-pressure air fluidization of contents	Air supply hose and material discharge hose	Compressed air system
Air-activated gravity-discharge hopper car	2,600 to 4,566	Integral vacuum pickup nozzle Hose or chute to feeders of vacuum or pressure systems	Aeration air supply hose Material discharge hose or chute Discharge outlet adapter connection Car breather vent	Aeration blower system Vacuum, V-P or pressure conveying system
Closed compartmented hopper car with integral vacuum unloading nozzles (at each hopper)	2,450 to 5,700	Hose connection to vacuum system	Material discharge hose	Vacuum or V-P conveying system
Pressurized tank car (comparmented)	2,400	High-pressure air fluidization of contents	Air supply hose and material discharge hose	Compressed air system
Standard covered hopper car	1,900 to 4,800	Gravity discharge with vibrator or car shaker assistance	Car shaker or vibrators Discharge chutes or vacuum pickup adapter connections	Vacuum, V-P, or pressure conveying system
Pressurized hopper car with aerated discharge outlet	2,980 to 5,700	Low-pressure air fluidization of contents (440 cfm at 15 psi)	Air supply hose (3 inch size) Material discharge hose (4 inch size; hoses may be made larger for high conveying rates)	4 inch size conveying line to a vented bin (larger for high conveying rates) Aeration blower system

TABLE 6–2
Types of Truck Trailers for Bulk Handling of Solid Materials

Type of Transport	Volume (ft^3)	Means of Discharging Cargo	Equipment Required To Discharge Cargo	Equipment Required in Plant for Conveying
Air-activated gravity-discharge hopper van	700 to 850	Integral aeration blower, conveying air blower, and pressure feeder at outlet connections	Material discharge hose	Conveying line to vented bin and a power supply receptacle
Pressurized container trailer	700 to 800	High-pressure air fluidization of contents (250 cfm at 25 psi)	Material discharge hose	Conveying line to vented bin
Combination gravity and pneumatic discharge tank trailer	700	Aeration and conveying air fluidizes powder (400 cfm at 15 psi); tilting of trailer causes flow of material to rear discharge outlet	Material discharge hose Aeration air supply hose	Conveying line to vented bin or to vacuum, V-P, or pressure conveying system
Pressurized tank trailer with aerated slope sheets	850 to 1,600	Low-pressure air fluidization of contents (400 cfm at 15 psi)	Material discharge hose	Conveying line to vented bin
Self-loading and unloading hopper trailer	700 to 2,100	Screw conveyor discharge into built-in pressure system	Material discharge hose	Conveying line to vented bin
Covered hopper trailer	700 to 800	Gravity discharge with vibrator assistance	Vibrators, discharge chutes, or vacuum pickup adapter connections	Vacuum, V-P, or pressure conveying systems
Standard van trailer	700 to 800	Bags or containers fork truck lifted to dumping station	Bag or container dump chute or hopper	Vacuum, V-P, or pressure conveying systems

ported only once, directly to the process, rather than pumped to a bin and then reclaimed.

The principal advantage of most bulk trailers is that they are equipped with an engine- or motor-driven blower or compressor and are, in effect, a portable pneumatic conveying system. Only a hose connection to a conveying line and, in some cases, a cable connection to a power supply receptacle is required to deliver its contents into a plant silo or bin. The railroad car is at a great disadvantage in this respect since it is unattended and subject to vandalism and neglect so that it cannot be fitted with an air supply unit, hoses or expensive unloading equipment.

The use of bulk-trailer engine-driven blowers may be limited to daylight hours because of the proliferation of antinoise laws. Where material must be delivered by bulk trailer during the night or early morning hours, it is advisable to install permanent electric-motor-driven blowers within sound-proof enclosures near the unloading site.

Labor Requirements

Bulk trailers are usually operated by the driver who doubles as an operator during filling and unloading his vehicle. He makes all necessary connections to discharge the lading into a plant storage system. No plant labor is required for other than routine weighing in and out on certified scales and the signing of delivery tickets. Railroad cars require plant labor to connect the car outlets to the plant conveying system and to tend the car during unloading. Depending upon the type of car, two men may be required to attach adapters to the car outlets and to open the car gates. When handling problem materials, the men may have to bar or poke the contents down toward the car outlets or move and fasten a heavy mechanical flow inducer to different parts of the car.

Unloading Rate

The time required for unloading a railroad car or trailer must be determined by test and will vary with the type of transport selected and the characteristics of the material handled. The discharge rate of a gravity unloaded transport is seriously affected by hold-up of some classes of material within the unit unless it is continuously attended. The discharge rate of air-activated and vacuum unloading types of transport is limited by the conveying rate of the conveying system to which it is connected and is affected only when the unit is low or nearly empty. Considerable time may be lost during cleanout operations in attempting to recover all residue in the transport. The discharge rate of pressure unloaded transports is limited by the built-in capacity of the air supply system and the operating pressure for which the tank section was designed. Standard practice is to design trailers and differential pressure railroad cars to operate at less than 15-psi pressure. This is the limit below which the A.S.M.E. Unfired Pressure Vessel Code need not be followed. This eliminates the need for inspection of the tank components and for special explosion insurance charges. Trailers and railroad cars are available that can operate at higher air pressures. These must be built to A.S.M.E. Code standards and must be inspected and so stamped. These vessels must be covered by explosion insurance. Obviously, high-pressure transports can deliver material at higher rates than low-pressure units for a given distance, or can deliver material over longer distances than the low-pressure transports, but their cost is greater than that of low-pressure units. Trailers, in particular, tend toward low-pressure operation so as to lower the vehicle weight and thus carry a heavier payload.

Demurrage

These charges are incurred due to inability to release railroad cars rapidly. This situation may occur because of a breakdown of the in-plant conveying system; use of cars as storage bins; lack of storage space to receive the lading; and shut-down of a plant due to a strike or other cause. Demurrage is incurred on railroad owned cars and generally incurred on privately owned cars. Charges are dependent on the demand by shippers for a specific type of car. Cars leased by shippers or by large users of bulk commodities may be retained for long periods or indefinitely without charges. However, they may be penalizing themselves by disrupting their shipping schedules and losing the mileage allowance that the cars can earn when on the road.

Transport Ownership

The transport may be wholly owned by the shipper, the receiver, a trucking company or a transport leasing company. Ownership carries with it the responsibility for maintenance, repair and control of the unit, the advantages of which must be determined for each specific application. In many cases, businesses prefer to divest themselves of ownership in order to charge the entire cost of receiving materials to operating expenses. In some cases, the number of deliveries required does not warrant ownership of the transports. In any case, contracts can be obtained for the delivery of a specified quantity of material from shippers using railroad or truck delivery depending upon the location of the source of material. The shipper would then bear all responsibility for maintenance of the transports or pass it on to a leasing company.

Railroad car leases generally contain a mileage allowance that is credited to the lessor by the railroads and the credit is passed on to the lessee. To the extent of rental, this allowance is paid in lieu of the railroad furnishing its own equipment. The lessee is charged with the responsibility of routing the cars so that loaded and empty mileage are equalized on each railroad. At the expiration of the lease, the lessee must pay for any empty mileage in excess of loaded mileage on any railroad.

RAILROAD CARS

Railroad cars are available to suit the handling characteristics of almost any commodity that can be pneumatically conveyed and to meet any unloading situation. Some cars are specifically designed for gravity discharge, but, by the attachment of suitable connections, can be adapted to discharge material into a pneumatic conveying system. Some cars are specifically designed to discharge material into a pneumatic conveying system; some are designed to become a part of a pneumatic conveying system—the supply bin; and some are designed to be, in effect, almost a complete pneumatic conveying system, needing only a source of air to activate them. A brief description of the various types of cars available and their applications follows.

Standard Box Car

The application of the common box car to pneumatic conveying appears remote, but these cars have been used for many years to haul grain in bulk as well as bagged material. To handle granular material, the doors are boarded up and material is blown into the car. As the level rises, additional boards are added until the material is near the eaves, then the car door is locked and sealed. Unloading progresses in reverse sequence—a door is opened and a manually guided vacuum pickup nozzle attached to a suction hose is inserted into the car above the top board. As the material level is lowered, boards are removed until the operator is able to enter the car and clean out the entire cargo.

To handle bagged material, the car is opened and unloaded in the conventional manner except that each bag is broken and emptied over a trackside dump hopper that discharges into a pneumatic conveying system for transport to the storage bins.

This car is ideal for the transport of deliquescent material such as salt, which cannot be stored in bins over a long period without bonding together into a solid mass. The car is filled as for granular material, but is emptied on demand by signaling to the operator who then proceeds to break up the pile so that it can enter the vacuum pickup nozzle. The wooden walls and floors of the car protect the salt from discoloration such as would occur with steel walls. The car actually serves as a day bin for some chemical processes.

The ready availability of this type of car and its low cost compared to other types of cars makes it worthy of consideration for dustless, granular cargoes, or for salt, as described above. Its greatest disadvantage is the need for continuous attendance to operate the pickup nozzle and handle the door closures.

Standard Covered Hopper Car

This type of car was originally intended to haul cement and other powdered materials affected by rain or damp weather. It is designed to be unloaded by gravity with the assistance, where required, of car vibrators. Vibrator attachment brackets are mounted on each slope sheet of the car. The operator must select the best location for the vibrator by observing the car interior for areas where material is hung up. Car shakers are available to rock the entire car, but railroad equipment builders frown on their use because of car-spring breakage. The discharge hoppers are fitted with a rack-and-pinion-operated slide gate and guides to which car connections can be made. The shorter cars have four discharge connections; each set of two straddling the rails athwart the car because of the center sill. Larger cars have six connections in sets of two athwart the car. The outlet locations are unfortunate in that their centerlines are practically over the centers of each rail and, on loaded cars, close to the rail thus making it difficult to attach gravity unloading chutes or vacuum pickup adapters to them. The most common attachment for gravity discharge is a canvas chute riveted to angle iron connection flanges that are clamped to the car outlets. This type of connection is commonly known as a "pair of pants" because each leg connects to a pair of athwart-the-rail car outlets that discharge into a common body between the rails that is connected to a discharge conveyor, or to a hopper supplying material to a pneumatic conveying system. Obviously, the conveyor or hopper must be located in a pit or trench below the rails. The fabric chutes eventually tear or hole-through, thus creating a cleaning problem in these underground vaults.

Above-the-rail adapters are available for making up car connections for direct gravity feed into a vacuum pickup pan that is hose-connected to a pneumatic conveying system. These unloading pans or adapters may be connected individually to the car outlet or may be skid mounted in pairs for attaching to two outlets for simultaneous discharge to the conveying system. Each pan is equipped with its own discharge nozzle and an air inlet nozzle arranged to pick up the material and direct it into the discharge spout. Here again, the close clearance between the track rails and car outlets may prevent insertion of skid-mounted pans or even of a single adapter pan. A heavily loaded car, or one with weak or broken springs, may have to be jacked up in order to attach the unloading pans.

When material is received from several sources, the arrival of cars equipped with four or six outlets may be anticipated. An unloading station equipped for handling cars with four outlets will require a car puller or hand jacks for spotting the six-outlet cars so that they will line up with built-in discharge chutes.

This type of hopper car is satisfactory for handling dustless, free-flowing materials. Dusty materials tend to spew material from open top hatch covers during an avalanche while unloading under suction. Fine powdered materials tend to hang up in corners and at knuckle points and form "cliffs" which must be barred down manually from atop the car using a 16-ft-long slice bar—a dirty, dangerous operation. Despite unloading problems, the low cost of this car makes it desirable for hauling large volume, low cost bulk cargoes.

Container Car

This type of car is intended to handle less-than-carload shipments of powdered materials. It consists of five or six air-activated pressure vessels mounted on a modified gondola car. Each container is essentially a 250-ft^3 blow tank using high-pressure air from an external source to activate the contents and blow the contents out into a conveying line. The tank is filled by gravity through a top-mounted inlet gate equipped with a weatherhood. It is discharged by connecting an air supply hose and a material discharge hose to the respective connections on the tank through the openings in the side of the gondola car and admitting air into the tank. The tanks are removable for storage within a plant. They are fitted with lifting lugs for attachment to an overhead crane. This type of car can handle free-flowing, granular materials, but its use is limited in view of the availability of similarly activated cars having larger capacity, fixed blow tanks. Its greatest usefulness would be in handling expensive or antiseptic materials requiring close quality control, since individual containers may then be rejected instead of rejecting an entire lading if any part of the lading does not meet specifications. Compressed air activation permits unloading regardless of weather conditions and simplifies making car connections. Any source of clean, oil-free compressed air may be used. In some cases, sufficient air for unloading may be available from existing air mains during off-peak periods which coincide with the car unloading schedule.

The containers may be safely stored in a yard or transferred to trucks and are adaptable to a variety of operating schemes. The containers, and the gondola car that carries them, are treated as one unit and railroad bulk rates apply.

Pressurized Tank Car (Compartmented)

This type of car has three or four cylindrical pressure vessels with conical aerated hopper bottoms mounted permanently on a special flat car. Each tank has a volume of 800 ft^3 and uses compressed air from an external source to activate the contents. The air-supply and material-discharge connections on each tank are valved and manifolded into common headers so that hose connections at the ends of the headers can be made by one man at one location. Unlike the container car, the tops of the tanks are equipped with interconnected platforms and ladders to reach them for access to the filling hatches.

The operating pressure is 40 psi—high enough to permit delivery of material to almost any point in a plant. The car can handle almost any material that can be pneumatically conveyed, including abrasive materials. The tanks can be coated to accommodate corrosive chemicals or foodstuffs (Figure 6–1).

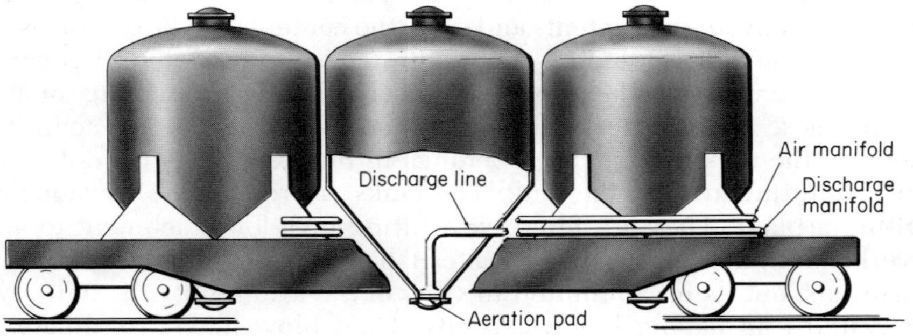

Figure 6–1. Typical design for a pressurized, self-discharging tank car for solids.

Air-Activated, Gravity-Discharge Hopper Car

This type of car represents a vast improvement over the standard hopper car in that it was designed specifically to handle fluidizable, dry, granular and powdered materials that can be conveyed pneumatically. The car is available in two sizes: 2,600 cubic feet and 4,566 cubic feet. The 2,600-ft^3 car is the most commonly used size (Figures 6–2 and 6–3). It is essentially a single compartment, airtight car with four hoppers; two on each side of the center sill. Each hopper is fitted with aerated troughs which pitch from each end of the car downward toward two discharge outlets at the center of the car. This location permits the car to be discharged at one location without the necessity of respotting it. The height of the outlet flanges above the track is at least 12 in. when the car is loaded thus permitting the insertion of various types of portable unloading devices beneath them. This location also favors gravity unloading through fixed chutes located between the track. The outlets can be fitted with special vacuum pickup nozzles for direct hose connection to pneumatic conveying systems.

An air-permeable fabric forms the sloping surface of the troughs and operates on the principle that aerated material flows to find its natural angle of repose. This angle is quite different than the angle of repose of the deaerated, or "dead," material. The aerated material never finds its angle of repose and continues flowing, since the slope of the trough is greater than that angle. Three inch size hose connections on piping manifolds, located on each side of the car, supply air from an external source to the plenum chamber of each pair of air-activated troughs on one side of the car, or can supply air to any individual trough by turning a 3-way valve at the hose connection. The discharge outlets are closed by an externally operated swing gate that is sealed from track dirt during transit by a bolted and gasketed cover plate. The amount of

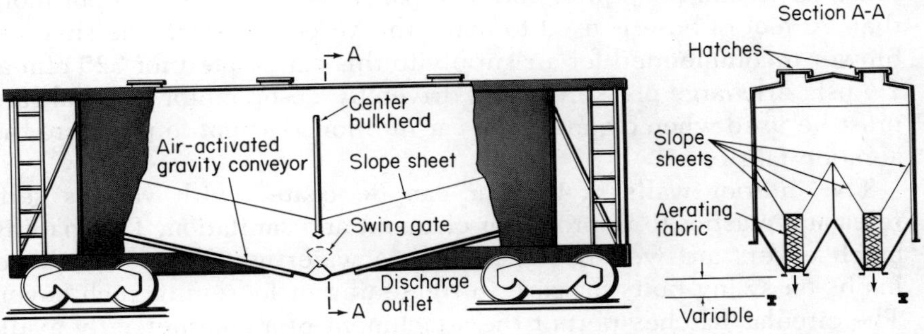

Figure 6–2. Sectional view of an air-activated, gravity-discharge hopper car.

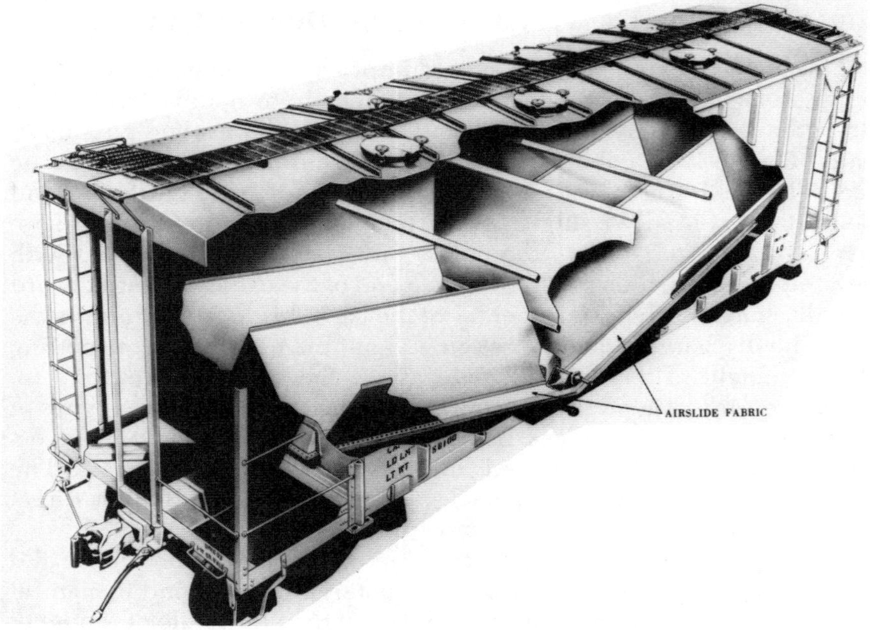

FIGURE 6–3. Cut-away view showing interior of 2,600-cubic-foot Air-slide® car. (GATX).

air required to unload the car depends on the fluidizing characteristics of the material being unloaded and is usually based on one-half car aeration since the air piping on each half of the car is independently manifolded. The airflow through the fabric varies from 3 cfm per square foot of trough at 1-psi pressure at the plenum chamber to as much as 28 cfm per square foot at 6 psi. This presumes that not more than 30 feet of hose is used to make the air connection. The smallest blower recommended for air supply to this car is rated for 325 cfm at 2½ psig discharge pressure and is driven by a 5-hp motor. Special care must be used when cleaning the car interior so as not to puncture the aeration fabric.

The interior walls of the car can be coated with various self-releasing plastics to improve car cleanout and sanitation. The loading hatch covers are well designed with a watertight seal, large hand knobs on swing bolts for ease in tightening or loosening each cover. The circular hatches permit the attachment of a commercially available breather vent filter designed specifically to fit the hatch and make use of its swing bolts to clamp the filter to the car.

The rate of flow of material from this car is limited only by the capacity of the unloading system to which the car is connected. The use of this car eliminates the necessity of operators entering cars to poke loose materials hanging up therein and improves operation of the unloading conveying system since the discharge rate is uniform and uninterrupted except during cleanout when hose and unloading connections are switched from one side of the car to the other.

Closed Multicompartment Hopper Car with Integral Vacuum Unloading Outlets

This type of car has been designed especially for handling dry granular and powdered materials that can be pneumatically conveyed. The compartments are made with vertical sides, steep slope sheets and no interior obstructions to cause bridging. There are no pockets, ledges or corners to trap lading and cause infestation. The side and slope sheets can be lined with various coatings so as to suit the properties of any type of commodity, to protect it from contamination or to release it for free flow toward the outlet (Figures 6–4 and 6–5).

The outlets are equipped with vacuum unloading pickup tubes for direct connection to a vacuum unloading hose on one side of the car and attachment of an air filter to a similar connection on the other side of the car. The rate of flow of material to the pickup tube is controlled by a rotating tube or gate over the material entrance port or by a sliding gate over the port, depending upon the make of car (Figure 6–6 and 6–7).

When the car is used as a storage bin, a special small diameter

Figure 6–4. Cut-away view showing interior of 3,500-cubic-foot Dry-Flo® car. (GATX).

FIGURE 6–5. Center Flow car. (Shippers' Car Line Division—ACF Industries.)

suction lance is inserted into the regular pickup tube to permit removal of material at low rates of flow.

The most modern multicompartment car is revolutionary in design since it eliminates the center sill and is braced like a pressure vessel. This design permits the location of all outlet hoppers on the centerline of the car—a feature not possible in cars with a center sill (Figure 6-5).

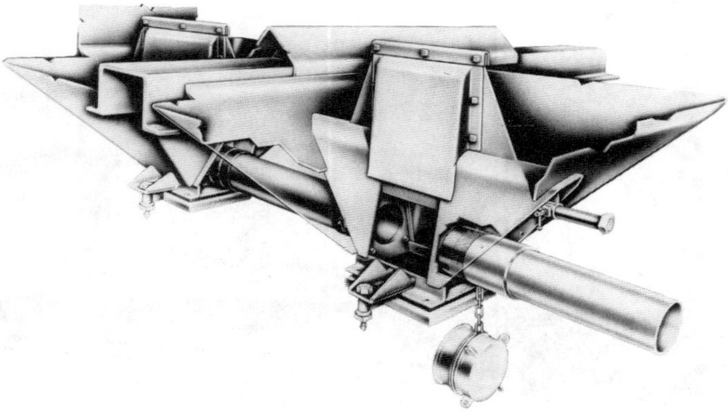

FIGURE 6–6. Pneumatic unloading nozzle for Dry-Flo® car. (GATX).

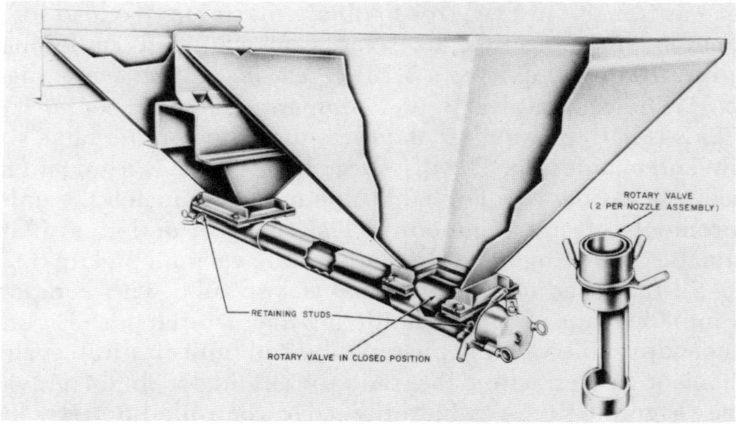

FIGURE 6–7. Pneumatic unloading nozzle for Dry-Flo Chem® car. (GATX).

FIGURE 6–8. Fluidized adjustable pneumatic outlet car. (Shippers Car Line Division—ACF Industries.)

The car is made in five sizes to obtain maximum payload for a wide range of loading densities. Sizes vary from a 2,971-ft^3 car for materials weighing 70 to 100 lb/ft^3 to a 5,701-ft^3 car for ladings weighing under 35 lb/ft^3. The cars have 2, 3, or 4 compartments with center-flow outlets. This is half the number of outlets furnished on the same size cars of conventional design. The discharge hopper size is fixed and flanged to receive a variety of interchangeable outlets to match the unloading requirements of any commodity. The types of outlets available for pneumatic conveying are: (1) An adjustable vacuum pickup tube (Figure 6–8); (2) an adjustable vacuum pickup tube with aeration pads built into the slope sheets for fluidizing the material (also Figure 6–8); (3) a nonadjustable vacuum pickup tube and built-in sliding gate and a mechanism for retracting the sanitary shield to obtain gravity discharge (Figure 6–9); (4) a butterfly-valve controlled outlet with aeration pads built into the slope sheets and with provision for attachment of an adapter for pneumatic unloading. The loading hatches of this car are located on the top centerline and are available in several designs to suit various loading arrangements. The centerline location of these hatches permits loading without the necessity of shifting a loading spout back and forth across the car. See Figures 6–8 to 6–10.

FIGURE 6–9. Gravity-pneumatic outlet. (Shippers Car Line Division—ACF Industries.)

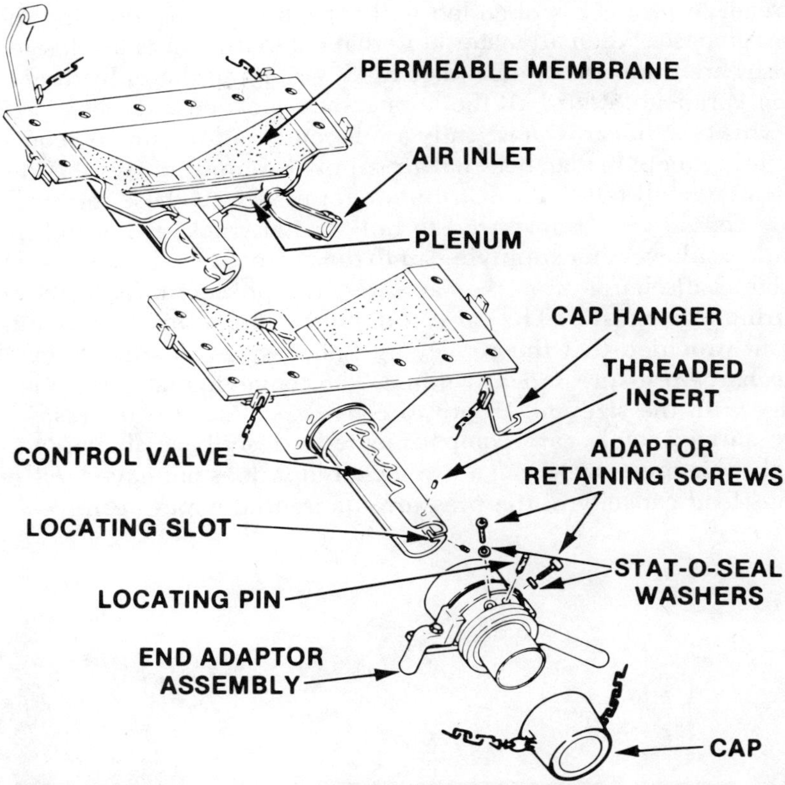

FIGURE 6–10. **Details of pneumatic outlet. (Shippers Car Line Division—ACF Industries.)**

PRESSURE DIFFERENTIAL HOPPER CAR WITH AERATED DISCHARGE OUTLET

This type of car is similar to the center-flow car but made in four sizes: 2,980-, 4,650-, 5,250-, and 5,701-ft^3 volume. It has the same pressure-vessel-type construction described above. However, it is equipped with porous stainless steel aeration plates in each outlet hopper. Each hopper aeration plenum chamber is fitted with an air-supply hose connection equipped with a metering valve and check valve cross-connected into the product line so that air may be supplied to that line at the same time that air enters the plenum. In operation, low-pressure air (about 5 psi) is supplied to the outlet plenum chamber until the entire car is pressurized; the product outlet valve is opened and the metering air valve is adjusted until a steady, unbroken stream of product is obtained without whipping or blocking the discharge hose. When all material in that compartment is discharged, the hoses are transferred to another full hopper and the unloading operation is repeated until all the hoppers are cleaned out.

This type of car is practically a self-contained pneumatic conveying system except that an external air supply is required. It is the ideal car for integrated rail-truck distribution terminals. At these terminals, the cars contents are transferred to bulk trailer trucks using the trailer's built-in blowers for supplying air to the railroad car (Figure 6–11). The product discharge from the car enters the product inlet to the trailer during this transfer. The car can deliver material to many points in a plant provided that the conveying line is sized to suit the available discharge pressure and air volume. The conveying rate, of course, will vary with the size and length of conveying line. In this respect, the pressurized tank car (compartmented) described above, with its higher air pressure has an advantage, but it does not have the tremendous load capacity of the pressure differential hopper car.

FIGURE 6–11. Transfer from pressure differential car to truck. (Butler Manufacturing Company.)

TRUCK TRAILERS

Bulk trailer designs are available that are prototypes of the various railroad cars described above.

STANDARD VAN TRAILER

This type of equipment is standard for handling bagged material or large bulk containers of modular design. The van may be applied in the same manner as railroad box cars.

The various types of bulk containers are discharged into a pneumatic conveying system in several ways. Bags are manually placed on a dump hopper equipped with rollers, a slitter, a screen, a vented hood, and a discharge hopper located above a rotary airlock feeder. Each bag is pushed over the slitter and the two halves of the bag lifted to dump the contents through the screen into the discharge hopper. The empty bags are shaken out and staked for disposal. Some types of bags are not slit, but are dumped, then dust is scalped off by a suction hood, and the bag salvaged for resale.

Sheet metal bins with side discharge doors are carried by fork truck, placed on a tilting station equipped with a screw feeder discharging into a rotary airlock feeder. The door side of the bin is tilted against a sealed hopper and the door latch tripped open. The screw feeder is started and the bin emptied. The tilting station may be scale-mounted to weigh out the contents and to detect the empty condition. Bottom-discharge and top-discharge bins are placed over a dusttight discharge hopper by fork trucks, inverted when required by the design of the bin, and material withdrawn by the rotary airlock feeder.

Large dodecagon-shaped cardboard containers are carried by fork truck and placed into a drum dump located over a discharge hopper. A vented hood is lowered and the drum tilted upside down to pour its contents into the hopper, from which it can be withdrawn by a rotary airlock feeder.

Collapsible rubber containers are handled by lift trucks of special design, or by monorail trolleys. The containers are removed from the van and placed in cone-type unloading cradles equipped with hydraulically driven arms. The containers are connected to the inlet spout of a rotary airlock feeder and, as the material is discharged, the arms squeeze the container into a funnel shape to facilitate emptying.

All of the above methods require the use of auxiliary powered lift trucks to handle the containers with all the attendant expenses connected with them. Gasoline-driven trucks are restricted to outdoor use. Propane-driven trucks require spare fuel tanks. Electric trucks require spare batteries and a battery charging station. All of these lift trucks require regular wheel changes and maintenance of the engine or motor.

Several of the container handling methods described above are definitely unsanitary. Bags always have surface dust that gets dumped along with its contents. Pallets for carrying large cardboard containers hold floor sweepings that get dumped into the hopper that receives the containers contents. Inverted bins can dump surface dust into the hopper that receives their contents.

Covered Hopper Trailer

This type of trailer is the equivalent of the standard covered hopper railroad car, but without its limitations, since only three or four discharge outlets are provided, located on the centerline of the trailer, high enough above the road to permit access for attaching unloading pans or chutes.

Pressurized Container Trailer

This trailer is equivalent to its railroad counterpart except that only two pressure tanks are used because of weight restrictions. These are mounted on a trussed frame in order to reduce the height of the trailer to road clearance limits. The tanks are built to A.S.M.E. Code requirements for operation at pressures up to 25 psi. Higher operating pressures would require heavier tanks with a consequent reduction in payload.

Air-Activated, Gravity-Discharge Hopper Van

This type of trailer is built to the same general design as its prototype railroad hopper car, but with less cubic capacity, in order to meet legal weight limitations (Figure 6–12). It is practically a self-contained pneumatic conveying system since it is equipped with a motor-driven blower for air supply and two blow-through type rotary airlock feeders; one under each discharge outlet, driven by transmission shafting connected to an electric motor. An equipment cubicle at the rear of the van contains the blower, motor-starting equipment and operating controls. A large, flexible electrical cable is used to connect the trailer's equipment to an external receptacle that is a part of the receiving plant's electrical distribution system.

Variations of this design omit the rotary airlocks and pressurize the tank for direct discharge into a conveying line. A return line from the

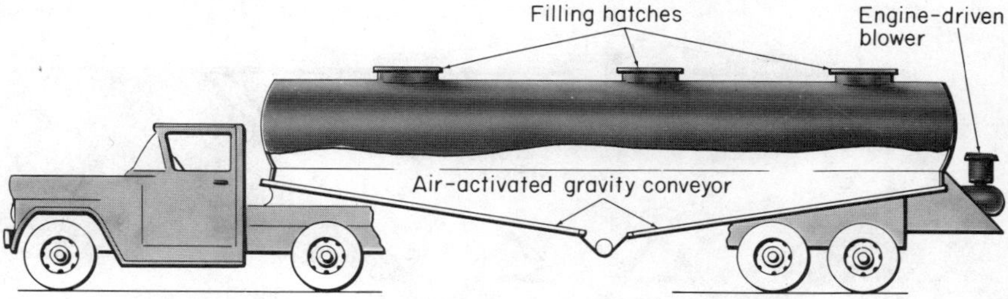

FIGURE 6–12. Air-activated, gravity-discharge hopper van.

receiving bin may be connected to a dust collector built into the trailer so as to form a closed-circuit, or loop, system. This arrangement is used on cargoes that require sanitary handling.

COMBINATION GRAVITY AND PNEUMATIC DISCHARGE TANK TRAILER

This type of trailer has no railroad counterpart and is no longer being manufactured, but a sufficient number of these are in service to warrant a description. The trailer is essentially a clean-bore, horizontal tank with an aeration pad built into its rear end. A pneumatic discharge tube, a mixing valve, and an air return tube are built into the tank. In operation, the front end of the trailer is elevated by a telescoping hydraulic hoist, thus causing the entire trailer unit to pivot on its rear wheels. The dump angle is 47 degrees, sufficient to permit gravity discharge of many types of materials when assisted by low pressure air activation through the air pads. The air percolates through the material and enters the return tube, which extends above the level of material when the tank is elevated. Air from the return tube picks up material at the mixing valve, which is located near the bottom of the elevated tank, and discharges it into the conveying line for transport to a silo. The angle of elevation is adjusted so as to ensure material feed to the mixing valve throughout the unloading period.

The available height required to permit maximum elevation of the trailer restricted its operation to such an extent as to force discontinuance of its manufacture.

PRESSURE-DIFFERENTIAL TANK TRAILER WITH AERATED SLOPE SHEETS

This type of trailer is built and operated like its railroad counterpart, but has the reduced capacity and height required to meet state laws. The trailer may also be used to discharge by gravity with the assistance

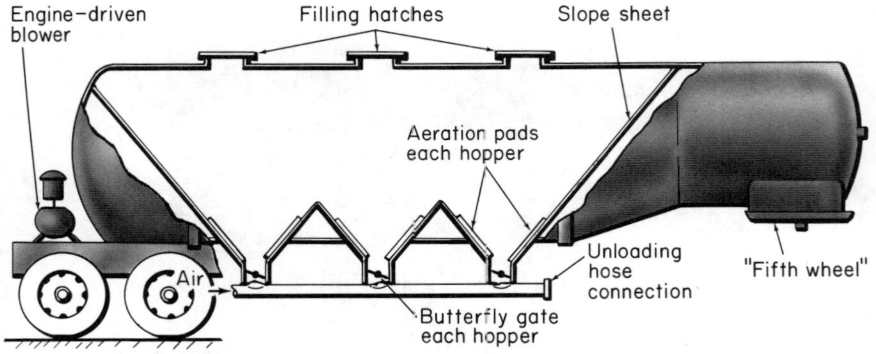

FIGURE 6–13. Pressure differential tank trailer—schematic view.

of air activation through the built-in air pads at each hopper. When used in this manner, the outlets can be fitted with vacuum pickup unloading adapters to feed material from the hoppers into an external pneumatic conveying system, instead of acting as a blow tank (Figures 6–13 and 6–14).

FIGURE 6–14. Nordic bulker. (Pullman-Trailmobile.)

Self-Loading and Unloading Hopper Trailer

The design of this type of truck varies considerably depending upon the ingenuity of the manufacturer. Essentially, the trailer is equipped with a dust collector and an exhauster that can draw material from a railroad car, or other source of supply, and deliver it directly into the tank compartment. Material may be reclaimed from the tank by feeding it into a blow-through rotary feeder using built-in screw conveyors or air-activated conveyors located at the bottom of the tank compartment. Air may be supplied to the feeder or exhausted from the tank by properly connecting a hose line from an engine-driven, positive-displacement blower.

TESTING TRANSPORTS

The large variety of transports available makes a selection rather difficult. However, when all the factors considered logically indicate a particular choice, it should be tested by actually shipping and unloading a shipment of the specific material under investigation.

Two courses are possible: (1) Observe the discharge of the transport at the plant of another user of the material to be handled (not a competitor!). The user's name can ordinarily be furnished by the material supplier. Or, (2) set up an unloading test for the material to be handled, following an over-the-road haul at least equal to the distance between shipper and receiving plant. The transport should be held in a yard for a long enough period to experience a variety of weather conditions before unloading. A test setup can be arranged in several ways—depending on the ingenuity of plant personnel, working in cooperation with representatives of the shipper and transport manufacturer. If the transport is of the pneumatically self-unloading type, all that is required is a pipe or hose line to discharge the material to a vented bin or into another transport that has a mechanical unloading device. If the transport requires an external air supply, a portable air compressor can be rented for the test and suitable precautions taken to keep the air moisture-free. If the transport has an aerated bottom, the manufacturer will furnish an aeration blower and connecting hose for the test, but venting and methods of handling the discharged material will again depend on the ingenuity of plant personnel. Several schemes for handling the material discharged from aerated or gravity-discharge transports include:

1. Use local team-track pneumatic conveying facilities and discharge the material pneumatically into small bulk containers that can be handled by existing plant equipment.
2. Use local hopper-car trestle facilities, and discharge contents of the transport by gravity into the small bulk containers.
3. Use the conveying equipment manufacturer's portable pneumatic unloading equipment on a rental basis to fill small bulk containers; or discharge via hose or temporary piping into existing bins.
4. Use mechanical conveyors to discharge the lading into the small bulk containers.

These precautions should be observed when arranging a test for a transport:

1. Where the shipper has not previously loaded transports that deliver his cargo into pneumatic conveying systems, he must be advised to screen all materials entering it so that lumps will not block the discharge spouts or piping.
2. In simulating a proposed conveying pipeline, design it with as few large-radius bends as possible, rather than with built-in kinks that would present the worst condition for testing. Failure to convey properly on such a test might kill further consideration of what could have been a money-saving installation but for a kink in the piping.

During a test, the unloading conditions and operations described in Table 6–3 should be observed and analyzed. They should be given consideration when designing a transport-unloading station.

The following additional observations should be made. Check the height of the lading from the eaves and compare it with the original height measured at the shipper's plant. This measurement should be available from the shipper's representative at the test site. The difference in height would be an indication of an increase in density caused by over-the-rail travel. On trucks, a cored out sample of the lading may be obtained and its density measured. After all the material appears to have been discharged and prior to final clean-out, examine the interior of the transport and note where the residue has accumulated. Observe the methods used to remove the residue, then re-inspect the interior to note the effect of final clean-out operations.

TABLE 6-3
Desirable Features and Conditions for Unloading Transports

Required Operations	Features and Conditions
Car and trailer unloading station	1. Should be in level area, protected from the weather when possible. 2. Spots should be well marked to center the discharge spouts. Use fluorescent paint when spotting is done at night. 3. Street or sidewalk hatch covers should be watertight and lightweight for handling by one man. Covers should be locked when in unprotected areas. 4. Steel blocks should be provided for chocking wheels. 5. Provide warning blinker lights or reflector stanchions to prevent collision by another vehicle. 6. Provide floodlights, where required, for night operations. Aim lights at the top of the car and at car outlet connections.
Connections to cars and trailers	1. Connections should be designed to eliminate the need for men working beneath the vehicle and in cramped spaces. 2. Attachments should be lightweight for handling by one man and should require a minimum of tools for connection to the discharge spout. 3. Hose connections should be snap-on or toggle clamp type with soft rubber gaskets. 4. Clear distance between discharge hopper of a loaded transport and road or track should be ample for insertion of unloading devices or attachments. 5. Transport discharge spouts should be equipped with easily operated gates above the covers so as to minimize powder spillage when making up connections. 6. Discharge connections should be flexible enough to compensate for the rise of the car on its springs as unloading progresses.
Access to loading hatches	1. Tops of transports should have grab bars at access ladders to assist operator in swinging onto vehicle. 2. Hatch covers on transports should be easily opened without the use of tools and should be fitted with watertight gaskets. 3. Platforms should have non-slip safety treads. 4. Top of transport should clear building canopies or cornices and yet permit passage of operator from one end of transport to the other.
Vent filters	1. Vent filters should be lightweight for handling by one man and accessible from top of transport without hoisting. 2. Filter should be easily attached to a loading hatch using the same clamping device used to secure the hatch cover. 3. Filters should be designed so that they may be cleaned after unloading, but before detaching so that dust is shaken down into the transport rather than onto plant grounds.

TABLE 6–3 (*Continued*)

Required Operations	Features and Conditions
Flow inducers	1. Transport should preferably empty itself without the use of external flow inducers or poking of material by operator through hatch covers. 2. Vibrators, where required, should be lightweight and easily fastened to one point on the transport most conducive to emptying. There should be no need to shift the vibrator from point to point. 3. Vibrators should cease operation whenever the discharge spout is blocked with material. 4. Vibrators should be operated by air or low voltage electricity. Portable cables and hoses should be attached to retracting reels for safety. 5. Vibrators should be noiseless, if possible, to eliminate neighborhood complaints, especially during night operations. 6. Vibrators should be sized with due regard to the detrimental effects of excessive vibration on the car structure.
Transport internals and clean-out	1. Transport interiors shall preferably be free of structural members and protuberances that may cause hangup of material. 2. Interior of transport should be coated with a releasing agent or plastic which will prevent adhesion of material to slope sheets. 3. All corners, valleys, and knuckle points should have large-radius curves to prevent retention of material at these points. Slope sheets should have a minimum angle of 45 degrees to the horizontal, but a slope of at least 60 degrees is preferred for self-unloading cars. 4. Aerating pads, blocks, troughs, or other aerating devices should retain a minimum of residue after transport is emptied. 5. Interior stiffeners, where required, should be designed to shed material and prevent hangup.
Rate of unloading	1. Rate of flow out of transport should be controllable to match conveying rate of in-plant conveying system or capacity of pipeline. 2. Flow within transport should be free from avalanching.

7 Plant Operations and Conveying Systems

Plant production-department operations that require consideration when contemplating a pneumatic bulk-handling system involve receiving, storage, and batching of material.

RECEIVING

Receiving operations involve a number of subfunctions that must be considered to insure the successful performance of the entire system.

WEIGHING

(1) The transport must be weighed in and out on registered scales at the shipping point or, preferably, at the receiving plant. (2) Allowance must be made for loads due to ice, snow, or rain. (3) A visual check of the empty condition through the loading hatch is required to prevent return of the transport with an excessive heel of material, especially on pressurized, self-unloading vehicles. (4) Materials that are required to be weighed after unloading, but before entering the silos require a considerable amount of space consuming equipment. The material must be handled in the following sequence: It must enter a receiver, an upper surge hopper, a tank scale, and a lower surge hopper before being discharged into a silo. Furthermore, the entire combination of equipment must be vented and an accounting made of the material scalped off by the vented air.

SAMPLING

(1) Cargo must be sampled to insure delivery into the proper bin and compliance with specifications. (2) Waiting time prior to unloading may have to be extended where stringent particle-size analysis must be met. (3) Arrangements must be made for disposition of a transport having a cargo of condemned material.

UNLOADING

(1) The type of operator required may be one of the following: The driver of a self-unloading transport; a raw-stock bulk-handling operator at a central unloading and storage station; or a part-time production worker, as required to meet the transport schedule. (2) The choice of controls required may be simple manual control where unloading and storage areas are adjacent to each other; semiautomatic control where more than one operator is available and unloading and storage areas are remote. (3) Unloading-station protection and facilities required call for determination and study of the following factors: need for roofing and housing for all-weather operations; access to system connections for hooking up transports; interferences due to movements of other transports in the same area; hazards involved if transport connections break or become disengaged; location of storage facilities for unloading equipment; and, finally, means of stopping unloading operation in case of trouble in some area remote from the unloading station. (4) Rate of unloading and total unloading time available are other factors that enter into the consideration.

STORAGE

FILLING

(1) Minimum storage volume required should be at least one and one-half times the capacity of the largest transport, in order to permit emptying the transport and still retain a heel of material. (2) The maximum storage volume needed requires a study of the following factors: reliability of transportation; distance from shipping points; effect of delays in receiving material due to weather, strikes or accident; area available for storage facilities; process demands; and reliability of the pneumatic conveying system.

LEVEL CONTROL

The following need consideration: (1) Manual or automatic control; (2) whether continuous indications or high, low, and intermediate indicators are required; (3) what protection is necessary to prevent over-filling bins; and (4) what method of checking inventory is to be used, i.e., weighing, visual, sounding tape, etc.

Venting

Factors to be considered include: (1) Need for a dust-collection system; (2) how collected dust will be disposed of, i.e., by return to source or degrading; (3) where the dust collector is located, indoors or out; (4) where discharge ducts for vented air are located; and (5) what hazards are involved if dust-filter bags should break or tear.

Emptying

This needs study of: (1) Rate of emptying; (2) means of emptying whether by gravity, by mechanical conveyors, or by electric, pneumatic, or mechanical flow inducers; (3) means of access for inspection and cleanout; (4) provisions for personnel protection during inspection; i.e., lockout of filling and emptying systems.

BATCHING

Reclaiming

(1) When contamination must be prevented, each ingredient should be handled in an independent conveying system. (2) Each product should have its own venting and dust collection system to prevent cross-contamination. The exception to this requirement is when the collected dust is to be discarded. (3) The type of receiver should be selected to suit the type of feed system on the processing equipment. The material may enter a feed hopper directly, under level control, and be vented through a loop system; or it may enter a surge hopper preceding a weight tank, and flow may be cut off by weight control; or it may enter a small cyclone and flow may be cutoff by stopping the conveying system feeder and allowing for the residue delivered by the clean-out airflow. (4) The type of reclaiming system is directly affected by the type of weigh system selected. A loss-in-weight batching system has a larger hopper than an equivalent net or gross weight system and it can be filled in one operating cycle, whereas the equivalent system may require many short cycle operations to meet the batching requirements.

Feeding to Process

(1) Feeding may be by gravity directly from feed hoppers, but level control is required to maintain low head in order to ensure uniform

product density and free flow without bridging of the material. (2) Feeding may be by gravity, directly into a weigh tank for net weighing, or into a bulk container for gross weighing. (3) Feeding may be into a proportioning system, in which case the type of feeder from the surge hopper to the weigh scale must be carefully selected to minimize overfill. The factors that must be considered are accuracy; the frequency of future formula changes; the number of existing formula changes; the possibility of contamination; the speed required to add material to the batch; the volume of material that must be handled; and the possibility of segregating the particles of material.

Weighing

(1) To minimize overfill, mechanical feeders such as screw feeders or rotary gates should have two speeds—a high speed and a dribble speed—followed by quick cut-off, using a gate or a brake on the drive motor. (2) The feed equipment should be designed to introduce the material in small increments—by using multiple-flight, fractional-pitch screw feeders, or rotary feeders with shallow pockets. (3) On pulverized, floodable materials, an auxiliary gate may be required after screw feeders or air-activated gravity conveyors and ahead of rotary feeders. (4) Provide flexible connections at all feeder entrances into a weigh tank and make these connections through the side of the weigh hopper so as to eliminate unbalanced vertical forces that may affect the weight. (5) Make all vent connections through the side of the weigh hopper in order to eliminate unbalanced vertical forces. This may be done by using flexible connections at the side outlet, an open draft-break connection, or by building air-permeable vent panels into the sides of the hopper. (6) The weigh hopper may build up a residue on its walls and allowance must be made for this. The amount of residue may be reduced by rounding all corners, eliminating shallow valley angles, coating the interior with self-releasing plastic or by using anti-static coatings; in short, by using the same principles as applied to the design of bins. (7) The possibility of underweight must be considered and provision made to jog a feeder or add materials manually to make up the loss. Overweight is a tougher problem to handle. It may require dumping the batch for rework or an on-the-spot formula change, if it exceeds tolerances set for the process.

TYPES OF CONVEYING SYSTEMS

Pneumatic conveying systems shown in manufacturers' bulletins and in engineering literature are variously described as high velocity systems, low velocity systems, dense phase systems, light phase systems, low pressure systems, high pressure systems, medium pressure systems, etc. Some of these terms may be used to differentiate between a particular manufacturer's lines of equipment, others are intended to differentiate one manufacturer's equipment from that of another manufacturer. There are no industrial standards available that define the terms used to describe various types of pneumatic conveying systems. These terms are so confusing and inexact that a reclassification of types based on the manner of combining the air and material appears desirable.

Three commercially available systems are recognizable: (1) Systems in which material enters a stream of air under either negative or positive pressure, or is induced into the stream of air by vacuum. These comprise the familiar so-called pressure systems, negative pressure or vacuum systems, and variations of them, such as the combination (V-P) and loop systems. (2) Systems in which air and material are intermixed simultaneously at the entrance of the conveying line. These systems resemble the first type, in which material enters an air stream, except that air and material are intimately intermixed in a special type of feeder at the entrance to the conveying line, resulting in a denser stream of material than is otherwise possible to obtain. (3) Systems in which air enters a mass of material to cause flow. These comprise the familiar pressure-tank bulk trailer systems and less familiar in-plant blow-tank systems.

The type of system selected for a particular installation is determined primarily by the properties and characteristics of the material handled and the necessity for preserving these properties and characteristics. (See Table 5–1, page 84.) Other factors that may determine the selection of a system and even supersede the primary requirements are as follows: (1) The available pressure range for long-distance conveying. (2) The functional requirements of the system, or flexibility in performing several functions, either separately or simultaneously. (3) The amount of accessory conveying equipment required to form a complete system. (4) The amount of control equipment required to operate the system. (5) The structural modifications required to install the equipment. (6) The power and utilities required to operate the system. (7) The overall cost of the system after consideration of all the above factors and receipt of quotations from vendors based on detailed plans and specifications.

The variety of functional operations required of a system may immediately determine the selection of one of these systems for a more detailed study. The functional operations that can be performed by pneumatic conveying systems are as follows: (1) Unloading a transport vehicle or a shipping container for delivery to a storage bin or directly to a process use bin. (2) Reclaiming material from a storage bin for delivery to a process use bin. (3) Distributing material from an unloading or process operation to multiple storage bins. Distributing material from a storage bin to multiple process use bins or to weigh hoppers. Distributing material from a processing unit to several other units for further processing. (4) Loading bulk transport vehicles or bulk containers. (5) Returning dust to storage.

The following detailed description of the three types of systems also presents an evaluation of those factors that affect selection and a description of the functional operations that each system can perform.

MATERIAL-INTO-AIR-STREAM-SYSTEMS

The material-into-air-stream type of system is the most versatile of the three types of systems and can handle a wide variety of materials over a wide range of conveying velocities. The velocities can be reduced sufficiently to handle many friable materials without breakage, and many abrasive materials without excessive erosion of piping and equipment. In addition, the low material-to-air ratio used permits handling of materials with a minimum of contact between particles. This makes the system ideal for handling a variety of particle shapes and sizes without blockage of the conveying line. The materials handled range from irregular lumps to chips, flakes, winged grains, hollow beads, granular and powdered solids.

The pressure range of these systems is limited by the available pressure ranges for the high-volume, positive-displacement blowers required for this type of conveying. Blower pressures are limited to a maximum of about 15 psig on positive-pressure systems. Blowers used as vacuum pumps are limited to a vacuum of about 13 in. Hg. The total range from 13 in. Hg vacuum to 15 psig pressure is available for combination vacuum and pressure (V-P) systems.

POSITIVE-PRESSURE SYSTEM

In this system a high velocity air stream is established in the conveying line by the air discharged from a positive-displacement or centrifugal blower through an injector fitting (Figure 7–1). Material is withdrawn from a storage vessel by gravity or mechanical conveyors and is delivered to a rotary airlock feeder. The feeder then discharges the material into the air stream, which transports it through a pipeline to a vented receiver. Discharge into several receivers is done, one at a time, by operating a diverter valve located at each receiver to shunt the flow from a main piping header into a selected receiver, or by

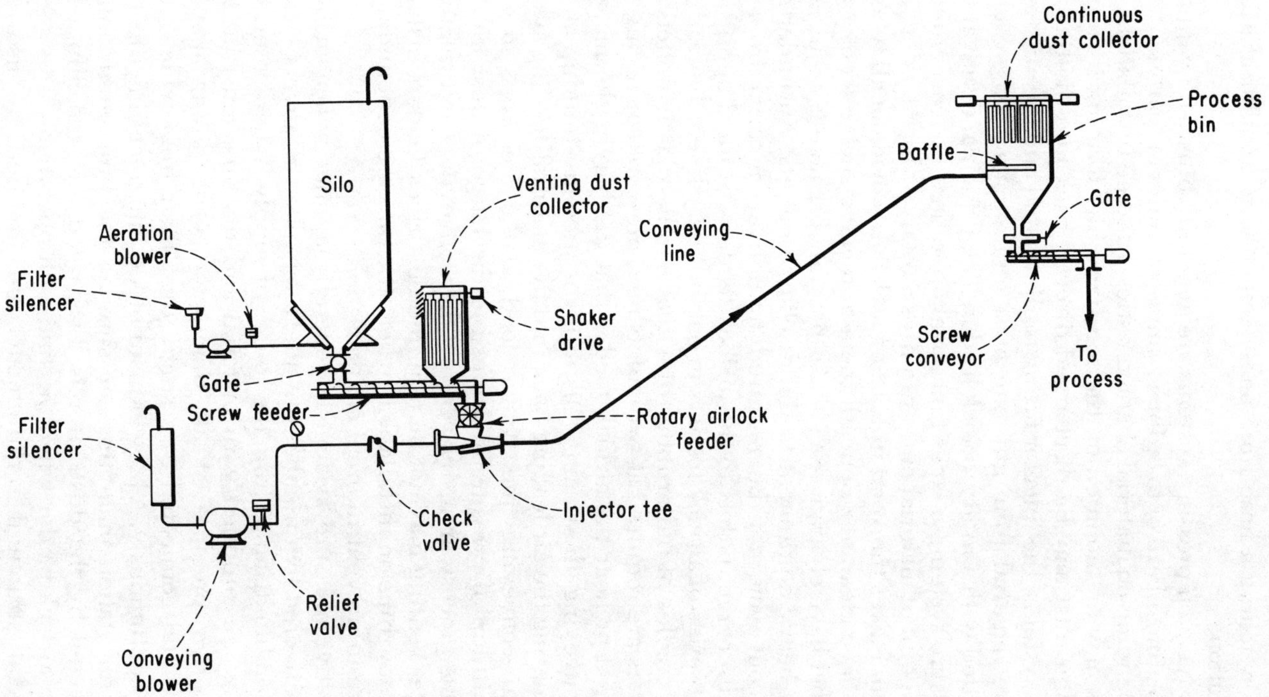

FIGURE 7–1. Positive-pressure, material-into-air system.

manually switching a hose from one connection to another at a hose selector station.

The receivers on pressure systems are generally bins with vent connections to atmosphere or to ordinary single or multicompartment dust filters, depending on material characteristics. In some installations, the dust filters may be mounted on the receiving bin, forming its top, so that the filter dust may be discharged directly back into the bin.

In some systems, the injector fitting is eliminated and the air stream can receive material from any one of several storage bins, each of which is fitted with a rotary airlock feeder. The principal disadvantage of the pressure system, so far as functional operations are concerned, is that every point of material entry requires a rotary airlock feeder. This requirement makes transport unloading awkward since not all types of transports have clearance beneath them to receive such a feeder and, for those that have clearance, a special portable carriage is required to attach this heavy equipment to the discharge outlets. An undertrack chute or conveyor may be required to deliver material from the transport discharge outlets to the airlock feeder. An ideal application for use of a series of airlock feeders in one delivery line is beneath the unloading cradles of bulk rubber bins that handle plastic pellets.

Dust filters are required at two points on a pressure conveying system: First, at or near the airlock feeder that discharges the material into the conveying air stream. This relieves pressure buildup at that point by venting the air leaking back into the material-feeding system from the air-conveying line. And second, at the receivers, so as to prevent buildup of pressure and possible discharge of dust into the surrounding process area, by venting the conveying air directly to atmosphere. Venting may not be required on airlock feeders installed directly below bins on installations where the bins are kept under dry air pressure to prevent internal condensation.

The positive-pressure system is generally applied when material must be delivered to a number of widely separated receivers located at a considerable distance from the source of supply. Where a series of storage bins are supplied with material and the bins must be kept under positive pressure at all times, the vent piping may be manifolded and connected to a single dust filter from which air is exhausted continuously. A blower manifolded into the filling lines at a hose selector station maintains pressure on the bins continuously. Dust collected by the dust filter is discharged into the filling line through an airlock feeder when the line is in operation.

No special structural requirements are imposed by use of a positive-pressure system other than the need for an underground transport unloading installation, where desired and location of bin venting dust filters.

The control requirements may be reduced considerably by eliminating diverter valves and using manual switching at hose selector stations. These stations can be set up to fill several bins from one point, but additional piping will be required. Hose connections can be used to cross-connect parallel pipe lines to perform several different functional operations using the same blower as an air mover. Obviously, where some of these operations overlap and cannot be staggered to make use of one blower, one or more additional blowers will be required.

The positive-pressure system has the inherent defect, common to all material-into-air-stream systems, of absorbing a considerable amount of power for moving air. In fact, the power required to supply air to the system far exceeds the power required to move the material to its destination mechanically. In many cases, the air lost by leakage of air through the airlock feeders may amount to 20 percent of the air supplied, if maintenance of the feeder is neglected.

Negative-Pressure System

This is the so-called vacuum system, in which a high-velocity airstream is established in the conveying line by air entering the suction side of a positive-displacement or centrifugal blower. Material is drawn directly into the conveying line by airflow into a pickup nozzle, or hopper. It may enter via a rotary airlock feeder supplied with material by gravity or by a mechanical conveyor. The airstream then conveys the material to a receiver-separator, from which the material is discharged through a second rotary airlock feeder to one or more storage bins. The conveying air passes through a dust filter and the vacuum producer, and then to atmosphere. This system is generally used where material must be picked up at each of several sources of supply for delivery to a receiver-separator at one location, or where it is preferable to have air enter the system at leakage points, rather than have dust leak out of the system at such points. See Figure 7–2.

However, although this system is recommended for pickup of material from several sources, it can also deliver material to several points of discharge; but it requires a considerable amount of additional equipment, as shown in Figure 7–3. Each point of discharge requires a receiver equipped with a discharge airlock feeder and a pair of diverter gates; one for connecting the receiver to the conveying main and one for connecting it to the dust return line.

The vacuum system is ideal for unloading bulk transport vehicles, many of which are equipped with pickup nozzles that require only a hose connection to the system for unloading. The material is drawn into the receiver-separator and is discharged by gravity into a turnspout

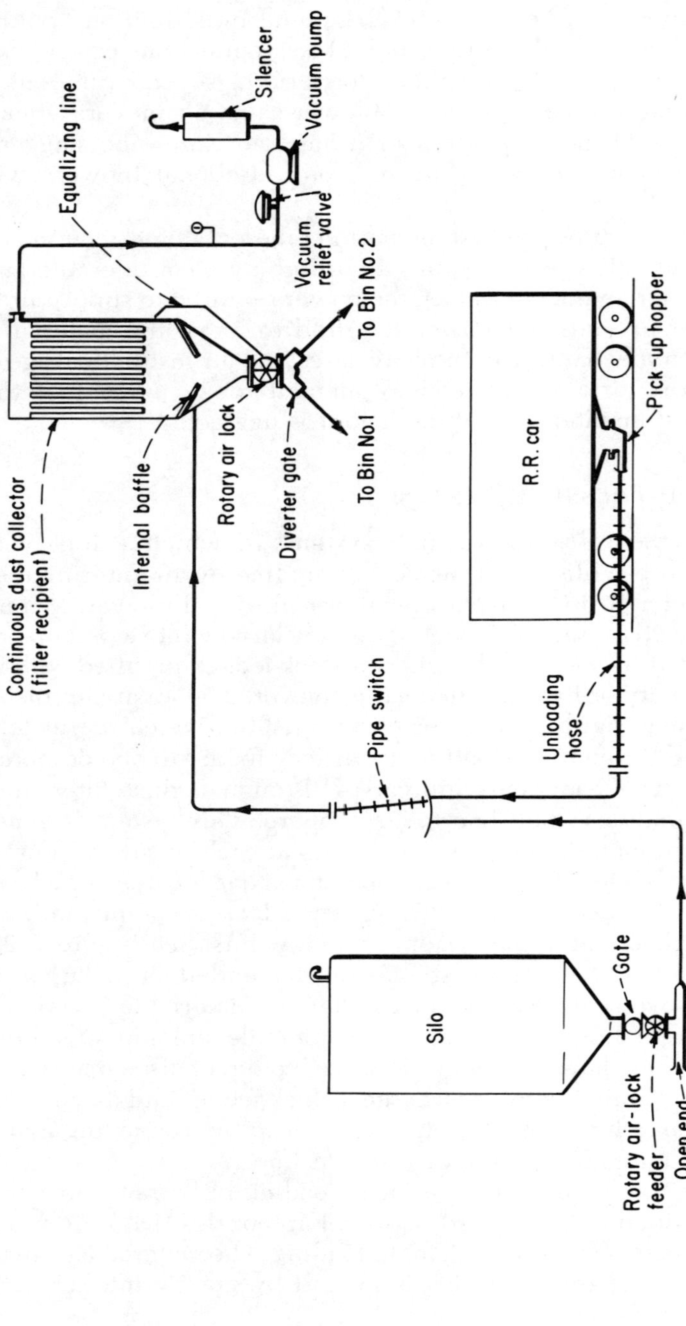

FIGURE 7–2. Negative-pressure, material-into-air system.

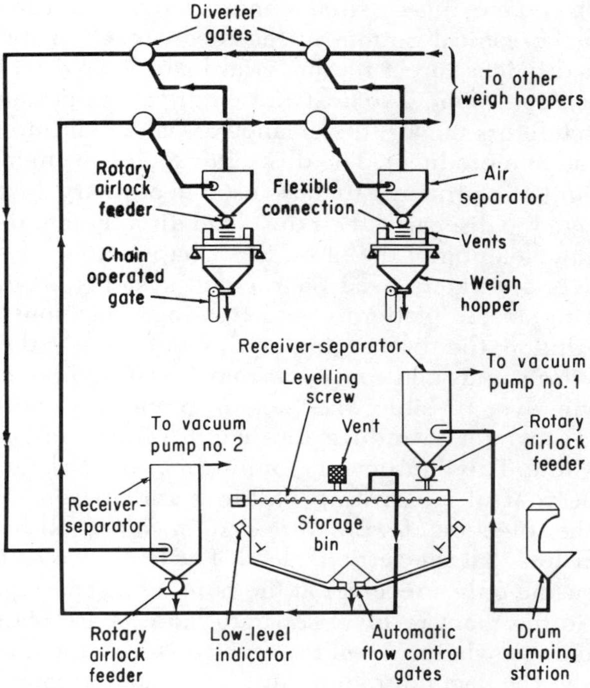

FIGURE 7–3. Negative-pressure, closed-loop system.

leading to any one of several bins, or into a mechanical or air-activated gravity conveyor system for transfer to any one of several discharge points.

Vacuum systems can be used to perform any one of several functional operations, one at a time, by installing diverter gates or hose selector stations at strategic locations so that piping may be interconnected with the equipment required to perform these functions. Thus one vacuum pump and receiver-separator may (1) unload a carload of material, (2) reclaim material from any one of several storage bins, (3) transfer material from one storage bin to any other storage bin, and (4) deliver material to a process area by simply installing a receiver above the use bin and running a dust return line from it to the receiver-separator.

Where several functional operations must be performed simultaneously, one or more additional vacuum pumps and receiver-filters are required.

Receiver-separators (so-called filter recipients), used on negative pressure systems are generally a combination of dust filter and cyclone separator. The conveying-air stream enters tangentially into an annular space in the conical bottom of the receiver or separator section. Here, internal baffles direct the material load downward toward the receiver discharge spout. The dusty air enters the upper, or dust filter section, which filters the air; the cleaned air is drawn into the suction line of the vacuum producer. The dust-filter section contains elements which are similar in principle to those used in ordinary dust collectors, but are arranged to discharge their dust load directly into the separator section during cleaning of the filter. See Chapter 10.

The receiver-separators used on vacuum systems are rather large units requiring access platforms, and lighting stanchions for several levels. In addition, the discharging equipment such as the rotary airlock feeder, turnspouts and conveyors may be housed beneath it. Furthermore, wherever possible, the vacuum pump is installed close to the receiver-separator to reduce the length of air suction piping as much as possible. This equipment grouping may be too heavy for installation on the roof of an existing building and in some cases, may determine the selection of a pressure system so that all heavy equipment is in an area that can carry the load. The loading on a roof may be reduced by using only a receiver at the point of discharge with a dust return line to the main receiver-separator and its associated vacuum pump, located elsewhere. When the receiver-separator is located atop a bin or tower, the vacuum pump must be located at ground level, or on a floor within an adjacent building. The length of vacuum piping must then be increased to suit the pump location.

A receiver-separator that requires shakers to clean the filter sections involves additional structural problems since all supports must be designed to resist impact loads induced by the moving mass of mechanism and dust-loaded filters.

The air and conveying lines on a vacuum system are of larger size than those used on pressure systems that require the same amount of free air for conveying, due to the decrease in density of air under vacuum.

The control requirements on vacuum systems may be reduced by using manual hose selector stations as described above for pressure systems.

The vacuum system has an inherent defect, common to all material-into-air-stream systems, namely, the amount of power just to move air through the system. There is no loss of air but rather an increase in the amount of air along a conveying line due to in-leakage under suction. Although the in-leakage is kept to a minimum in order to prevent entrance of rain, some of the equipment such as diverter gates and airlock feeders are designed to have air-swept bearings in order to keep them free of dust. This design feature generally distinguishes vacuum system equipment from pressure system equipment where other methods must be used to keep bearings dust free.

Combination System

This system uses a combination of the positive- and negative-pressure systems to perform two functional operations simultaneously—namely, pickup of material from each of several sources of supply and discharge to each of several delivery locations. Another use is where vacuum unloading of transports is required, along with pressure delivery to several locations. The material is drawn directly into a negative-pressure system and taken out by a receiver-separator discharging the air to a vacuum producer as described above. However, the vacuum producer then discharges the filtered conveying air into a positive-pressure system. Here, the material discharged from the negative-pressure receiver separator enters a rotary airlock feeder and is discharged into the positive-pressure conveying line. See Figure 7–4.

The complex pneumatic conveying systems seen in many industries are actually groupings of several combination systems; and the number of functional operations that can be performed is extended considerably. The factors that determine selection of positive- and negative-pressure systems also apply to the combination system. Analysis and careful study of the required functional operations will indicate which type of system will predominate, but each blower should be made to serve at least two functions, if possible.

Closed-Loop System

A variation of the positive-pressure system is the closed-loop system. This is essentially the positive-pressure system described above, but with the air or conveying gas vented through a pipeline back to the suction of the air mover instead of to the atmosphere. Figure 7–3 shows another form of closed-loop system operating under negative rather than positive pressure. Here material is picked up from one or several points by a conventional vacuum system and delivered by a closed-loop vacuum system to any of several weigh hoppers.

Recycle arrangements are used in several situations: (1) Where the material or conveying medium must be free from moisture and contamination; (2) when an inert gas must be used for conveying a hazardous material; (3) when the vented air cannot or need not be passed through a fabric dust filter; or (4) when the dust load at the receiver is very light

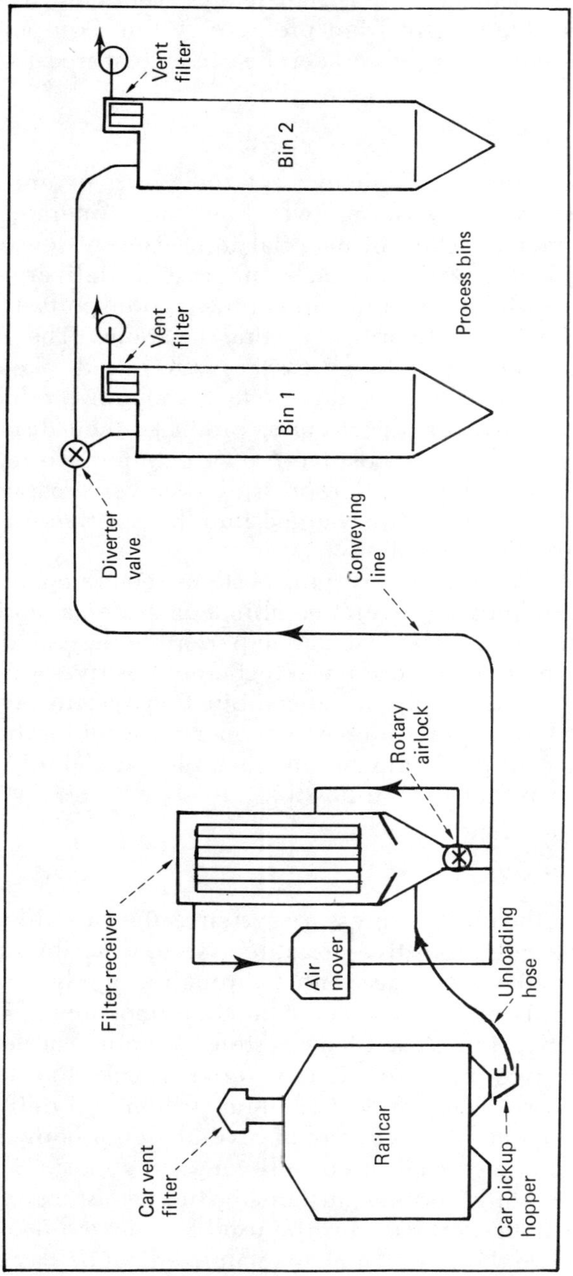

FIGURE 7–4. Vacuum/pressure system unloads transports and delivers material to bins.

but must be contained in a closed-circuit system. Additional equipment, not normally used in other types of conveying systems, is required for loop systems that use inert gas for conveying. The system becomes even more complex when the conveying gas is nitrogen. All the inert gases are produced by combustion of natural gas or propane in a pressurized furnace. The products of combustion are cooled in a heat exchanger, and moisture, one of the products of combustion, is removed in a separator. This assemblage of equipment is called an inert gas generator. When a lower moisture content is desired, the gas is compressed or refrigerated and passed through a desiccant dryer that is continuously reactivated. The inert gas is piped to the inlet side of the conveying blower through a pressure regulating valve or a regulator controlled by an oxygen analyzer. A fabric line filter is installed at the blower discharge to collect dust from the products of combustion and broken down particles of desiccant. (See Chapter 13.)

When nitrogen is required for conveying, the inert gas must be passed through a chemical absorption system for removal of the carbon dioxide. The CO_2 is purged to atmosphere and the nitrogen is discharged through a compressor, or, it is refrigerated and passed through a desiccant dryer to lower the moisture content in the same manner as for inert gas.

The equipment used on inert gas conveying systems must be gas-tight and all material must enter and leave the piping system through rotary airlock feeders.

When the loop system is used for dust return, a continuously cleaned filter must be used at the inlet of the conveying blower with dust discharged through an airlock feeder, into the delivery line.

AIR-MIXING SYSTEMS

Systems in which air and material are intermixed in a special feeder are, in effect, positive-pressure systems. They operate at relatively high pressure and low air-to-material ratios. From the discharge point of the feeder onward, they resemble the positive-pressure systems described above, except that pipelines are smaller and dust-filter size requirements are reduced.

Two types of special feeders are in common use. One employs a special rotary feeder that receives material directly from a shallow surge hopper fed by a mechanical or air-activated conveying system (Figures 7–5 and 7–6). Air is supplied to the end covers of the feeder from a high-pressure blower and enters each side of a rotary feeder pocket just after the filled pocket passes beyond the material inlet from

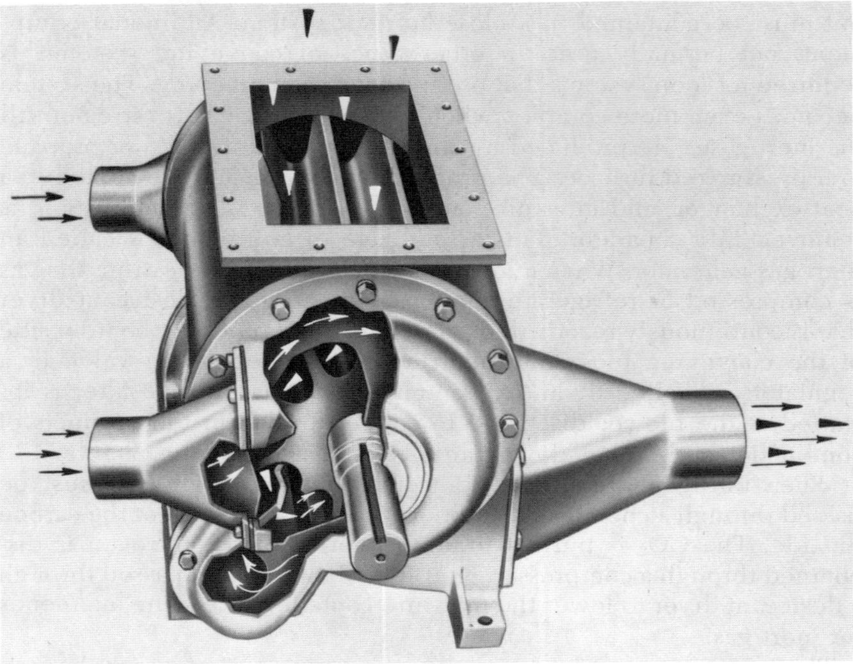

FIGURE 7–5. Interior view of air-swept, double-entry feeder valve. (CEA-Carter-Day Company.)

the supply hopper. At the entrance to the conveying line, the dense, high-pressure mixture is blown out and through the conveying system. Residual air in the feeder pocket is vented off into the surge hopper by a pre-vent connection on the feeder so that the empty pocket is ready to receive another charge of material. This type of feeder is very sensitive to the head of material in the surge hopper. Too high a head of material prevents effective venting; consequently, conveying may cease completely.

The other type of special feeder (Figures 7–7 and 7–8) uses an impeller screw with gradually decreasing pitch which receives material from a hopper supplied by gravity, or by a mechanical or air-activated conveying system. This "choke" screw compacts the material as it is forced through a barrel housing and check valve into a mixing chamber. The mixing chamber is fitted with several nozzles that are supplied with high-pressure air from a sliding-vane or centrifugal compressor, and directed toward the entrance to the conveying pipeline. The entering material drops directly into the stream of air

FIGURE 7–6. External view of air-swept, double-entry feeder valve. (CEA-Carter-Day Company.)

and is conveyed through the line. This impeller screw is supported in an overhung position by a pair of external bearings; it therefore requires a balancing chamber and weights at its discharge end in order to reduce vibration caused by wear.

A more recent design of impeller screw is supported by bearings at both ends which virtually eliminates vibration and, hence, screw damage when operating under no load or light load. The discharge of the screw is at right angles to the shaft axis and compression is effected by reverse pitch flights opposing the impeller flights. The discharged material enters the same type of mixing chamber as in the overhung shaft design. See Figure 7–9.

Air-mixing systems are limited to conveying pulverized and powdered materials, due to the high material-to-air ratios used and their dependence upon the fluidity of aerated material for successful conveying. These systems are used for conveying semi-abrasive material relatively long distances both with a minimum of air and in a dense mixture that tends to reduce pipeline wear. The low air-to-material

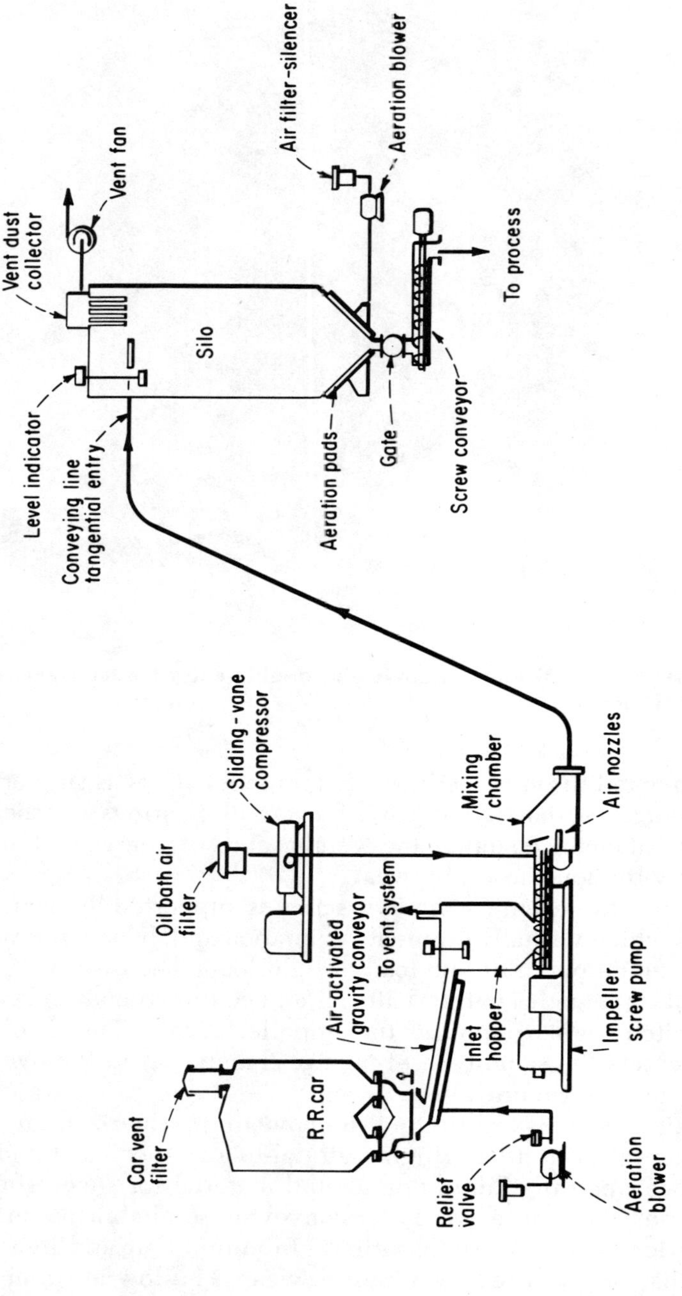

FIGURE 7-7. Air-mixing system.

FIGURE 7–8. Fuller-Kinyon air-mixing system. (Fuller Company/GATX.)

ratio used makes this type of system useful for handling material that must be kept sanitary and free from airborne spores.

In the impeller-screw type of feeder, pulverized materials are required in order to develop a compression seal at the discharge end. This prevents blow-back from the pressure zone.

The maximum operating pressure on air-mixing systems using the special rotary airlock feeder is 15 psig. This limit is determined by the pressure ranges of available low-volume positive-displacement blowers when they are compounded, that is, used in series to boost the pressure. Systems using the impeller screw feeder may operate at pressures up to 40 psig. Compressor pressure during operation is normally 4 to 6 psig above line pressure. This difference is the pressure drop across the air injection nozzles located in the mixing chamber at the discharge end of the impeller.

The air-mixing systems are essentially pressure systems and can deliver material to any one of several discharge points through diverter gates. However, since material and air are mixed at only one point, the feeder's versatility is limited. All material must be supplied to the pump hopper by gravity, or, from a remote area, by an auxiliary system of mechanical or air-activated gravity conveyors. The hoppers must be located in a pit or in an underground vault in order to unload a bulk transport vehicle.

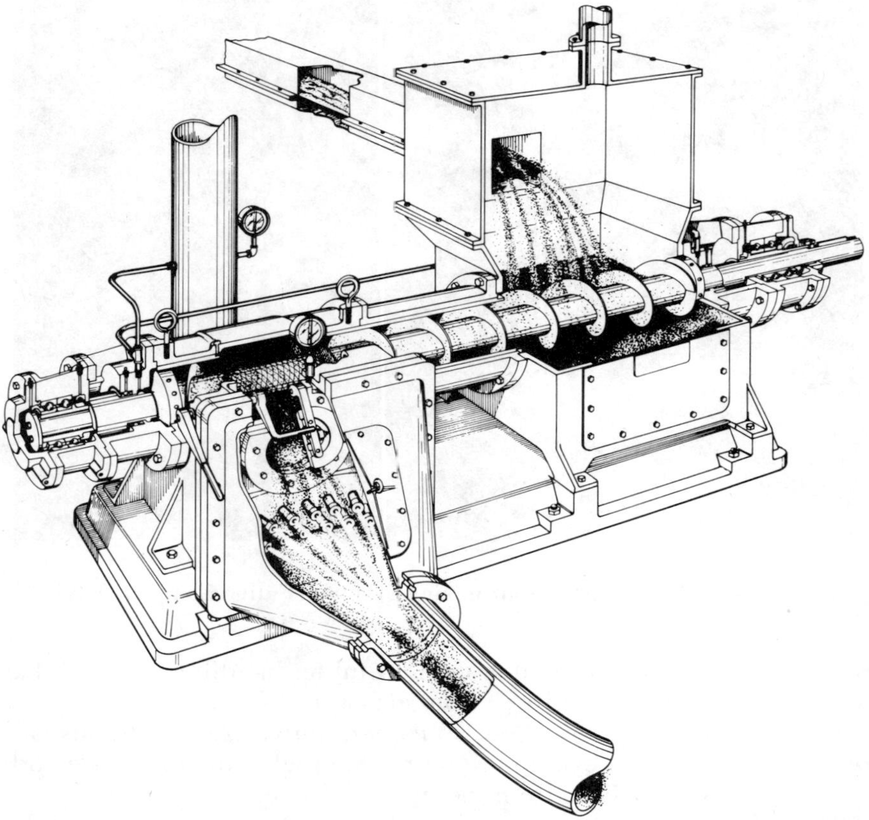

FIGURE 7–9. Modern Type "M" design of air-mixing system. (Fuller Company/GATX.)

The rotary airlock feeder type of system loses a considerable amount of air through the feeder due to the high pressures used for conveying. However, the high conveying rates obtainable more than compensates for this power loss.

The impeller screw type of system absorbs a variable amount of power in the air compressor depending upon the conveying line loading. The line loading, in turn, affects the conveying line pressure. The impeller screw absorbs power due to the compression of material as it is forced through the seal against the conveying line pressure. When no material enters the screw, the power is low. When highly aerated material enters the screw, it can seriously reduce the efficiency of the impeller screw pickup. Power is also reduced and the material may not even enter the conveying line.

AIR-INTO-MATERIAL (BLOW-TANK) SYSTEMS

Systems in which air enters a mass of material to cause flow are exemplified by several commercially available in-plant blow-tank set-ups. In general, these comprise a cylindrical pressure vessel with a conical bottom, which has the following features and attachments: (1) A large-diameter material-inlet valve operated by an air cylinder at the top of the vessel and fitted with an outsize pressure-seal ring gasket to insure seating. (2) An air-cylinder-operated vent valve and vent port at the top of the vessel to release displaced air during filling. (3) A level detector at the top of the vessel, weigh scales, or spring-mounted supports, for determining when the tank is full of material. (4) Nozzles or aeration pads along the sides and near the bottom of the tank for admission of conveying air from a blower or high-pressure air receiver in such manner that the material is nearly fluidized. (5) A powder-sealed discharge pipe with or without an auxiliary booster air nozzle to augment the fluidizing air and assist in transporting the material through the conveying line. (6) A control panel for sequential operation of valves to empty and fill the tank on an intermittent automatic cycle, or manually, as required. Several types appear in Figures 7–10 to 7–12, inclusive.

A high-pressure air compressor sold with the equipment supplies air to an air receiver that discharges to the blow tank through a solenoid valve. Air may be supplied by a blower discharging directly into the blow tank, or from an existing plant's compressed-air system if sufficient extra capacity is available and the air is clean and oil-free.

Starting with an empty blow tank, operation is as follows (see Figure 7–13): The vent and material-inlet valves open, and material enters until the tank is filled, as detected by the level detector, by a relay actuated by settling of the tank on its mounting springs, by a reed switch at a pre-set weight on a weigh scale or by a pneumatic load cell. Displaced air is vented off into the material-feed hopper or into the conveying line. The level detector or weight-operated relay actuates electrical controls to close the vent and material-inlet valve and open the air-supply solenoid valve. Air from the air receiver, or blower, enters the blow tank, percolates through the material and builds up pressure until the highly aerated material starts to flow through the conveying line. The conveying-stream flow rate and the blow tank and air-receiver pressure vary throughout the conveying period, as shown by the typical set of curves in Figure 7–14.

These curves show the performance of a free-feeding system in

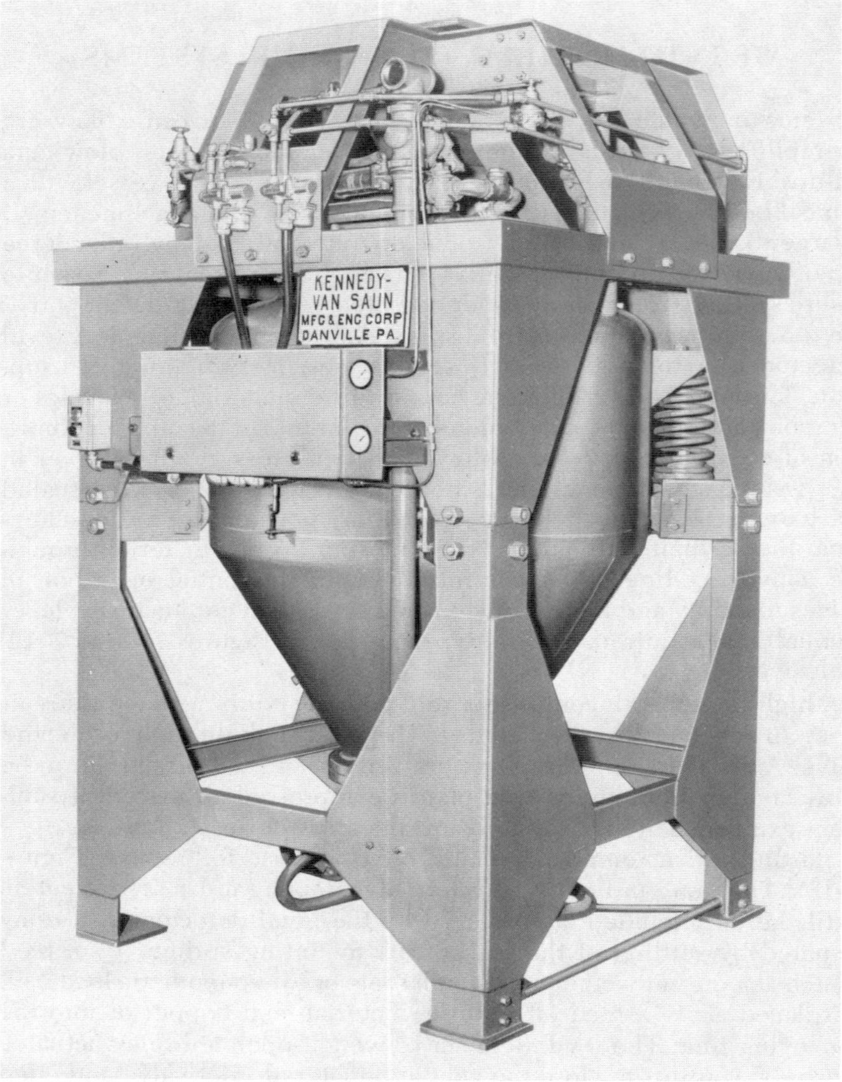

FIGURE 7–10. Kennedy automatic pneumatic transport pump. (Kennedy Van Saun Corporation.)

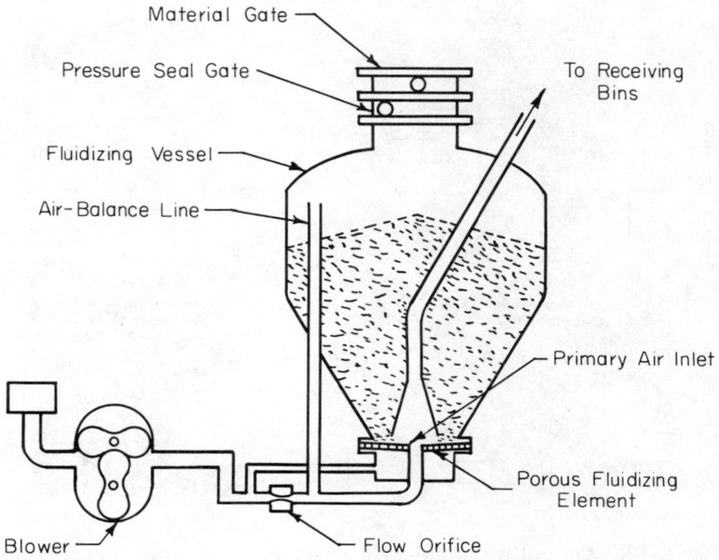

FIGURE 7–11. **Low-pressure blow tank. (The Ducon Company.)**

which the material-to-air ratio is low at the start and finish of the discharge cycle, since the air flow is high during those portions of the cycle.

Figure 7–15 shows the performance of a controlled-feed system in which a constant material-to-air ratio, hence, constant velocity of discharge, is maintained by metering air into the tank in proportion to the loss in tank weight as detected by a load cell that supports the tank.

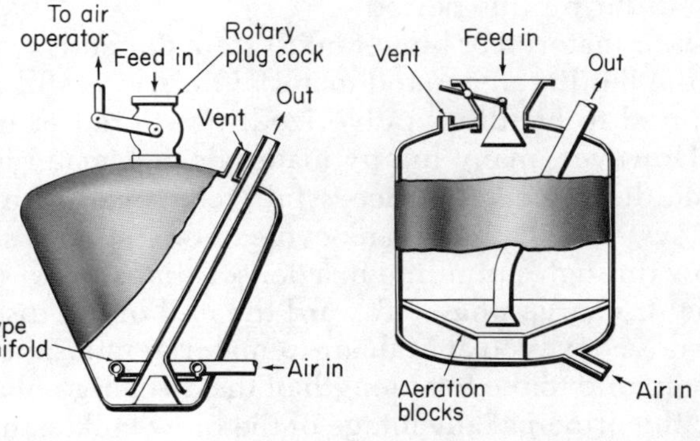

FIGURE 7–12. **Variations in blow-tank design: *left*, high pressure; *right*, low pressure.**

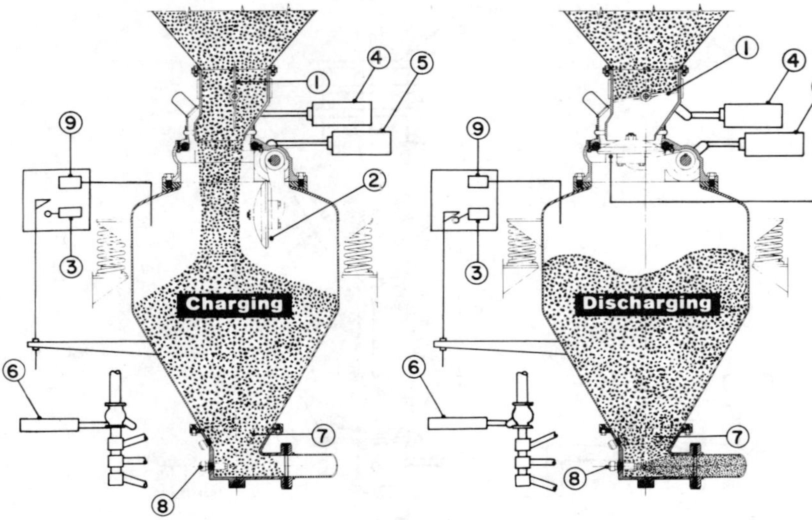

FIGURE 7–13. Operating cycle of a high-pressure blow tank. (Kennedy Van Saun Corporation.)

When all material has left the tank and only air is flowing through the conveying line, a low-pressure switch (or a weight-operated relay that detects the rising of the tank on its mounting springs) actuates electrical controls to both close the air-supply valve and open the vent and material-inlet valves to start the next cycle. Meanwhile, as material flows into the blow tank, pressure in the air receiver is increased to normal conveying pressure by the air compressor. On blow tanks that receive air from blowers, the blower must be stopped or its discharged air bypassed during this period.

The air-into-material or blow-tank system depends to a considerable extent on the fluidity of aerated material for successful conveying and is thus limited to handling pulverized, powdered or mixed granular material. However, many lumpy materials and materials that are not normally fluidizable can be successfully conveyed by a high-pressure blow-tank system. The blow-tank cycle of operation is such that material can flow through a pipeline in a dense, almost solid stream through most of the discharge phase. Toward the end of the discharge period, the material is conveyed at high air-to-material ratios, but this period is so short compared to the total length of the discharge phase that it does not negate the principal advantage of the blow tank, namely, its ability to convey material at very low velocity. The low velocity of the conveyed stream makes the blow tank ideal for conveying powdered

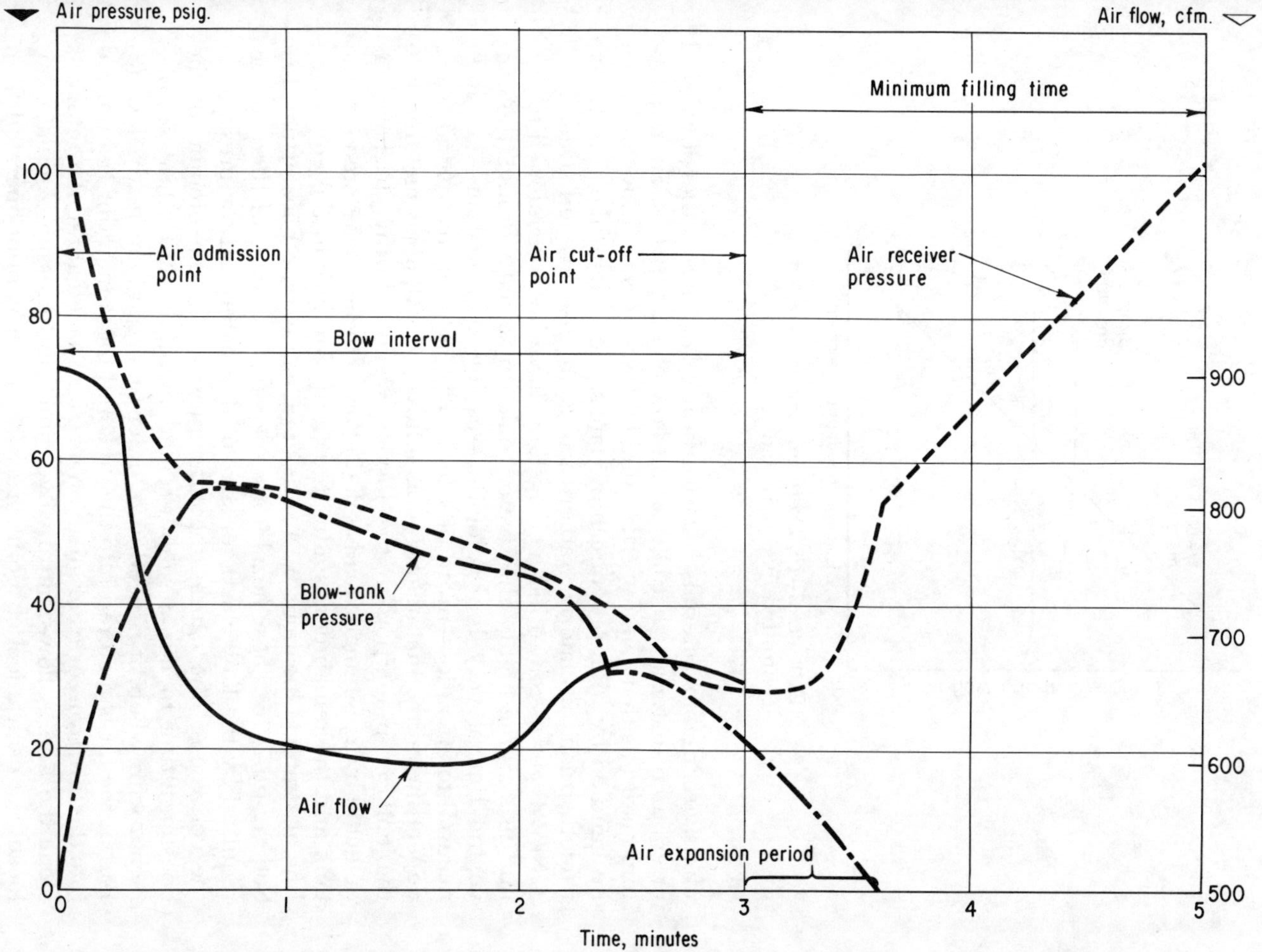

FIGURE 7–14. Record of air flow and pressure during the operating cycle of a high-pressure blow tank.

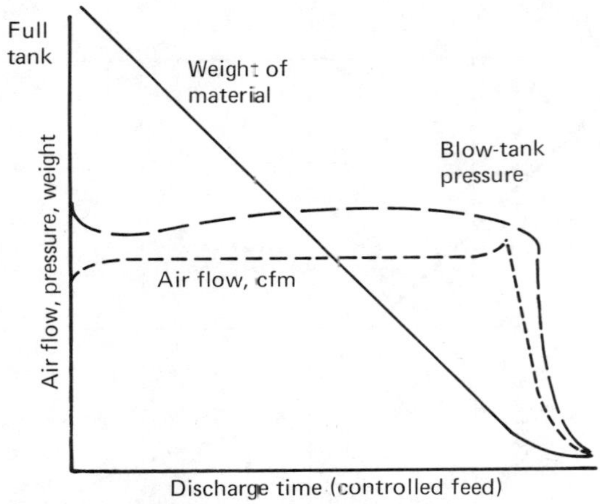

FIGURE 7–15. Performance curves of a controlled feed blow tank.

abrasive or friable materials, but it can be used to convey other powdered or granular materials that have good flow characteristics. Some problem materials that cannot be handled by low-pressure blow tanks can be handled by high-pressure blow tanks, more than likely as an extruded stream of material, rather than as an air-conveyed stream.

New types of controlled-feed systems have been developed to permit discharging highly abrasive or friable materials through the conveying line at very low velocities—as low as 150 fpm—and high material-to-air ratios—up to 500 to 1. This is done in several ways, depending on the manufacturer, as follows: (1) All air is metered into the tank in proportion to the loss in weight during discharge as detected by a pneumatic load cell. (2) Part of the air is metered into the tank and the remainder is admitted peripherally into the conveying line at strategic locations by a series of Coanda-type booster nozzles. See Figure 7–16. These nozzles are spaced at intervals of 10 to 40 ft as required by the flow characteristics of the material. Low pressure in the conveying line causes the high pressure in the plenum of the booster fitting to flex a rubber-sealed manifold and admit supplementary air where it is needed in order to maintain a steady flow. (3) Part of the air is metered into the tank and the remainder is admitted into the line by a series of injection nozzles spaced at regular intervals along the line. A flow-control valve, filter, and pressure regulator are located at each injection point. Each regulator is pilot-operated from the next upstream booster point in order to maintain a uniform pressure drop along the line. (4) Part of the air is metered into the tank and the remainder is admitted into an internal tracer pipe that is built into the conveying line and has air jets or perforations distributed along its length. The material is introduced into the line as a series of solid plugs. The tracer air jets build up pressure between the plugs in order to maintain their flow toward the discharge point.

Pilot-plant testing is definitely required for materials intended for use in these systems. Systems fitted with booster nozzles or jets will require initial field setting of air flow at each entrance or injection point even after installation until uniform flow is attained.

Although the blow-tank cycle is an intermittent one, a continuous discharge may be effected by installing a pair of blow tanks, arranged to operate alternately. While one tank is discharging material, the other would be receiving material. The fact that operation is intermittent does not limit the capacity of a blow-tank system. The range of tank volumes and operating pressures available is such that any required discharge rate can be met. Volumes range from one to 400 cubic feet and pressures range from 10 to several hundred pounds per square inch, depending upon the application.

The circumstances that would require the use of two blow tanks in parallel are as follows: (1) where headroom is limited and a single,

FIGURE 7–16. Blow tank used as a blender. Booster fittings at discharge and on transfer lines at manual switching station. (Dynamic Air Incorporated.)

large volume tank cannot be accommodated; (2) when the capacity of the material inlet feeder is too small to meet the total cycle time requirement for the use of one tank; (3) when required discharge velocities are so low as to increase the total cycle time beyond the limit for use of one tank; (4) when a standby blow tank is required in order to ensure material supply to a continuous process. In the first three circumstances, the tanks would be operated alternately, so that only one of them is discharging into the conveying line at any given time.

Air-into-material systems can operate at a variety of pressure ranges. Most low-cost systems are limited to a maximum design pressure of 15 psig in order to be free of A.S.M.E. Unfired Pressure Vessel Code requirements, but systems can be designed for pressures as high as required to convey over very long distances. Most high-pressure tanks are designed for use with the 100 psig compressed air systems that are installed for general utility service in most plants, but the conveying pressures are usually lower than this due to the nozzle effect of air entering the tank during the discharge period.

Blow-tank systems have the same disadvantage as air-mixing systems, namely, the fact that all material must be supplied by gravity or by a system of auxiliary conveyors. Delivery can be made to any one of several discharge points as in any pressure system. However, a blow-tank system has one great advantage over all other systems in that all materials entering it could be weighed by simply mounting the tank on a weigh-tank scale. A fairly accurate volumetric check of quantities delivered may be made by counting the number of discharge cycles, and estimating the volume of material in the tank by checking the level at the material cut-off point and calculating the volume from the tank dimensions.

The numerous designs of blow tanks available makes selection difficult. Some are patterned after the common, pressure-differential, bulk trailer; some are high-pressure tanks with valves and controls attached to the tank; and some are highly integrated transport pumps with all valves and controls built-in. Many blow tanks require a high-pressure air compressor and air receiver unit for air supply and some are specifically designed to use air directly from the discharge of a positive-displacement blower.

The following factors can serve as a checklist of functional requirements for blow-tank installations and provide the design details necessary to meet these requirements.

Protection

(1) Blow tanks operating at pressures higher than 15 psig should be constructed in accordance with the A.S.M.E. Code for Unfired

Pressure Vessels, or a State or Municipal code based on the A.S.M.E. Code. (2) Blow tanks operating at less than 15 psig, though not requiring A.S.M.E. Code testing or inspection, should be constructed in accordance with that Code. (3) Safety valves set to relieve pressures in excess of the safe working pressure of the tank should be mounted directly on the tank. Its size is determined by A.S.M.E. Code requirements.

Filling

(1) The material-inlet valve should close against its seat after material flow is cut off by an external butterfly gate or batching valve. This prevents trapping of material between valve and seat with consequent leakage of air. (2) The material flow into the tank should be cut off at the height corresponding to the tank's working volume as detected by a level probe or sensor; by a weight-actuated switch, on scale-mounted tanks; or, by a position-stop relay, on spring-mounted tanks. Do not depend on the setting of a time delay relay to cut off material flow into the tank since this flow can vary greatly and may cause overfilling of the tank with consequent failure of the material-inlet valve to seat tightly. (3) Vent the tank directly using a vertical, self-cleaning riser discharging into a dust filter or bin that is under suction. Avoid kinks or bends in the vent line that can become blocked with powder. (4) Provide flexible connections between piping and spring- or scale-mounted tanks to allow for settlement of the tank when filled. (5) Provide an opening in top of the tank for visual check of material level. (6) Provide controlled flow of material into the tank wherever layout permits interposing a feeder between the blow tank and a supply bin. (7) Avoid the use of butterfly type valves as the material-inlet valve on high-pressure blow tanks. Air leakage around the valve stem will increase with time. If the butterfly valves are selected, use two in order to prevent the blowback of air. Interconnect their controls so that operation of the lower one is delayed to allow the upper one to close first and act as a batching valve. (8) Locate the air-cylinder operator for a material-inlet valve outside of the tank. Leaking packing on internal air cylinders may draw in materials at certain pressure differentials and foul the control valves.

Air Supply

(1) Provide an air compressor, a blower, or some source of air, large enough to meet the conveying cycle requirements including the use of auxiliary air nozzles for transport or tank cleanout. (2) On air-compressor units, provide an aftercooler and some other form of air

dryer to insure a supply of dry air into the blow tank. (3) Where an air receiver is required, size it to recover full pressure before the material-inlet valve closes. (4) Provide an air-line lubricator on the air-supply lines to the air-cylinder valve operators. (5) Provide a check valve in each air-supply connection to the blow-tank transport and activator nozzles in order to prevent entrance of material into these lines. (6) Make certain that available air pressures are sufficient to clean out a plugged conveying line. (It may be impractical to clear solidly plugged lines with compressed air.)

Discharge

(1) Provide a valve in the discharge line when handling materials that are difficult to fluidize, or to control the flow of materials that are too easily fluidized. (2) Vent the discharge bins through reverse air blast or mechanically cleaned dust filters. (3) Size the discharge line to convey the maximum amount of material with the minimum amount of air at an operating pressure differential corresponding to the maximum working pressure in the blow tank.

Access for Servicing

(1) Provide an access manhole of generous proportions in the blow-tank shell for servicing internals such as air nozzles, aeration pads, conveying elbows, level sensors, and material-inlet valves. (2) Provide externally mounted activator nozzles with built-in check valves on high-pressure tanks that do not use aeration pads.

The selection of a pneumatic conveying system requires a thorough knowledge of equipment details since material properties and characteristics influence the design of equipment to an extent not fully appreciated by the average purchaser. (See Table 5–1, p. 84.) Some types of equipment may be used for several types of conveying systems. A detailed study of equipment appears in Part II of this volume.

FUTURE SERVICE

If, at a later date, the materials entering the blow tank may change from a fluidizable material to one that is not fluidizable, the blow tank may still be used. However, certain designs of tanks can use air more efficiently than can others for nonfluidizable materials. These designs are as follows, in order of preference: (1) Bottom discharge tanks. (2) Top discharge tanks that have a built-in nozzle that discharges air directly into the entrance nozzle of the conveying line. All other types will require higher air pressure or greater air volume than originally used in order to simulate fluidization. The material may not fluidize at all and may have to be discharged as an extruded stream of material. In such cases, the system may have to be replaced by one that is more suitable for the flow properties of the material.

CLASSIFICATION OF PNEUMATIC CONVEYORS

The advantages and disadvantages of the various classifications and types of pneumatic conveying systems is summarized in Table 7–1.

TABLE 7–1
Classification of Pneumatic Conveyors

Classification and type of system	Advantages	Disadvantages
Material-into-air	1. Handles widest range of materials, including irregular shapes.	1. Uses lower material-to-air ratios than other systems. 2. Requires larger equipment than other systems—air movers, pipelines, dust filters, etc.
Positive-pressure	1. Delivers material through a single pipeline to several discharge points. 2. Uses smaller pipelines than comparable vacuum systems. 3. Air leakage is outward so that moisture can not be drawn into equipment.	1. Requires an airlock feeder and blowback vent filter at each material entry point. 2. Requires large dust filter at each material discharge point. 3. Transport unloading requires use of fixed or portable airlock feeder installation.
Negative-pressure (vacuum)	1. Transports easily unloaded using pickup hoppers or nozzles. 2. Material can enter a single pipeline from several sources of supply without venting. 3. Material can enter line using airlock feeder, controlled-feed tank, or pickup hopper.	1. Material can be delivered only to a single discharge point. 2. Dust filter and airlock discharger required at the discharge point. 3. Moisture and outside air can be drawn into equipment. 4. Requires larger piping and equipment than positive-pressure systems, due to lower-density air. 5. Air mover and receiver-filter requires location atop silos or building roof.
Negative-pressure (vacuum) with dust-return loop	1. Delivers material to several discharge points. 2. Dust is returned to a single, conveniently located dust filter for re-entry into conveying line or silo	1. Requires a collector and an airlock discharger at each discharge point. 2. Requires diverter valves in the dust return and conveying lines at all but one discharge point.
Combination vacuum/pressure	1. Permits pickup of material from transports by vacuum and simultaneous discharge to each of several discharge points by pressure. 2. Dust from vacuum-side receiver is delivered into conveying line.	1. See above for respective portions of system—vacuum or pressure side. 2. Requires larger-capacity and -horsepower air mover than for the vacuum or positive-pressure system. 3. If air mover breaks down, two plant operations are affected—unloading and reclaiming.
Closed-loop, positive or negative pressure	1. Permits use of inert gas for conveying, with minimum make-up requirement. 2. Permits re-use of conveying air where air is dried or filtered to be free of contamination. 3. Eliminates the use of diverter valves on dust-return line at each discharge point.	1. Same as for combination vacuum-pressure system.
Air-into-material	1. Uses highest possible material-to-air ratio compared to other systems. 2. Requires smaller equipment than other systems—air movers, pipelines, dust filters, etc.	1. Operation is intermittent. 2. High pressure systems (over 15 psig) require ASME-Code-constructed vessels. 3. Surge hopper is required for rapid filling if auxiliary feed capacity is limited.

Classification and type of system	Advantages	Disadvantages
Air-into-material (continued)		
	3. Low velocity, dense mixture permits handling abrasive and friable materials. 4. High-pressure units and systems permit delivery through very long pipelines. 5. Delivers material to several discharge points using a single pipeline. 6. Each batch can be weighed before being conveyed.	4. A dust filter is required for venting surge hopper and blow tank. 5. Material must be delivered to blow tanks at higher than pneumatic conveying rates, due to limited filling time between discharge cycles.
Free-feed blow tank	1. Bottom-discharge units can handle lumpy, non-fluidizable and fluidizable materials. 2. Bottom-discharge units can be emptied completely when handling sanitary materials.	1. Material-to-air ratio is variable—high at start and finish of each cycle. 2. Top-discharge units leave residue in bottom of tank that may cause contamination of material.
Controlled-feed blow tank	1. Material-to-air ratio of discharge is uniform. 2. Conveying-line velocity can be reduced to the slug flow limit. 3. Booster nozzles may be installed in the conveying line to maintain dense, low velocity flow.	1. Use is limited to fluidizable, powdered or mixed-granular materials.
Air-mixing	1. Material-to-air ratio is high, but not as high as in air-into-material systems. 2. Conveying is continuous. 3. Delivers material to several discharge points through a single pipeline. 4. High pressure permits delivery through long, small-diameter pipelines.	1. Use is limited to fluidizable, powdered or mixed-granular materials. 2. Can not handle friable material. 3. Material must be delivered by metered auxiliary conveyors. 4. Feed hopper must be level-controlled and vented to dust filters. 5. Feed hoppers must be maintained full when handling abrasive material.
Feed-screw into air nozzles	1. Requires smaller auxiliary conveyors than blow tanks. 2. Requires lower headroom than blow tanks.	1. Subject to erosion and loss of capacity when handling abrasive material. 2. Requires skilled maintenance, as for high-grade machinery. 3. Must run fully loaded when handling abrasive material—wear increases when idling. 4. Requires high-pressure, sliding-vane or reciprocating air compressor. 5. Compressors require cooling water. Oil separators must be installed ahead of air nozzles.
Air-swept, double-entry rotary feeder	1. Can be made with cutting vane to handle tacky material such as sugar or detergents.	1. Pressure limit is 20 psig, requiring low-volume, positive-displacement blowers in tandem. 2. Capacity varies with head of material in feed hopper.

8 Vendor Selection and Procurement Specifications

SEARCH FOR A VENDOR

There are several ways of judging a prospective vendor of pneumatic conveying equipment so that the purchaser may be certain of receiving the full cooperative effort required for a successful pneumatic conveying system. However, the selection of prospective bidders from among the many manufacturers of this type of equipment comes first, and the search may start by perusing in three areas.

A Directory of Manufacturers

This will list many names, with the prominent display of the names of large, old-line companies. It also gives some indication of the financial standing of these firms, but gives no clue to the services, know-how, or equipment they have available to handle specific material.

Advertising in Trade and Technical Journals

The advertisements will generally take several forms. They may cite a particularly successful installation for a specific material; they may list available services or material handled; they may show actual photographs of a system, but without any indication of how much of the photographed equipment was furnished by them. Also, they may bruit the names of large corporations, again without indicating the quantity or kind of equipment sold to these corporations.

A marginal vendor may capitalize on a few successful conveying installations of relatively simple design. Another vendor, offering "systems engineering," may have supplied a conveying system of simple design that, with a few simple electrical interlocks, was integrated with process equipment. Other vendors have installed some automatic systems and now offer "automated conveying systems." Selection under the above circumstances is difficult, but look for vendors who have handled a specific material; who cite an installation that appears to meet the conveying requirements; whose claims appear to be limited to pneumatic conveying; and who manufacture a line of conveying equipment.

Articles in Trade and Technical Journals

These also take several forms. They may have been prepared by the vendor's sales or engineering personnel to describe a successful installation or to introduce a new line of equipment. These are broadly descriptive of methods, but generally lack the kind of detailed information essential to a real knowledge of pneumatic conveying. This the vendor reserves for himself; it is his stock-in-trade. Some articles describe the troubles encountered on a particular system by a purchaser who has discovered, too late, that a pneumatic conveying installation is a cooperative effort between purchaser and vendor. Much can be learned from this type of article about the pitfalls to be avoided, but very little specific or detailed information is offered on how to avoid them.

Still another type of article describes with many mathematical formulas the experiments conducted on a specific phase of pneumatic conveying or the conveying of a specific material. Such articles are generally very informative, but the application of this information is extremely limited and of little value in designing a practical conveying system intended for everyday production service. Here, again, the guidelines set down for vendor selection from advertising should hold.

NARROWING THE FIELD

Reviewing the various sources of information will limit the number of vendors who appear to be qualified to handle a prospective installation. The field may be narrowed considerably, and many vendors eliminated completely, by submitting to them a detailed specification prepared on the basis of information contained in this book. This specification may be prepared in tabular form giving basic information with details amplified in a separate procurement specification (Figures 8–1 and 8–2). The tabulated data should be grouped under several general headings as follows:

Process Specification. This will cover conveying rate, the services required for the conveying system and whether they will be performed independently or simultaneously, the type of system preferred, and the degree of control required.

XYZ CHEMICAL COMPANY

EQUIPMENT DATA SHEET

PNEUMATIC CONVEYING SYSTEMS FOR DRY BULK MATERIALS

MH 3405 PAGE 1 OF 2

PROJECT TITLE		DATA SHEET NO.
PROJECT LOCATION		DATE
PREPARED BY	REVISIONS	DATE

QUANTITY	ITEM NO. AND TITLE
SERVICE CONDITIONS	QUOTATIONS SHALL BE IN ACCORDANCE WITH XYZ CO. PURCHASE SPECIFICATION NUMBER

PROCESS SPECIFICATION

MATERIAL CONVEYED

CONVEYING RATE

SERVICE

1. UNLOAD TON HOPPER CAR(S) ON TRACK· CAR SPOT(S)
   AND DELIVER TO
2. TRANSFER TONS OF FROM TO
3. OTHER

OPERATIONS ☐ 1. ☐ 2. ☐ 3. ARE TO BE PERFORMED ☐ SEPARATELY ☐ SIMULTANEOUSLY

TYPE OF SYSTEM

☐ PRESSURE ☐ VACUUM ☐ PRESSURE-VACUUM ☐ TO BE RECOMMENDED BY VENDOR

CONTROL

☐ FULLY AUTOMATIC ☐ PUSHBUTTON START-STOP OF MOTOR DRIVEN EQUIPMENT ☐ AUTOMATIC POSITIONING OF VALVES AND GATES PRIOR TO START-UP

☐ AUTOMATIC SHUTDOWN AT HIGH LEVEL IN ☐ AUTOMATIC START-UP AT LOW LEVEL IN

☐ DUAL CONTROL FROM STATIONS AT WITH MASTER CONTROL AT

☐ CONVEYING TO STOP WHEN THE FOLLOWING CONDITIONS OCCUR

☐ OTHER –

MATERIAL SPECIFICATION

SPECIFIC GRAVITY		SETTLED SPECIFIC WEIGHT		AERATED SPECIFIC WEIGHT	
1.	2.	1.	2.	1.	2.
SIEVE ANALYSIS	RETAIN %		THROUGH %		
SIEVE SIZE	1.	2.	1.	2.	
80					
100					
200					
325					
AVERAGE PARTICLE SIZE SMALLER THAN 325 MESH	MICRONS / INCHES	MICRONS / INCHES	MICRONS / INCHES	MICRONS / INCHES	

ABRASIVENESS

SAMPLES ☐ FORWARDED HEREWITH ☐ SHIPPED SEPARATELY

FIGURE 8–1. Typical process and material specification sheet.

XYZ CHEMICAL COMPANY

EQUIPMENT DATA SHEET

PAGE 2 OF 2

MH 3408

PNEUMATIC CONVEYING SYSTEMS FOR DRY BULK MATERIALS

ELECTRICAL SPECIFICATION

MOTOR DRIVES	TYPE; SERVICE; PHASE; CYCLE; VOLTS
	STARTERS FURNISHED BY
CONTROLS	SERVICE; PHASE; CYCLE; VOLTS
	PANELS - N.E.M.A. TYPE
	RELAYS ☐ INTEGRAL WITH PANEL
	☐ SEPARATE MOUNTING PLATE FOR INSTALLATION IN
	CONTROL TRANSFORMERS FURNISHED BY
	INTERLOCKS ☐ AS RECOMMENDED BY VENDOR ☐ OTHER –
	TERMINAL MARKINGS ☐ VENDOR STANDARD ☐ XYZ CO. STANDARD; ALARM HORN ☐ INTEGRAL WITH PANEL ☐ FURNISHED SEPARATELY BY

NOTE: CONTROL EQUIPMENT EXTERNAL TO PANELS WILL BE FURNISHED BY XYZ CHEMICAL COMPANY

MECHANICAL SPECIFICATION

FILTER RECEIPIENTS	SERVICE ☐ CONTINUOUS ☐ INDOOR ☐ INTERMITTENT ☐ OUTDOOR ☐ WEATHER ENCLOSURE AT BOTTOM ☐ INSULATION REQUIRED
	FILTER MATERIAL
	SERVICE PLATFORMS FURNISHED BY
VACUUM PUMPS OR BLOWERS	SERVICE ☐ INDOOR ☐ OUTDOOR
	BASE ☐ CONCRETE ☐ STRUCTURAL STEEL ☐ FURNISHED BY VENDOR ☐ FURNISHED BY XYZ CHEMICAL COMPANY
SILENCER	☐ MOUNTING LUGS FURNISHED BY VENDOR ☐ MOUNTING LUGS FURNISHED BY XYZ CHEMICAL COMPANY ☐ LUG LOCATION DRAWING BY VENDOR ☐ LUG LOCATION DRAWING BY XYZ CHEMICAL COMPANY

DIVERTER GATES AND VALVES	DUTY	SERVICE		OPERATION	
		INDOOR	OUTDOOR	MOTOR	MANUAL
	1.				
	2.				
	3.				
	4.				

CAR UNLOADING HOSE
SECTIONS . LENGTH PER SECTION TYPE OF HOSE CONNECTION

RECLAIMING SCREW FEEDERS FURNISHED BY ☐ VENDOR ☐ XYZ CHEMICAL COMPANY

PROTECTIVE FILTERS FURNISHED BY ☐ VENDOR ☐ XYZ CHEMICAL COMPANY ☐ NONE

MODIFICATIONS TO PROCUREMENT SPECIFICATION NUMBER SP-1304 (NOTE: THE FOLLOWING PARAGRAPHS HAVE BEEN MODIFIED - SEE PURCHASE ORDER SPECIFICATION FOR MODIFICATIONS)

FOR REMARKS SEE ATTACHED SHEET

FIGURE 8–2. Typical electrical and mechanical specification sheet.

Material Specification. This will include all known properties and flow characteristics of the materials to be conveyed.

Electrical Specification. This will include tabulation of available electric-current characteristics for power and control, type of enclosures required for starters and panelboards, interlocks required, terminal-marking standards, and a list of equipment that will be supplied by the purchaser.

Mechanical Specification. This will include basic data concerning each piece of major mechanical equipment such as indoor or outdoor service, continuous or intermittent service, type of foundation or base, duty of diverter gates and valves, air, electric motor or manual operation, and a list of equipment supplied by the purchaser.

PREPARATION OF A PROCUREMENT SPECIFICATION

The tabulated data regarding conveying requirements should be supplemented by a Procurement Specification describing the services, responsibilities, and limitations that will govern the relationship between the purchaser and vendor.

This specification should cover the purchaser's requirements as completely as possible and, where details of procedure are unknown, should solicit the vendor's suggestions for modification of the design with the proviso that the modification or design deviation must be specifically defined in the vendor's quotation. The general headings of such a procurement specification would include the following:

General. A brief statement as to the intent of the specification; a statement of the mutuality of interest in the project; a statement as to responsibility for performance of the system and all its components after mutual agreement on the type of system to be installed; a reference to the purchaser's flow diagram or to the necessity for one to be prepared by the vendor.

Vendor's Quotation. A requirement that the vendor list and describe each piece of equipment he proposes to furnish for a completely integrated and operable conveying system. Descriptions should be accompanied by such catalog cuts, drawings, sketches, etc., as are required to show the construction and operation of the equipment and the assembly of the equipment to form the proposed system, so that the equipment may be evaluated for conformance with requirements.

Service Requirements. Specific requirements for each system and reference to accompanying documents such as: (1) a tabulation of equipment data sheets; (2) standard purchase-order form; and (3) purchaser's preliminary drawings. The purchaser's preliminary drawings should show conveying-line lengths, runs of piping and number of bends from air- and material-intake end of the line to points of delivery, as well as equipment locations and elevations above grade. Runs of conveying line shown on these drawings may be modified by the vendor within the limits imposed by adjacent structures, provided the modification results in improved operating characteristics, increased efficiency of the system, or decreased cost of installation in the opinion of the purchaser's engineers.

Drawing Procedure. A detailed description of the purchaser's and vendor's responsibilities for interchange of drawings, and a list of specific types of drawings required from the vendor. The purchaser should furnish the vendor with drawings of buildings, structures, and other possible interferences, showing the locations, grades, and elevations required for the proper preparation of the vendor's layout drawings. The purchaser should be responsible for accuracy of dimensions of existing buildings, interferences, etc., shown on his drawings. Any conflicting or missing dimensions should be called to the attention of the purchaser's engineers by the vendor for clarification, and proper dimensions should be supplied to the vendor.

The vendor should prepare and submit drawings as follows.

Layout drawings of piping and equipment. Piping drawings should be complete to the extent that they may be turned over to a piping contractor for installation without modification or additions by the purchaser.

Detailed assembly and wiring diagrams of control panels, relay panels, sequence timers, or other factory-wired assemblies.

Schematic wiring diagram for the entire electrical system, including controls that are required to integrate the mechanical equipment into a complete pneumatic conveying system.

Outline dimension drawings of all equipment. Drawings should be certified accurate by the vendor when equipment is to be bolted to supports or foundations that are installed by the purchaser.

Shop drawings of equipment, with operating mechanisms sufficiently detailed to enable the purchaser's personnel to understand the operation of the equipment and the manner of assembly, so that equipment may be properly operated and serviced.

A composite electromechanical schematic functional diagram for the entire system, showing the relationship of electrical equipment to mechanical equipment. The drawing should be sufficiently detailed to show operation of the entire system and to enable electrical operations to be correlated with mechanical operations by maintenance personnel.

Installation. A statement as to who will install equipment. If the purchaser elects to install the equipment, the vendor should be required to: (1) mark all equipment by painting item numbers thereon for field identification; marking on units requiring field assembly should be painted on the knocked-down components of the unit; (2) match-mark piping, structural steel, equipment parts, etc., by painting in such a manner as to facilitate ready identification and assembly for field erection.

The purchaser should list here all the associated equipment that he will furnish to complete the installation—such as motor-starting equipment; electrical conduit and wiring not included in factory-wired or fabricated assemblies; sheet-metal chutes to bins; piping supports; instrument tubing not a part of factory-assembled panels; and mechanical feeders or conveyors, and their drives, not normally furnished by the vendor.

Service Engineer. The vendor should furnish the services of a qualified engineer to: (1) assist in startup of the system; (2) instruct the operators in proper operation and maintenance of the system; (3) correct defects that may be revealed during startup; (4) make adjustments to equipment to secure optimum performance. These services should be offered for a period as recommended by the vendor, with charges stipulated for traveling and living expenses. However, if services are required beyond this period for the purpose of correcting defects revealed during startup, all expenses for the service engineer should be borne by the vendor.

Performance Guarantee. The vendor should guarantee the complete system and component parts to perform the bulk-handling operations described in the specifications, and at the specified conveying rates. Failure of the equipment to conform to the warranty should be cause for returning the equipment or any part thereof to the vendor, with full refund of the purchase price of the equipment so returned. However, the vendor should be granted a reasonable time to correct any defective conditions at his own expense.

Material and Workmanship Guarantee. This is standard with practically all vendors on equipment supplied by them, and lasts for a period of one year from date of shipment. However, since construction extends over a long period of time, it may be more than a year before the equipment can be placed in operation. The purchaser should therefore extend the guarantee to suit his specific conditions.

Equipment Details. This should include a detailed description of every piece of equipment required to form the complete conveying system. The requirements should be based on functional considerations given in this book with due regard for the peculiar properties and characteristics of the material to be conveyed. These requirements may have to be modified at a later date in order to suit the limitations of the equipment supplied by the selected vendor.

ANALYSIS OF BIDS

The replies to a quotation request, which includes a detailed specification such as outlined above, will eliminate all but the most responsible vendors. A bid analysis must then be prepared that will compare the bids, item for item. This may appear complicated when different types of systems are offered, but a breakdown of the bids into functional equipment groups can simplify the process. An investigation of intangible factors will require an interview with each vendor, during which answers to the following questions can be obtained:

1. Can the vendor furnish a complete conveying system integrated electrically and mechanically?
2. How much of the required equipment does he fabricate in his own shop?
3. Do individual pieces of equipment that are offered meet the specifications exactly? The vendor should point out the absence of any important detail or feature that is required to handle the type of material to be conveyed.
4. Has he handled this type of material before? If so, can he cite a particular installation without revealing your competitor's confidences?
5. Does he have his own laboratory for analyzing material characteristics, and a pilot plant for testing the conveying of the material?
6. Can he offer constructive suggestions to improve the proposed conveying system?
7. Does he have a service department composed of trained engineers or technicians, or do his salesmen double as service engineers?
8. Can he meet the guarantee requirements?

Even the most reliable vendor may balk at some of the purchaser's requirements, especially when bidding on a system for conveying a problem material or a material he has never handled. The circumstances that may bother the vendor are as follows:

1. Handling a semi-abrasive material. The purchaser may have to make tests to convince the vendor that equipment can be guaranteed for a stipulated period, or he may have to assume all risks of the equipment wearing out prematurely.
2. Handling a tacky or sticky material. Here again, the purchaser may have to make tests to convince the vendor that the buildup of material or blinding of dust filters will not occur—or assume the risk himself.
3. Handling a friable material. The purchaser must set tolerances on the amount of degradation he will permit.
4. Location of dust filters outdoors. Many vendors will guarantee the performance of dust filters only if the purchaser can keep the temperature of material entering the filter above the dew point of the conveying air.

After selecting a vendor, the purchaser must advise him as to how conveying rates for acceptance of the system will be determined. This stipulation in the contract will permit the purchaser to invoke the terms of the performance guarantee if rates do not meet expectations. In order to invoke this clause, and in line with the concept that the installation of a pneumatic conveying system is a cooperative undertaking, the vendor's service engineer must be included on the purchaser's startup team. This engineer should be briefed on the interrelationship of the conveying system with the plant production processes and then be permitted to check out all equipment supplied by his firm prior to startup.

ENERGY CONSIDERATIONS

The energy needed to operate the system must be assessed (1) to ensure that the most efficient system is chosen, and (2) to make an accounting of the utility requirements that must be furnished—electric power and compressed air.

Determination of air horsepower. The system with the lowest overall energy cost per ton of material handled is the most efficient. The principal energy demands in a pneumatic conveying system arise from the furnishing of conveying air, the venting of this air through the dust filters, and the cleaning of the filter fabrics. Since air movers of one or more types are involved, the simplest standard of measurement is the air horsepower developed by the air mover. This may be determined for each type of air mover from its air-inlet capacity at standard conditions (14.7 psia and 70°F) and its discharge pressure.

The actual brake horsepower necessary to develop the required air horsepower depends on the volumetric and mechanical efficiencies of the air mover. These efficiencies vary with the type of air mover and its drive arrangement, and they are difficult to obtain. Because efficiency differences among air movers are too small to substantially influence a selection, air horsepower usually suffices as a standard of comparison. However, as an aid in estimating power requirements, the following efficiencies may be used: reciprocating compressors, 76 percent; sliding-vane compressors, 67 percent; positive-displacement blowers with lobed impellers, 65 percent; positive-displacement blowers with meshed screws, 70 percent; and fans with radial blades, 64 percent.

In any air mover used for pneumatic conveying, the volume of air entering the pipeline must equal the weight rate of flow required to maintain the material-to-air ratio of the conveying system at standard conditions. This is most important for vacuum conveyors and air movers at high altitudes. The volume of air entering the air mover should be checked by using the following formula:

$$V_1 = \frac{1{,}545WT_1}{144MP_1} \tag{8–1}$$

where V_1 = volume at air-mover inlet, cfm
W = weight rate required, lb/min, standard air
T_1 = absolute temperature at inlet, °R
M = molecular weight = 29 for air
P_1 = absolute pressure at inlet, psia
1,545 = universal gas constant

Positive-displacement blowers having lobed or meshed screw impellers compress air adiabatically, since no cooling is provided. Theoretical horsepower for these units may be calculated by the formula

$$\text{hp} = 0.0153P_1V_1\left[\left(\frac{P_2}{P_1}\right)^{0.283} - 1\right] \tag{8-2}$$

where hp = horsepower
V_1 = volume at air-mover inlet, cfm

P_1 = inlet pressure, psia
P_2 = discharge pressure, psia

Reciprocating and sliding-vane units compress air isothermally, since their cylinders are water- or air-cooled. Dust filters cleaned by reversed jets may be assumed to obtain their air supply from reciprocating compressors. Theoretical horsepower for these units may be calculated by the following formula:

$$\text{hp} = \frac{144}{33{,}000} P_1 V_1 \ln \frac{P_2}{P_1} \qquad (8\text{–}3)$$

where hp = horsepower
V_1 = volume at air mover inlet, cfm
P_1 = inlet pressure, psia
P_2 = discharge pressure, psia

The use of the above formulas may be circumvented by reference to Table 13–2, which shows the theoretical air horsepower required to compress 100 cfm of air at standard conditions to various discharge pressures via isothermal or adiabatic compression. The inlet volume in hundreds of cfm, multiplied by the tabular value for the discharge conditions, is the total air horsepower.

The table below was developed for use with positive-displacement blowers having lobed or meshed-screw impellers that operate as vacuum pumps. These horsepowers take cognizance of the increased volume, hence lower density, of the air at the inlet nozzle of the vacuum pump. They are based on the total volume of standard air drawn into the inlet of the conveying system. This volume in hundreds of scfm, multiplied by the tabular value for the required vacuum, is the total air horsepower. At more than 1,000 ft above sea level, Eq. (8–2) for adiabatic conditions should be used.

Vacuum, in.Hg	Theoretical hp per 100 scfm air
5	1.19
6	1.46
7	1.78
8	2.09
9	2.39
10	2.77
11	3.14
12	3.50

Fans and low-pressure centrifugal blowers are considered to affect the specific volume of the air passing through them only slightly, if the ratio of total pressure increase to initial pressure is less than 0.01. Therefore, the pressure and vacuum tables for horsepower do not apply to these units. Air horsepower becomes

$$\frac{\text{Total inlet air, scfm} \times \text{pressure differential, lb/ft}^2}{33{,}000}$$

Since fan-pressure differential is usually stated in inches of water pressure (one inch of water is equivalent to 5.19 lb/ft^2) the air horsepower becomes

$$\frac{\text{Total inlet air, scfm} \times \text{pressure differential, in.}H_2O}{6{,}356}$$

DETERMINATION OF ENERGY PER TON

From the above data, the total air horsepower needed to convey, to exhaust the dust filters, and to clean the filter fabric can be tabulated. Given the conveying rate in tons per hour for a continuous system, the energy per ton becomes

$$\text{Energy/ton, hp-h} = \frac{\text{Total air horsepower}}{\text{Conveying rate, tons/h}}$$

In blow-tank systems, conveying is intermittent and the energy per ton becomes:

$$\text{Energy/ton, hp-h} = \frac{\text{Total air horsepower} \times \text{discharge time/charge, min}}{\text{Tons/charge} \times 60}$$

The equipment supplier should furnish the basic information for the above calculations plus the brake horsepower of all other motor-driven equipment such as dust-filter plenum drives, reverse air blowers, discharge feeders, exhaust fans, and valve drive motors. In addition, the supplier should state the compressed air requirements of fabric-cleaning devices and air-cylinder drive motors (for diverter and shutoff valves).

II EQUIPMENT & CONTROLS

9 Airlock Feeders and Gates

Airlock feeders and gates are used to effect the discharge of powdered or granular material from equipment operating under a pressure differential. On equipment such as dust filters and receivers operating under a negative pressure, material cannot be discharged against atmospheric pressure. The backflow of air entering at the discharge connection prevents the outflow of material to such an extent as to completely fill the filter or receiver.

On feeders that must deliver material from bins at atmospheric pressure into a conveying line that is at a higher pressure, the air at the higher pressure will tend to blow through the feeder clearance spaces and pockets and prevent material from entering the conveying line. The various types of gates used to prevent the backflow of air are, in order of sealing ability, the rotary airlock feeder, the double-door discharge gate, and the flap or dribble valve.

ROTARY AIRLOCK FEEDERS

The principal device for metering material into or out of a pneumatic conveying system continuously with a minimum loss of conveying air is the motor-driven, rotary airlock feeder (Figure 9–1). This unit is essentially an air-sealed star wheel with material buckets and air seals designed to suit the characteristics of the material to be handled (Figure 9–2). The design for a dustless, granular material will be quite different than that for a very fine, dusty material. See Figure 9–3. Therefore, a thorough investigation must be made to insure that the design details of an airlock feeder are compatible with the characteristics of the material that it will handle.

The design details also affect the size of the feeder, since air leakage from the high- to low-pressure side of the unit will cause a fine, fluidizable material to expand and decrease in density upon entering the pockets. Irregularly shaped material may require more time to fill the pockets and require slowing down of the feeder. Both these conditions

FIGURE 9–1. Motor-driven rotary feeder and motion switch. (Fuller Company/GATX.)

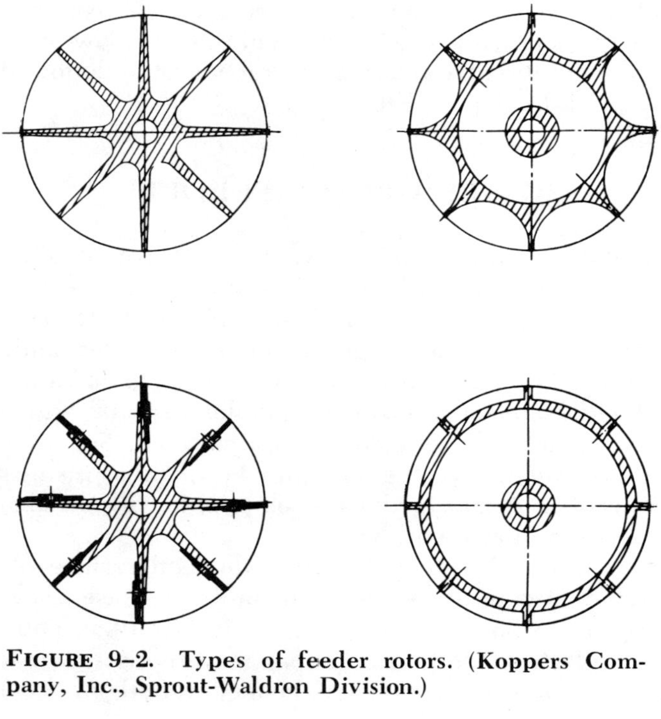

FIGURE 9–2. Types of feeder rotors. (Koppers Company, Inc., Sprout-Waldron Division.)

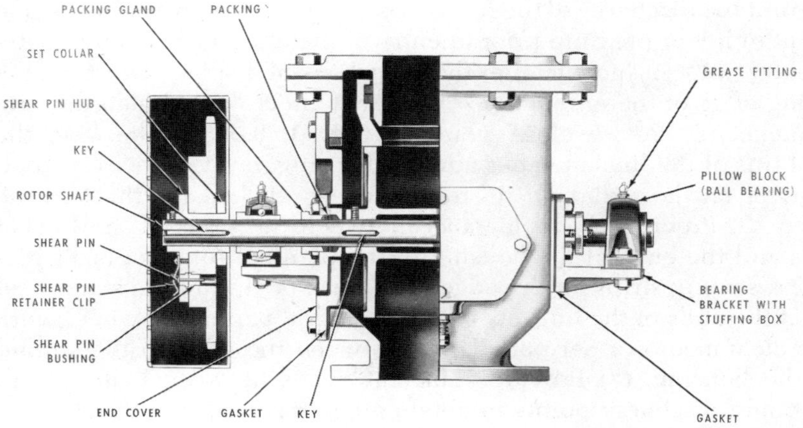

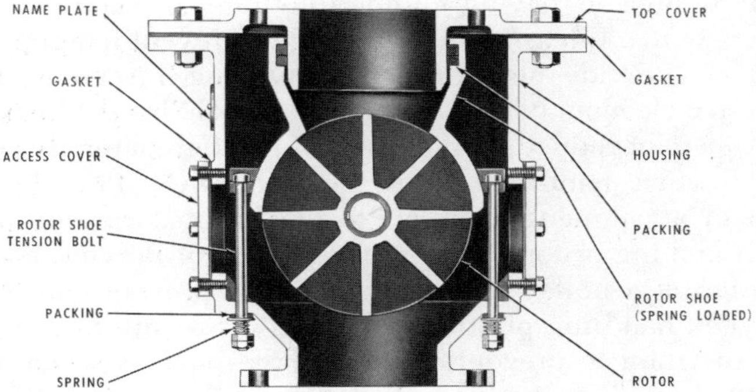

FIGURE 9–3. Rotary seal feeder suitable for abrasive materials. (Detroit Stoker Company.)

require that the rotor size as well as speed be changed to accommodate unexpected changes in flow volume.

The following checklist of functional requirements and design details to meet these requirements should guide the user in his selection of a rotary airlock feeder.

REDUCE INTERNAL AIR LEAKAGE

These are the steps necessary to minimize leakage from the high to the low pressure side of the feeder. (1) Use a pre-vent pipe connection to permit the discharge of the high-pressure air in the empty buckets to a point of lower pressure prior to entry of material; or baffle the material entry point in such manner that blowback of displaced air from the leading edge of the empty bucket will not affect flow of material into the bucket. (2) Provide close clearance (0.003 to 0.005 in.) between the radial tips of the bucket vanes and the housing; or fit frictionless sealing strips on the radial tips to reduce direct air leakage through the feeder. (3) Provide close clearance between the sides of the bucket vanes and the ends of the housing in the plane of rotation; or fit frictionless sealing strips on the edges of vanes; or shroud the star wheel so that the ends of the buckets form an integral part of the wheel, with close clearance or a peripheral seal between the edge of the shroud and the housing. (4) Provide sufficient vanes between material entrance and discharge points to obtain a labyrinth sealing effect.

Fill Pockets Completely

(1) To handle sticky or compacting materials, provide the star wheel with shallow buckets having rounded bottoms and corners. (2) Coat the buckets with a releasing-type coating to prevent hangup of sticky materials. (3) Provide blow-through conveying-line connections for most effective cleanout of problem materials. (4) Size the buckets with due allowance for reduction in density of aerated materials or for partial filling when handling lumpy materials. (5) Provide arch or lumpbreaker attachments at inlets to feeders that receive materials capable of arching or forming lumps in bins. (6) Provide raked inlet throat design or a pellet valve attachment to prevent overfilling the pockets when handling plastic materials. (7) Provide forward-curved vanes on the rotor to prevent overfilling the pockets when handling plastics. (8) Provide a hardened and ground tool steel cutting blade that projects above the other rotor vanes radially and at the rotor ends to remove built-up sticky materials that adhere to the housing. Such a blade must be braced by at least three struts in order to prevent chatter. (9) Back-bevel the tips of rotor blades or vanes to prevent smearing and binding that may be caused by tacky materials. (10) Provide air-blast cleanout jets in end plates of the housing that are directed into the bottom of the rotor pocket in its discharge position. This will ensure complete removal of tacky material.

Prevent External Air Leakage

(1) Provide packing boxes on end covers deep enough to receive at least three rings of packing when handling very dusty material at high

pressure differentials. (2) Provide lantern rings with connection of an external air supply in addition to packing on the end covers where the shafts protrude, in order to insure a purging-air flow into the feeder and to prevent dust blowout. (3) Provide purge-air pressure or garter-spring-loaded lip seals on the end covers where shafts protrude, for sealing dusty material at low-pressure differentials. (4) Provide external purge-air connections on end covers of shrouded star wheels in order to keep the space between shrouds and end-covers free of material buildup.

Provide for Operation and Maintenance

(1) Provide tell-tale space and a shaft slinger between the packing box and outboard bearings, in order to detect powder leakage and prevent the entrance of bearing lubricant into the feeder, or entry of blown-out material into the bearing. (2) Provide a renewable shaft sleeve for the entire length of each packing box that is fitted with a gland follower. (3) Provide doors or plates to permit access to the feeder interior for checking clearances, replacing seal strips, inspecting for buildup, and removing jammed material. (4) Provide overload or jam protection such as a motion switch, shear-pin hub sprocket, controlled torque coupling, pivoted motor drive or torque arm drive overload release. (5) Provide a drive mechanism with means of making small speed adjustments during start-up in order to obtain rated capacity. (6) Provide for access to end of the shaft and a center bore for checking speed of rotation. (7) Attach the drive motor to the feeder housing in such a manner that increase of belt or chain tension will not affect internal clearances by distorting the housing. (8) When handling abrasive materials, hard-coat the rotor-blade tips with colmonoy or some other hardfacing material. Then grind the rotor to size, and hard-chrome-plate the interior of the housing.

The rotary airlock feeder is usually one of the pneumatic conveying system components requiring frequent inspection and adjustment, therefore, it should be located for easy accessibility and removal for servicing without the need for erection of scaffolding. In order to insure continuity of some processes and to eliminate time-consuming field adjustment—especially in bad weather on outdoor installations—it may be desirable to retain a complete feeder including drive as a spare unit, since most drives require special ratio gearmotors. This is definitely required on units that are equipped with a cutting blade, since both rotor and housing must be remachined when the blade is worn. Adjustable cutting blades and wear strips are prone to misalignment in service owing to loosening of bolts. These will require frequent inspection in order to maintain proper clearance. Wear strips will eventually abrade and air leakage will insidiously increase until major conveying problems occur unless a regular program of inspection and replacement is adopted.

Ancillary features required to insure successful operation of rotary airlock feeders are: (1) adequately sized dust collectors and air vents on the inlet side of feeders delivering material into positive-pressure systems, to prevent pressure buildup and discharge of dust from equipment: (2) pre-vent piping connected to the low-pressure point in such a manner as to be self-cleaning and not block up, and (3) external air supply to lantern rings or purge chambers available at higher pressure than exists within the feeder. This should be supplied through a solenoid valve that is open only while the feeder is in operation.

DOUBLE-DOOR DISCHARGE GATES

This type of gate is less common than the rotary gate since its action is intermittent. It is used principally to handle high-temperature, abrasive, or easily degraded products that flow out of a hopper in a steady stream. It should not be applied to full hoppers that would tend to flood the gate and hold the doors open. It is ideal for application on dry dust collectors and cyclones having a steady dust load. Basically, the gate consists of two counterweighted or counterbalanced door assemblies mounted in tandem, one above the other, and seating against an outlet port. The space between the doors forms the airlock. The doors may be gravity-, motor-driven-, cam-, or air-cylinder-operated. Designs may vary considerably depending upon the differential pressure that must be overcome.

In the gravity operated gate, the weight of material on the upper gate overbalances the door, material is dumped into the airlock as the pressure is equalized and the door reseals itself against the upper, or inlet, port. The lower door then becomes overbalanced and drops open. The air pressure is equalized and material falls through the outlet port to the downstream equipment. The maximum differential across this type of gate is about 5 inches of water.

In the motor-driven, cam-operated gate, the doors are opened alternately at a speed dependent on the gearmotor drive ratio. The doors do not sense any weight of material; therefore the doors open whether or not there is a supply of material above them. A spring, or the counterweights, slam the doors to the closed position to obtain positive seating. Obviously, any material buildup on the door or seats will cause the gate to leak air and thus restrict the outflow of material. The maximum differential across a motor-operated gate is about 20 inches of water.

A modification of the double-door gate has a set of split doors located above the upper door that act as batching doors to limit the flow through the gate. In addition, the space between the sealing gates is fitted with a cam-actuated relief valve for equalizing the pressure in the airlock. Before the upper door opens, the relief valve opens to the suction chamber from where the material is being drawn and pulls a vacuum in the airlock space. After the upper door is resealed and before the lower door is opened, the relief valve admits atmospheric air to the airlock space so that when the lower door trips, the material can drop out of the gate. This type of gate can seal up to a maximum differential pressure of 10 inches of mercury (about 5 psi). See Figures 9–4 and 9–5.

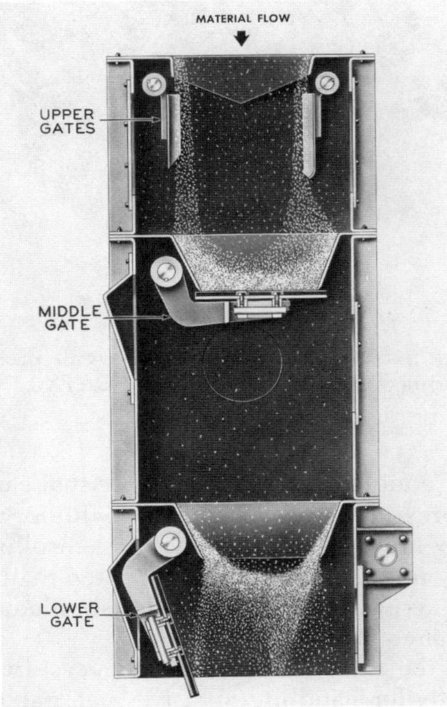

FIGURE 9-4. Internal view of gate discharge lock. (Fuller Company/GATX.)

FIGURE 9–5. Gate discharge lock with pressure-equalizing valve. (Fuller Company/GATX.)

Another type of double-door discharge gate that can be operated at high differential pressures uses separate air cylinder drives to actuate each door. The operation of the cylinders is controlled by a remotely located electrical cam timer that can be adjusted to obtain a variety of timing cycles and actuate a number of gates simultaneously. Pressure differentials can range from 3 to 6 psi.

The housings of the gates can be made from cast iron with hard cast iron flaps and seats for handling abrasive materials, or they can be all-welded units made from carbon or stainless steel plate, or cast aluminum for non-abrasive materials.

The following checklist of functional requirements and design details to meet these requirements should guide the user in his selection of a double-door discharge lock.

Unrestricted Material Flow

(1) Contour the body design to eliminate corners or shoulders where material could hang up. (2) Valve doors should retract out of the path of flowing material. (3) Inlet and outlet ports should be a straight chute design rather than a tapered transition piece. (4) The chamber between the upper and lower valves should be sized so that material cannot back up and interfere with the door during normal operation. This volume should be at least four times the maximum operating rate when handling materials that could flood this space. (5) Provide shallow, self-cleaning spaces around the lip of the door seats so that material cannot pack in the annular space between body and door seat, if powder backs up that far.

Provide for Operation and Maintenance

(1) Provide outboard bearings to prevent entry of dust into the bearing or contamination of product by lubricant. (2) Provide renewable valve seats and easy access for renewal. (3) Provide an access door large enough for removal or adjustment of the sealing doors. (4) Provide simple adjustments for repositioning the sealing door on its shaft relative to the driving cam position. (5) For use with abrasive materials, provide purge air connections at bearing seals.

Reduce Internal Air Leakage

(1) Provide self-seating valve doors. The door should pivot in order to lie flat against the seat. (2) Make the gate volume large enough to prevent overloading. Preferably, select a gate with a linkage that provides gravity free-flow flushing of an overload. (3) Provide sharp edges at the lip of each door seat in order to prevent buildup of material. A reversed bevel on the lip makes that edge self-cleaning, because high-velocity air scours the seat as the door closes.

Prevent External Air Leakage

(1) All pivot shaft bearings of doors should be of the sealed type, on gates operated under air pressure. When handling very abrasive powders, a packing gland should be provided on the material side of the bearing with an air-purge chamber between gland and bearing.

Special attention should be directed to the driving mechanism of the double-door discharge lock. Motor-driven, cam-operated doors limit movement to the normal operating cycle since timing is fixed. Discharge locks with independent air-cylinder drive on each door permit (1) timing to be programmed and varied to suit operating conditions, (2) wiring the air cylinder solenoid valves to selector switches so that both doors can be opened simultaneously, flushing through lumps and material when equipment is being cleaned out, and (3) greater spacing between the doors to hold larger volumes of material than can be accommodated by a standard gate lock.

In some applications, a double-gate airlock may be preferable to a rotary airlock since the former has the capability of flushing material through the gates for equipment cleanout.

The maximum differential pressures for double-door discharge gates are as follows:

Type	Pressure
Gravity-opened, spring-returned	5 in. of water
Motor-driven, cam-operated	20 in. of water
Individual air-cylinder-operated	3 to 6 psi
Motor-driven, cam-operated with pressure-equalizing valve between doors	7 to 15 psi

FLAP VALVES

Flap valves are sometimes called trickle, or dribble valves, because of their action. These valves are essentially a discharge box fitted with a weighted fabric flap or a counterweighted metal flap held against a metal seat by the operating suction of the equipment to which it is attached, Figure 9–6. As the weight of the collected dust increases, it eventually overcomes the seating force of the vacuum, opens the flap, and is discharged. Specially formed flexible, flat-walled sleeves that are held closed by suction are actually flap valves since they depend on the weight of material to force itself through the sleeve.

This type of valve may be used to discharge dust on equipment operating under very low suction pressure. In many cases, an extension of a dust hopper chute may be required in order to obtain sufficient head of material above the flap to make its weight effective. The best (and most logical) application for this type of valve is on multicompartment dust filters where each compartment is isolated during

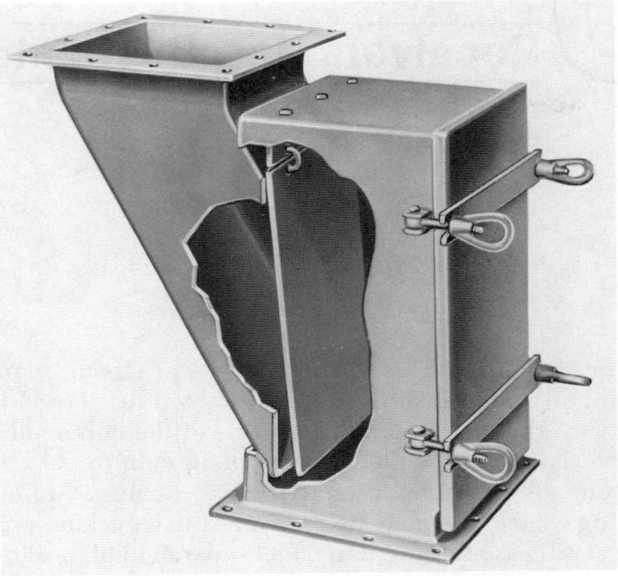

FIGURE 9–6. Trickle valve. (The Ducon Company.)

cleaning and the suction is considerably reduced. The dust will then flow out freely during cleaning, but the flap will seal tightly when normal operating suction is restored.

Flexible sleeves that are held closed by suction on the inlet side cannot be discharged into other equipment maintained under negative pressure. Any blockage of the inlet by lumpy or sticky material will permit the outlet suction to close the sleeve tightly and prevent the discharge of any material.

10 Receivers and Dust Filters

RECEIVERS

The recovery of material from the conveying-air stream requires separation of air and material. This may be effected in a type of receiver designed to suit the dusting characteristics of the collected material. Dustless material may be collected in a plain cylindrical tank with a hopper bottom and a vent opening to the atmosphere. Granular material containing coarse dust may be recovered in a cyclone-type collector having a scroll case designed to effect separation of most of the dust by centrifugal force with the air vented to atmosphere directly or through an exhauster. Fine, dusty powders require additional equipment for their recovery. They may be collected in several ways: (1) In a cyclone-type collector with the air vented to a fabric filter; (2) in a fabric filter designed to handle the dust load; or (3) in a combination receiver-filter unit that attempts to combine in one piece of compact equipment the functions of a cyclone-type receiver and a dust filter.

In all types of receivers, material is discharged through an air-sealed gate designed to prevent entrance or loss of air depending on whether the receiver is under suction or pressure. The type of gate used depends to a large extent on the pressure differential across it and, in increasing order of pressure differential, the gate may be a trickle valve, a double-gate airlock feeder or a rotary airlock feeder.

The high conveying velocity and the uniform dust load in material-into-airstream types of pneumatic conveying systems permits effective separation of the material in cyclone-type receivers. The collection efficiency of a cyclone depends on the cyclone shape and proportions and the mechanical forces these impose on the airstream and on the airborne dust particles. The effect of these forces depends on the size and distribution of the particles, their specific gravity and the particle shape. Each design of cyclone has built-in characteristics that may be evaluated from examination of the manufacturer's efficiency curve. Typical curves are shown in Figure 10–1. These show the relatively high efficiencies obtainable in this type of equipment. However, any increased loading will increase the total discharge of dust through the

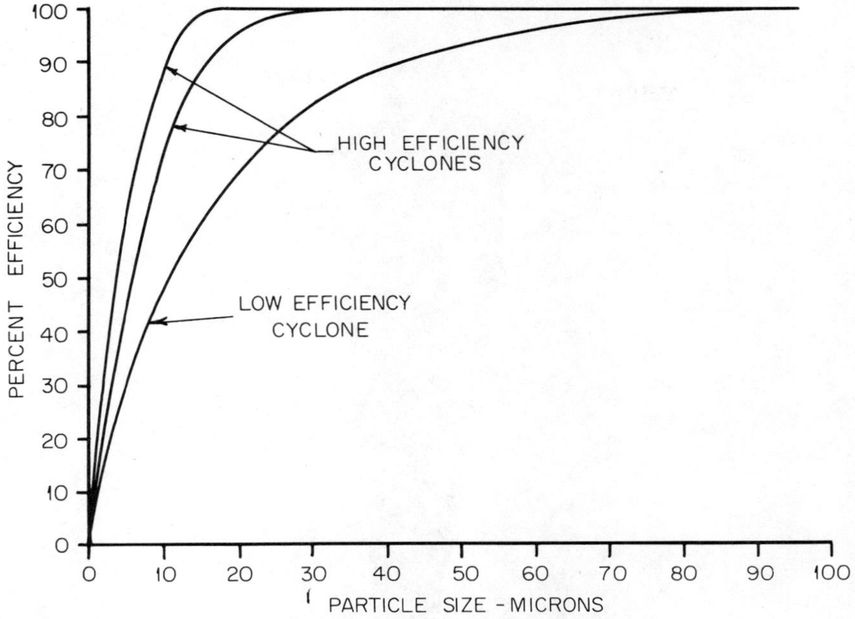

FIGURE 10–1. Typical cyclone efficiency curves.

outlet even though the overall collection efficiency increases. The design of cyclones for use on negative-pressure conveying systems may include a vertically-mounted motor-driven exhaust fan on its roof, forming an integral part of the unit (Figure 10–2).

The use of a cyclone ahead of a fabric dust filter is the ideal way to effect separation, since the dust loading on the fabric is thereby reduced. The cyclone and its discharge gate also seals the receiving bins so that they are not subjected to conveying line pressure. However, since this type of separation requires additional equipment, several alternate methods are used to reduce the capital investment in conveying equipment. The system requirements and material characteristics will determine which of these alternates should be selected.

POSITIVE-PRESSURE SYSTEMS

(1) The conveying stream can be discharged directly into receiving bins and the air vented to atmosphere. This method is not recommended for dusty material. (2) The material can be discharged directly into bins vented to the atmosphere via an intermittent type of dust filter that is under positive pressure during conveying. This method is

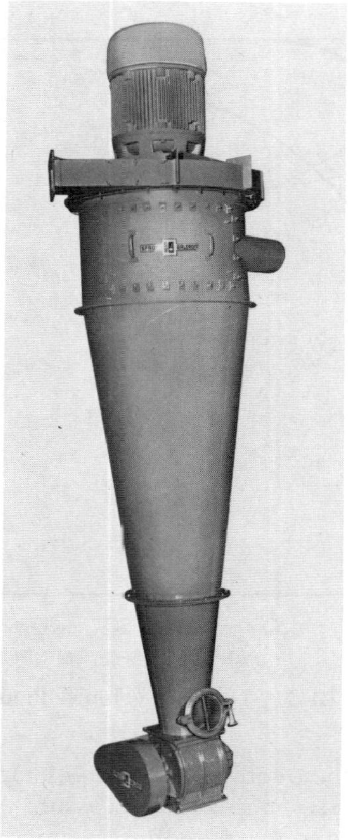

FIGURE 10–2. Cyclone-type receiver. (Koppers Company, Inc. Sprout-Waldron Division.)

used when the conveying cycle is of such short duration that the dust filter is not overloaded and can be cleaned during the shutdown period. (3) Finally, material can be discharged directly into bins vented to a continuous type of dust filter equipped with an exhaust fan.

NEGATIVE-PRESSURE SYSTEMS

With this method, material can be discharged into a combination receiver-filter unit that is cleaned continuously during the conveying period (Figure 10–3).

FIGURE 10–3. Combination receiver-filter. (Fuller Company/ GATX.

DUST FILTERS

One of the most important subjects in pneumatic conveying is dust filtration. Dust filters are required whenever air must be discharged from a conveying system handling dusty material. They are used to control dust nuisance and to recover valuable dust. Sizes may range from relatively small bin-venting filters to large multicompartment filters that receive dust from an extensive exhaust system used to vent bins, hoppers, conveyor housings and leakage through airlock feeders. The material presented here is sufficient in scope to permit intelligent discussion of filtration problems with manufacturer's representatives.

All dust filtration in pneumatic conveying work is accomplished by passing the dust-laden air through a filter fabric in the form of cylindrical tubes, or envelopes. As the air passes through the fabric, the dust is retained on the surface and in the interstices of the yarn strands, building up a filter cake that itself acts as a filter medium. In order to reduce the resistance to airflow so as not to affect the conveying rate, the filter cake must be dislodged periodically. The method of cleaning the fabric forms the basic difference between the various makes of filters. The greater the area of filter cloth provided for a given dust loading on a specific type of filter, the longer it will take to build up a filter cake that will affect the airflow.

Filter Fabrics

In comparing the area of filter fabric offered by different vendors, one should recognize the different methods of cleaning the fabric. The criterion should be a constant pressure drop across the fabric for a specified airflow rate and a specified dust loading rather than the obsolete "air-to-cloth ratio." If checked dimensionally, this term is seen to be the face velocity of the air through the effective area of the fabric for a given type of filter. The air-to-cloth ratio only has significance when comparing the performance of a particular manufacturer's line of filters when handling different materials. The cleaning method determines whether a dust filter is suitable for continuous or intermittent conveying service.

A filter fabric must have the following properties in order to function properly:

Porosity. It must be sufficiently porous to permit a satisfactory flow of air. The porosity of a fabric is generally stated as the airflow in cubic feet per square foot of cloth as determined by the Frazier test, specified in A.S.T.M. Standard No. D-737, "Standard Method of Test for Air Permeability of Textile Fabrics." The Frazier permeability alone is no guide to fabric selection since materials having the same permeability may not have the other characteristics required to successfully filter a specific material. Also, the nature of the filter cake formed on the fabric surface may reduce airflow.

Mechanical Strength. It must resist the tensile forces caused by operating pressure differentials; by mechanical shaking during cleaning and by pulsing during reverse airflow. It must withstand abrasion where it is clamped to tube sheet ferrules, where supported on metal grids and where subjected to the impact of the filtered materials.

Solids Retention. Its construction must be open enough to prevent the solid accumulation of fines in the interstices of the yarn, yet close enough to prevent them from blowing through the fabric. This may seem anomalous, but it is the filter cake on the surface of the fabric that acts as a filter medium and screens the interstices.

Corrosive Chemicals. It must resist attack and weakening due to chemical action between the fibers and the filtered materials especially when moisture may be present due to condensation.

Heat. It must resist heat in some process applications of pneumatic conveying. Each fabric has a definite temperature limit beyond which it will tend to disintegrate.

Cleaning. It must have a surface texture conducive to easy cake discharge. This, too, may seem anomalous since the surface must also retain the cake. However, the forces involved during cleaning are usually airflow in the opposite direction to normal airflow in addition to mechanical shaking. The fabric should have a high rate of electrostatic charge dissipation so as to shed charged dust particles.

Dimensional Stability. The fabric must resist stretching and shrinking that would affect its premeability. Many untreated fabrics tend to increase in permeability after laundering.

Fabric Construction

A knowledge of the fiber properties and construction of the fabric will provide some indication as to how well these requirements can be met. However, final selection should be made by test and an economic study that weighs first cost against probable maintenance expense. All fabrics will eventually blind when handling fine materials, or materials that crystallize or polymerize when embedded in the fabric, and it is better to replace all cloth in a filter at regular intervals, determined by experience or periodic permeability tests, than to operate until airflow is drastically reduced and other problems are created.

Filter fabrics may be felted or woven. Felted fabrics are relatively thick and are composed of fibers which are compressed under high pressures. The pressure enmeshes the fibers, forming a strongly bonded sheet of material with complex labryinthine interstices through which air can pass in three directions. Synthetic monofilament fibers cannot form a strong bond between themselves and are usually pressed over a woven sheet of skrim fabric for strength. Wool fibers are naturally barbed and have the stronger bond. Most felts can absorb moisture due to capillary action between fibers even though the fibers themselves are moisture resistant. A simple test for relative absorption of moisture is to stand equal lengths of fabric on edge in a jar with a thin layer of water in the bottom. After several minutes withdraw the samples and note the height to which the water has been drawn up in each sample.

The thickness of felt provides maximum dust impingement. The many changes of direction due to lateral air flow traps the smallest particles. Ultraviolet light inspection of the cross section of used felt fabrics which have handled dust containing fluorescent materials, and microscopic examination of the cross section of felts handling other types of dust shows that the particles rarely penetrate through more than one-third the felt thickness. However, there is a reduction in air permeability, but not as much as would occur in a woven fabric with the same quantity of trapped dust per square inch of filter area. Felted fabrics are used for maximum product recovery, where filtered air is recirculated or discharged into a work space. They are primarily used in collectors with a high air-to-cloth ratio (high face velocity).

Woven fabrics are composed of twisted yarns woven into geometric patterns with various spacings between the yarns and with a surface finish designed to retain or shed a filter cake depending upon the application. The porosity of the fabric depends on the type of fiber, the tightness of twist and size of yarn, the type of weave (geometric pattern), the tightness of weave or thread count, and the type of surface finish. Correct coordination of all these factors for a successful filter application cannot be done by abstract analysis; the fabric must be tested in the specific type of dust collector considered suitable for a particular application on the basis of operational and economic factors. Woven fabrics have a low ratio of weave openings to yarn area and require a surface filter cake to retain the dust. This limits the face velocity of the air to low values [1.5 to 3 (ft/min)/ft^2] in order to prevent dust from blowing through the openings and destroying the filter cake.

The porosity of woven fabric is so dependent on construction that some familiarity with textile terminology appears desirable. The yarn may be made from (1) staple—short fibers of cotton or synthetic material—which must be drawn into parallel strands and twisted into yarn by spinning, or from (2) continuous filaments of synthetic material manufactured by extrusion of a spinning solution through a finely perforated nozzle then spun into yarn. The degree of twist of the yarn affects porosity; a tight twist resists penetration by dust particles and a loose twist permits retention of dust or "blinding" of yarn. Filament yarns are stronger than staple yarns. Spun yarns have more flexibility than filament yarns and more resistance to flexing.

Weaving is the interlacing at right angles of a series of yarns to form a textured fabric. The basic weaves are plain, twill and satin. In the plain weave, each crosswise or filling yarn passes alternately over and under one warp or lengthwise yarn. Each warp yarn passes alternately over and under each filling yarn. This type of construction makes the tightest weave, but by varying the thread count, the weave may be made as open and porous as required. In the twill weave, the filling yarn passes under one warp yarn then over two or more warp yarns. In each succeeding row the filling thread moves one warp either left or right thus forming the diagonal lines which identify a twill weave. Twills have fewer interlacings than the plain weave and, depending on the thread count, greater porosity. In the satin weave, the filling yarn passes over one and under four or more warp yarns in one row and under one and over four or more warp yarns in the next row. This weave has even fewer interlacings, spaced widely, but regularly. It has a smooth surface and greater porosity than the other weaves. Cotton fabric in this weave is known as cotton sateen. Regardless of weave, the size of yarn and the number of warp and filling yarns per inch (thread count) greatly affect porosity. A low count will permit higher airflow rates at low pressure drops whereas a high count will trap more fine particles and produce a higher pressure drop across the fabric (Figure 10–4).

The woven fabric may have various finishes. Cotton fabrics may be preshrunk to maintain dimensional stability. They may also be napped in order to hold a filter cake. (Napping is the process of scratching the cloth surface to raise the fibers and form a soft pile.) Synthetic fabrics may be heat-set to obtain dimensional stability and a smoother surface with uniform porosity. All fabrics may be silicone sprayed to obtain abrasion resistance, improved cake release or low moisture absorbency.

Table 10–1 shows the properties of commonly used filter fabrics. Many other fabrics are available for special applications; selection of these should be made only after determining that the more common fabrics are not suitable for the proposed service and after discussing the application with a reputable firm of filter media specialists.

FIGURE 10–4. Weaves used for dust-filter fabrics. (WestPoint Pepperell, Industrial Fabrics Division.)

A typical example of the unpredictability of fabric performance are the following specifications for the original tube material furnished by a filter manufacturer and for a replacement material furnished six years later by the same vendor for the same filter and for the same service.

	Original	Replacement
Fabric manufacturer	ABC Co.	XYZ Co.
Material	Filament Orlon	Staple Orlon
Thread diameter	0.0085 in.	0.024 in.
Thread count	70 × 68	90 × 60
Weave	Twill	Satin
Permeability (Frazier)—cfm/ft²	20 to 25	43

Despite the apparent dissimilarity of these fabrics, they gave the same performance. In this case, filtration and pressure drop were evidently a function of the type of filter cake built up on the fabric while in service, rather than a function of the type of fabric selected.

TABLE 10–1
Properties of common filter fabrics

Fabric	Temperature limitation, °F	Chemical resistance: Acids	Chemical resistance: Alkalis
Cotton	180	Poor	Good
Wool	200	Good	Poor
Nylon	200	Poor	Good
Polypropylene	200	Excellent	Excellent
Orlon	260	Good	Fair
Dacron	275	Good	Fair
Dynel	160	Good	Good
Glass	550	Excellent	Poor
Nomex	400	Fair	Good
Polyethylene	165	Good	Good
Teflon	400	Excellent	Excellent

Relative ratings: Excellent, Good, Fair, Poor.
Source: Publications of the Industrial Gas Cleaning Institute, Inc., Alexandria, Va.

The fabrics are formed into filter bags having configurations to suit the collector. The bags may be cylindrical tubes open at both ends for attachment to top and bottom tubesheets using band clamps or snap rings, and fitted with anticollapse spreader rings to permit the released dust to fall out. Or the bags may be cylindrical tubes closed at the top to permit suspension from a system of shaker bars, and open at the bottom for fastening to a tubesheet. Or they may be cylindrical tubes closed at the bottom and open at the top for installation over a wire retaining cage suspended from a tubesheet.

The bags may be oblong: in one version, closed at one end, open at the other for slipping over a retaining cage (like a pillowcase); in another version, closed at one end for top suspension, with several cylindrical openings at the lower end for clamping to a tubesheet.

In any bag design, the following details should be considered:

Thread used to sew the bag should be of the same material as the bag, or compatible with it. The open ends of the bag, which are attached to tubesheet ferrules by clamps or snap rings, should have a reinforced cuff. All hems and seams should have two rows of stitches, with edges folded or treated to prevent fraying. Spreader rings should be enclosed in a pocket that is doubly reinforced inside and outside the bag.

The suspension end of hung bags should be doubly reinforced and folded over to form a loop for hanging on a wide suspension hook, or fitted with a metal grommet if hung from an eye bolt or hook bolt. Cuffs of oblong bags that are hooked over support pins should be fitted with metal grommets at the hook locations. Where bags must be grounded, a copper wire should be sewn the full length of the bag, with sufficient lead to clamp to a tube ferrule.

FILTER CLEANING

The shape of the filter element, its method of support, and the normal direction of airflow through the filter will determine the method of cleaning. Normal airflow will hold the dust cake against the fabric. Stopping the flow of air will tend to release the dust cake. Simultaneously flexing the element, or reversing the airflow, will ensure release of the cake so that it can drop into a hopper.

Flexing can be induced (1) mechanically, by motor-driven crank mechanisms that oscillate closed-end tube elements, or (2) pneumatically, by reverse airflow for flow-through or closed-end tube elements.

On flow-through elements that collect dust on the inside surface, spreader rings may be used to form nodes for more-effective reverse-air flexing and to prevent collapse of the elements. (A collapsed element would prevent the dust from falling into the hopper.) On closed-end tubular elements that collect dust on the inside surface, a top-suspension tensioning device is depended on to prevent collapse of the elements during cleaning and to hold their surfaces taut to form the dust cake during normal airflow.

Envelope, oblong, or tubular elements that collect dust on their external surfaces are generally placed over metal retaining cages or grids that keep the fabric taut so as to retain the dust cake and maintain the internal open area required for clean-air flow. During cleaning, reverse airflow flexes the fabric away from the retainer grids; restoring normal airflow whips the fabric against it.

On collectors where normal airflow through the fabric is not stopped, an extra-high reverse airflow is needed to dislodge the dust cake. This may be done by using high-pressure air jets or separate air-reversal blowers.

Cleaning Methods for Baghouses

Commercial baghouses use one or more of the cleaning methods previously discussed. These methods may be further described as follows:

Shaking only is used on unit filters having woven bags where service can be interrupted long enough for cleaning—usually 3 to 4 min. The bags may be (1) envelope type, mounted over a metal spacer, with the closed end connected to a motor-driven shaker grid and tensioning bar, and the open end fastened to a clean-air plenum wall; or (2) tubular type, with the closed end suspended from motor-driven shaker bars, and the open bottom end fastened to a tubesheet for the entry of dust-laden air. The cleaning cycle may be initiated by (*a*) a timer set for a specific cleaning period, (*b*) a pressure-differential sensor, or (*c*) a manual stop button connected to a time-delay relay. In this method, the exhaust fan stops, allowing the internal pressure to reach atmospheric, and the shaker motor operates for a preset interval. After cleaning, operation of the dust collector may be restored automatically or manually, depending on the service.

Shaking with reverse airflow is used with either type of bag described for unit filters. However, for continuous service, the filter elements must be distributed among two or more compartments, and the exhaust fan must remain in operation during the cleaning cycle. Each compartment is fitted with its own shaker mechanism and a set of reverse-air dampers. Each set is comprised of an outlet damper, ducted to the exhaust fan, and a smaller damper opening directly to atmosphere. A program timer permits cleaning the bags in each compartment successively, with the time interval between cleanings adjusted to suit the service.

During cleaning, the timer activates the reverse-air damper drive to close the clean-air outlet damper to the fan, and simultaneously opens the atmospheric-air inlet damper. Suction from the active compartments draws air through the bags in the reverse direction. Limit switches make sure the dampers are correctly positioned and start the shaker mechanism. After the shake period, the reverse-air damper drive opens the clean-air outlet damper and closes the atmospheric-air inlet damper, reactivating the compartment. During the cleaning cycle, a

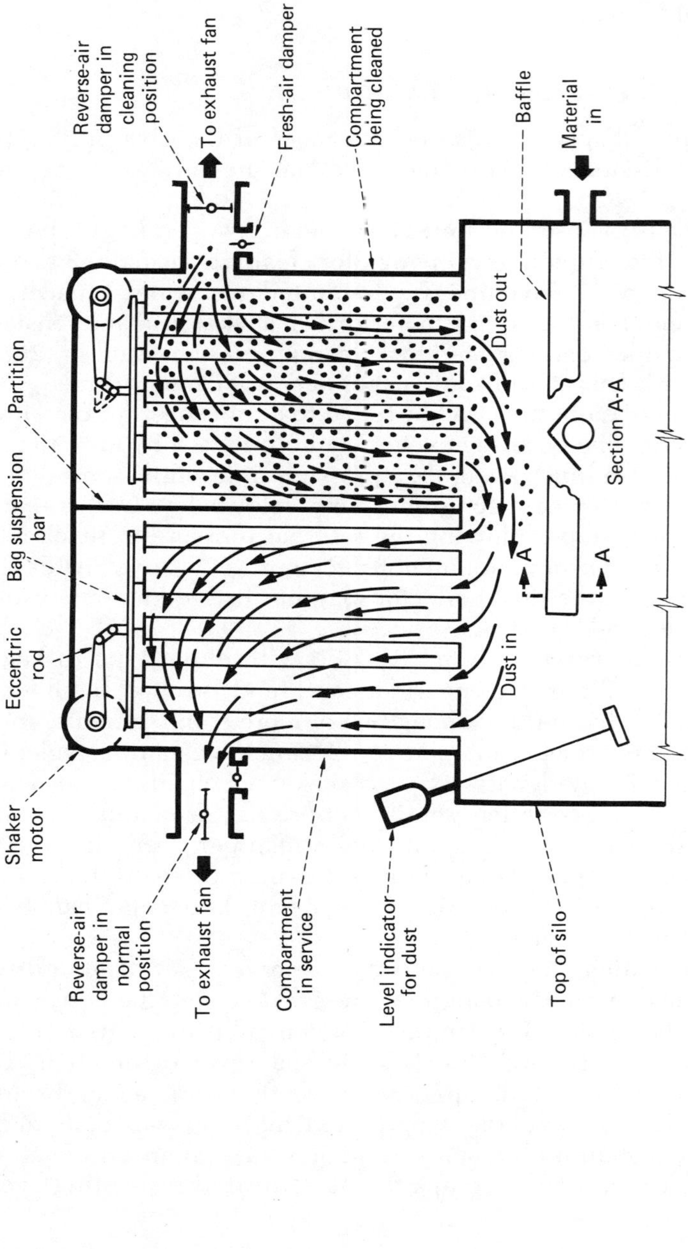

FIGURE 10–5. Dust collector cleaned by bag-shaking and reverse air-flow.

certain amount of dust is picked up by the reverse airflow and deposited on the bags in the active compartments.

Figure 10–5 shows a two-compartment filter used for venting a silo receiving material from a pneumatic conveying system. Figure 10–6 shows the variation in bin pressure during the cleaning cycle for this collector.

Reverse-airflow cleaning is used for all types of filter elements (tubular or envelope, woven or felt materials, dust inside or outside of the bag, elements distributed between several compartments or all in one housing). There are many arrangements for creating reverse airflow—all intended to enable continuous service. Several common arrangements are:

1. A compressed-air nozzle at the clean-air discharge of each bag injects high-pressure air into the tube in the reverse direction from normal airflow in pulses of short duration ($^1/_{10}$ to $^1/_{25}$ sec) for systems having closed-end tubular bags supported on metal retainer cages suspended from the clean-air plenum tubesheet. The pulses are controlled by external timers that operate special air-supply solenoid valves, each of which serves a group of bags. The time between pulses is adjustable, but it is factory-set for the intended service. See Figure 10–7.

2. A variation of this method is to divide groups of bags (including the clean-air plenum) into zones by means of internal partitions. The clean-air plenums are manifolded into an outlet duct via poppet valves. Each zone is equipped with a single pulse valve that supplies compressed air to the group of bags. During the cleaning cycle, the

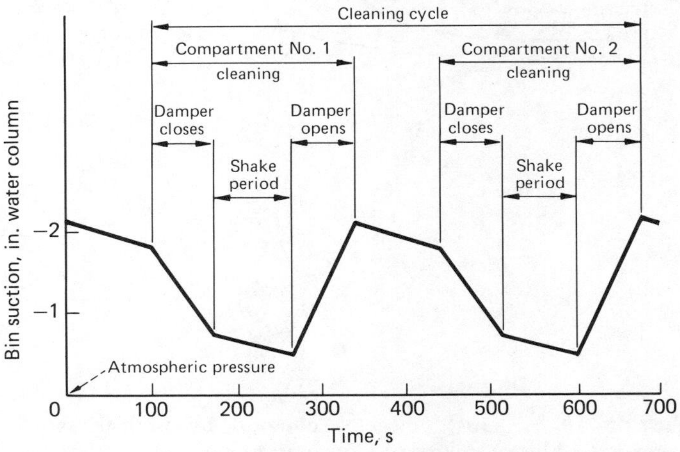

FIGURE 10–6. Variation in bin pressure during cleaning of a two-compartment filter.

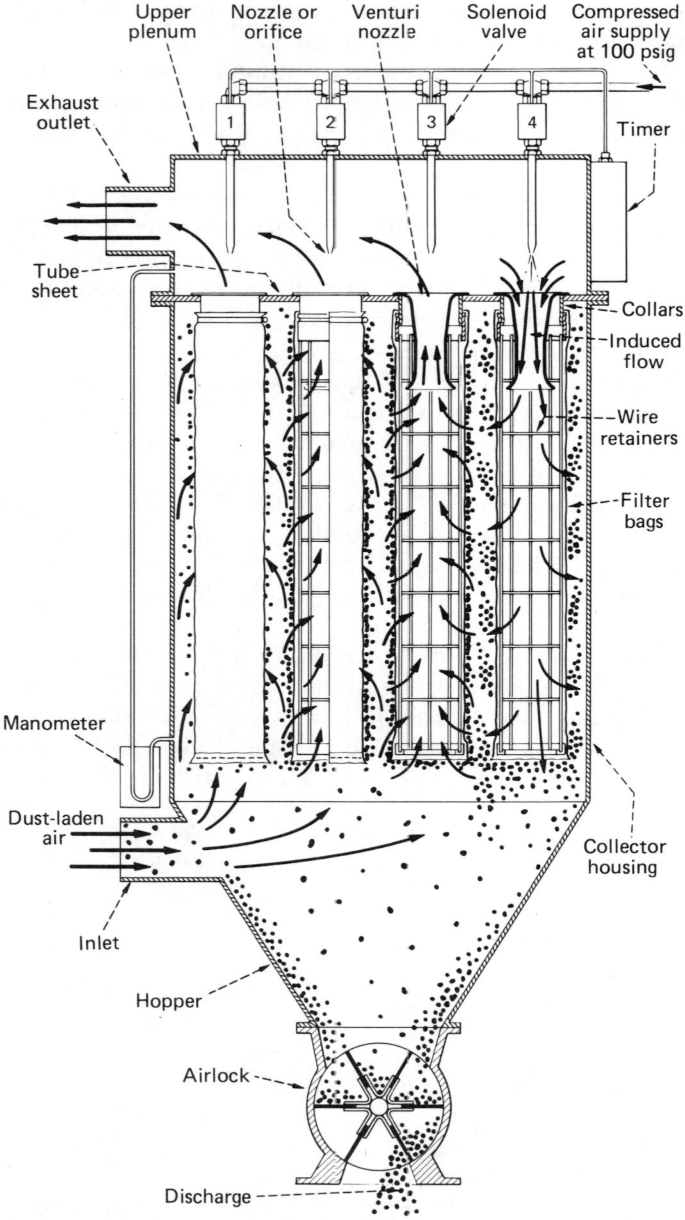

FIGURE 10–7. Dust collector cleaned by high-pressure reverse air blast. (Mikropul Corporation.)

poppet valve in the zone to be cleaned closes, stopping airflow through the zone. The pulse valve opens for 0.1 sec, admitting a burst of air into the plenum to clean the bags. The poppet valve then automatically reopens, putting the cleaned zone back onstream. This operation is repeated in each successive zone until all bags in the collector have been cleaned.

3. A motor-driven plenum chamber receiving air from a separate blower delivers reverse air to closed-end oblong bags that are supported as in the previous arrangement but not zoned. The plenum rotates and blows reverse air into the clean-air discharge end of each bag as its nozzles travel over the bags. Timing is determined by the speed of rotation of the plenum chamber. See Figures 10–8, 10–9, and 10–10.

4. A counterweighted frame carries a series of blowrings that encircle each filter tube. The frame is driven slowly up and down by chains supported from overhead shafts and sprockets. Air is continuously supplied to the blowrings by an external blower or fan connected by a

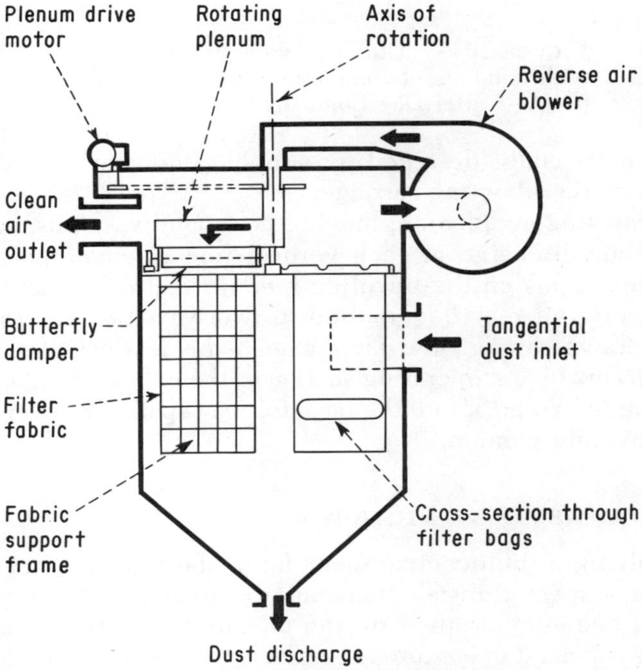

FIGURE 10–8. Dust collector cleaned by low-pressure reverse air blast.

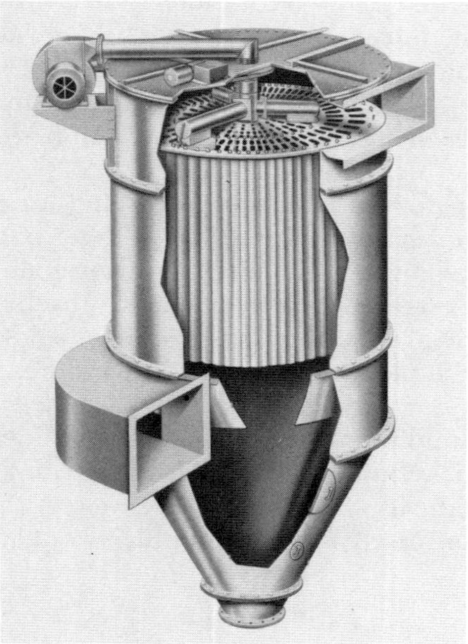

FIGURE 10–9. Cut-away view of dust collector with rotating plenum for reverse-air cleaning. (CEA—Carter-Day Company.)

hose, or a specially designed air-supply connection to a manifold mounted on the blowring carriage.

5. A traveling air plenum, moving horizontally, is positioned over the clean-air discharge of each vertical line of envelope-type filter bags. Diaphragms on the traveling plenum seal off adjacent rows of bags. Reverse air may be supplied in two ways: (*a*) atmospheric air may be drawn through the filter bag in the reverse direction from normal airflow by the operating suction of the exhaust fan, or (*b*) clean air may be blown back into the bags from a separate blower mounted on the traveling plenum.

Multicompartment Filters

In applying a multicompartment filter, the total area of cloth selected for a specific installation must be increased to allow for the outage of one compartment during cleaning. If a two-compartment dust filter is placed in continuous service, the pressure drop across the filter during cleaning of one compartment is more than doubled since the total airflow and its dust load must pass through the remaining

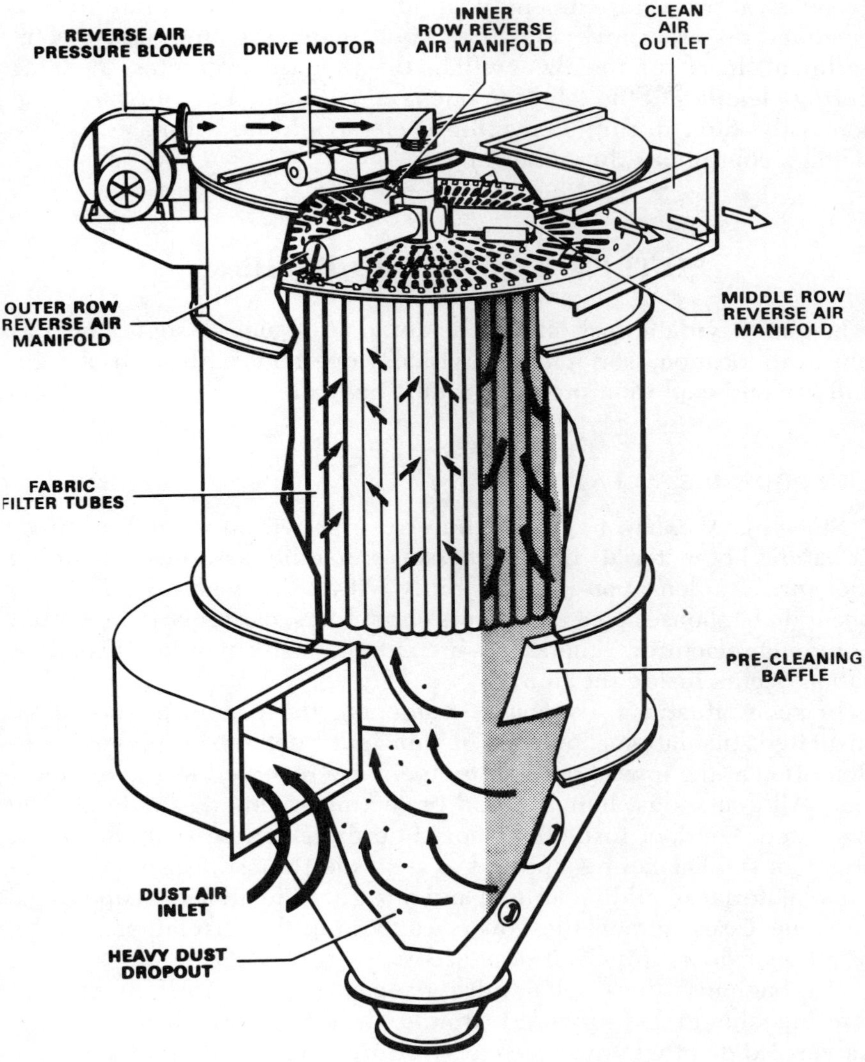

FIGURE 10–10. Operation of dust collector cleaned by rotating plenum. (CEA-Carter-Day Company.)

compartment, having only one half the total cloth area, and through its associated ducts and dampers. The pressure drop through the fabric varies directly as the airflow, but the pressure drop through dampers, ducts and tubesheets varies almost as the square of the airflow. Figure 10–6 shows the variation in bin suction pressure during the cleaning cycle of a two-compartment dust filter receiving material from a positive-pressure conveying line. The greater the number of compartments in a filter, the lower will be the pressure drop across the filter during cleaning of the fabric in one compartment. The pressure drop across the filter during a cleaning cycle should be estimated before sizing a continuous duty filter unit.

FILTER DESIGN REQUIREMENTS

When specifying a dust filter, consideration of such design details as physical location, servicing provisions, operations, dust discharge, and fire and explosion protection must be made.

Outdoor Installation

Since many exhaust systems draw air from within warm buildings, or handle hot materials from processes, or handle materials containing moisture, condensation is a possibility within the baghouse. This can occur on baghouses located indoors or outdoors, particularly in winter. Some manufacturers' guarantees are automatically avoided if condensation occurs inside the unit.

In such situations (especially outdoors), the baghouse should be insulated. Insulation should be of sufficient thickness to prevent condensation at the lowest outdoor temperature recorded in a given location. All doors, dust hoppers, and projecting stiffeners should also be insulated. Outdoor insulation should be protected with aluminum or stainless steel jacketing, applied so as to shed water. Flashings of the same material should be formed and fitted around all access openings. In some cases, removable covers containing the insulation may be fitted over doors to permit easy access.

If a baghouse does not require a separate enclosure, a watertight housing should be provided around dust hoppers, dust-discharge valves, and other dust-recovery equipment located below the baghouse to permit access for servicing during bad weather. Hinged or easily removable weathertight hoods should be furnished to protect all external operating mechanisms.

All electric conduit to timers and actuators mounted on the baghouse should be watertight. Pneumatic actuators, solenoid valves, shaking mechanisms, and air piping to them must also be insulated or traced, to prevent condensation of moisture and its subsequent freezing, which will prevent operation of the accessories.

If baghouses handling hot gases are shut down for the weekend, they become especially susceptible to condensation problems that prevent start-up. Such units should be equipped with an independent heater and a pressurizing fan designed to maintain the normal operating temperature within the unit during shutdown. A gravity-closed damper on the discharge of the main exhaust fan will permit retention of the heated air. Baghouses operating under pressure should have a weathertight cover or an automatic backdraft damper at the air outlet to prevent intake of cold air during shutdowns.

Access for Servicing

The location of access doors is principally determined by the method of grouping the bags and attaching them to their supports. The closed-end tubular bag and its metal retainer suspended from the clean-air plenum tubesheet is the most common design because it permits the closest grouping of bags in a given space. Access is through hinged doors located entirely across the clean-air plenum, atop the collector. The bag and its retainer are simply dropped through the tubesheet and fastened to it by clamps, snap rings, or twist locks, using the bag material as a gasket. Internal air-nozzle piping is made to swing out of the way or to be easily disconnected. Or the clean-air plenum is subdivided and valved so that each subdivision can be isolated and pulsed by a single solenoid valve.

Where the bags must be serviced regardless of weather conditions, the clean-air plenum can be designed as a walk-in housing requiring only one door, but only on collectors fitted with an air nozzle over each bag. Where bags and retainers are clamped from below the tubesheet, the doors must be located on the side walls of the unit so that bags are within 30 in. of the opening to allow proper takeup of clamps.

If adjacent walls or equipment interfere with proper location of a side door, the available doors must be used, necessitating the removal of a group of bags in order to reach those farthest from the door.

Bags suspended from shaking bars, with their open ends fastened to a bottom tubesheet, are usually grouped on each side of a narrow walkway within the clean-air compartment that is reached through a single door in the wall. During rebagging, the most remote bags must be attached and tensioned. Then, work must progress toward the walkway.

Envelope-type bags on metal spacer frames are installed in horizontal rows, with their open ends clipped to the vertical slots of a clean-air plenum, and their closed ends hooked to a support frame. Several rows of bags are located in tiers, one above the other. All bags in a compartment are accessible from two doors—one leading to the clean-air plenum, the other leading onto a walkway in the dusty-air chamber.

Long, large-diameter bags that are open at both ends are usually fastened to a bottom tubesheet that is accessible through a side door at the bottom of each compartment, and to the top tubesheet that is accessible through one or more doors in the roof of the dusty-air inlet plenum. The side doors lead to walkways between the bags. Some designs limit the number of rows of bags on either side of the walkway in order to improve access for clamping and to avoid damage to the bags.

All doors should be loosely hinged and clamped or latched in such a manner that all surfaces of the gasket are sealed airtight. Gaskets should be closed-pore sponge rubber (40 Durometer hardness) that is cemented into a confining recess around the door or around its frame. The rubber and its cement should be compatible with each other and with the dust to which they are exposed. Latches or clamps should be so designed that they can be opened manually by operators without using a wrench or special tool. The frames of all doors should extend beyond any external insulation so that they can be flashed properly. The inside surface of vertical or sloping doors on the dusty side of the collector should be flush with adjacent walls in order to prevent dust buildup on ledges.

Platforms, railings, and ladders should be furnished for reaching all elevated equipment and all access doors. These should be built in accordance with applicable safety codes. Rooftop equipment should be accessible for servicing from within the peripheral handrailing. Ladders should have safety cages starting 7 ft above floors or platforms, and should be equipped with grab rails and, for greater safety, with narrow, nonskid treads in place of rungs. Platforms should be made from serrated grating so that dust cannot accumulate and slipping is prevented.

Servicing Adjustments

Tubes on shaker-type units should be suspended from support grids by means of eye bolts or threaded rods with sufficient takeup for adjustment of bag tension. Tube fasteners to tube sheet ferrules or grid supports for open-end tubes may be of the toggle-clamp or snap-lock type, when operators are not permitted to use tools. Program timers should have a supply of change gears, potentiometers, or some other device to permit shortening the cleaning cycle below that recommended by the vendor. Jog buttons should be provided to permit operating the shaker motors and dampers while the system is shut down. These will assist in checking motor operation and in releasing the dust load when tubes are overloaded due to system troubles.

Operation

Operate filters under negative pressure at all times in order to prevent the discharge of dust to process areas, or the pressurizing of bins or receivers on positive-pressure conveying systems. Use air-cylinder motors for damper operation on units with short shaking or cleaning cycles. Where the length of shaking cycle permits, give preference to electric-motor damper operators, which, despite their longer period of response, are more reliable. Baffles should be furnished to distribute the dust load evenly across the filter. Where baffles are not practicable, inlet connections should be arranged to reduce entering dust velocities so as to prevent impingement on interior parts, on the shell of the filter, or on filter bags. Impingement on fabric has the same effect as high face velocities and the affected portion of the fabric will become blinded, thus reducing the effective cloth area.

Do not interconnect a pneumatic conveying system receiver with a central exhaust system dust filter. Provide a separate dust filter to suit the conveying system requirements, otherwise, troubles may occur on the central exhaust system from overloading due to the difference in cleaning cycle time required for the two services.

To reduce dust load and to make effective use of fabric cleaning methods, the following rules should be observed in system design:

Where material enters a silo or bin, locate the filter as far from the entrance point as possible.

Elevate the bottom of the filter-bag housing above the top of the silo by adding an additional section of casing below the bag housing.

Add an additional section of casing between bag housing and hopper on combination receiver-filters.

Use the fabric with the best dust-releasing characteristics for the material handled.

For abrasive materials, install a polyurethane or ceramic-lined scalper ahead of the fabric filter.

On bins with vent filters, keep the high-level indicator as far below the bottom of the fabric as possible.

Install an air-volume chamber or accumulator in the compressed-air header ahead of each large dust filter or close to each group of small dust filters.

Reduce the velocity into the dust filters by installing an entrance transition with a 14-degree diverging angle. (Sudden transitions *do not* lower velocity).

Also, a detailed knowledge of filter design parameters is essential if these units are to be properly applied to a pneumatic conveying system.

PROTECTION AGAINST FIRE

When hazardous materials are handled, protection must be provided to eliminate or minimize hazards. Some materials may require only fire protection; others, fire and explosion protection.

The best procedure for determining the hazards in handling a particular dust with unknown characteristics is to submit test samples to the fire-insurance underwriter's laboratory or to a commercial testing laboratory that has facilities for conducting:

1. *Combustion test*—Ignition by match flame or Bunsen burner to determine whether the material is self-extinguishing, or whether the flame will spread.

2. *Spontaneous-combustion test*—Exposing material to high temperature for a prolonged period (several hours). The material is contained in a crucible immersed in a hot salt bath. Moist materials should be tested at their usual moisture content, and again after drying.

3. *Autoignition-temperature test*—Same as the preceding test, but raising the temperature continuously to determine the temperature at which ignition occurs. Evaporation of water from a moist sample may exclude air, so that the autoignition point could be on the high side.

4. *Dust-explosion test*—Material is placed in a bomb having a variable vent opening, mixed with compressed air, and then ignited by an electric spark. The pressure developed is measured and reported for two conditions: (*a*) unvented, and (*b*) for a specific ratio of vent area to bomb volume. Materials known to be explosive should also be tested for rate of pressure rise in addition to the pressure developed from an explosion.

Where possible, any baghouse subject to fire or explosion should be located outdoors and away from important equipment.

Fire protection may be provided by a built-in sprinkler system. The sprinkler heads would be rated about 25°F below the temperature limit of the bags if no other protection were available; and 100°F above this limit, if early-warning protection were installed. The sprinklers would be connected to a building sprinkler system. The heads are usually located where they can cover most of the bag area. Additional heads may be needed to cover the dust hoppers. Because of difficulty in placing heads in some baghouses, assistance of the fire underwriter's engineering department should be obtained.

Opening of the sprinklers would be indicated by the water-flow detector and alarm on the building sprinkler system. However, a local alarm to warn system operators may also be required. The local alarm would be a fire detector within the baghouse—the detector being set at the same temperature as the sprinkler head. Provision must be made to dispose of the water released by the sprinklers. When the baghouse is outdoors, a nonfreeze sprinkler system will be needed. A dry system, using air in the sprinkler lines, is expensive and requires considerable maintenance. A system filled with a nonfreeze solution may foul the bags when the heads are released. Hence, the bags may require replacement even if they are not burned.

Several alternatives are available for sprinklers:

1. An inert-gas (Halon) injection system may be installed. When actuated, it will exclude or reduce the oxygen content in the baghouse so that combustion cannot be sustained. The sensor is a rate-of-rise temperature detector, set at the temperature limit of the bags, that operates a control unit to electro-explosively rupture a diaphragm to release the inert gas into the baghouse. The gas storage vessels are located outside the collector. If the system is actuated, there is no mess to clean up except char from the fire. The control unit can also initiate other functions such as closing dampers, stopping fans, and sounding alarms.

2. A steam-smothering or a water-deluge system that is actuated manually or by a rate-or-rise temperature detector uses open nozzles, located in the same manner as sprinkler heads.

3. A manually operated deluge system can be devised for collectors having reverse-jet air manifolds by connecting a water header to the air manifold pipes at the ends opposite the air header. When water is turned on, the air pulses speed up wetting of the bags. The timer can be temporarily switched to pulse all nozzles simultaneously, and pulse time could be increased to obtain faster water distribution.

Protection against Explosion

Explosion protection may be provided by installing an inert-gas suppression system or explosion doors.

The suppression system is similar to that for fire protection, except that a highly sensitive pressure detector initiates action of the control unit. The pressure setting is determined from data for the rate of pressure rise obtained during the bomb test. The baghouse may be isolated so as to contain the inert gas without loss by the use of fast-acting inlet and outlet dampers, or by shutting down the exhaust fan and rapidly closing an outlet damper.

The inert-gas suppression system detects the pressure rise caused by an explosion and acts in milliseconds to stop pressures from reaching values that would rupture the bag housing. Because of the short duration of an explosion and the dependence on proper functioning of all control equipment, a backup system of explosion doors may be desirable.

Explosion doors are generally installed in baghouse walls on the dusty-air side. They may be hinged doors with special, spring-loaded, quick-releasing latches. They may be rupture panels made of frangible materials such as aluminum foil, lead sheets, polyethylene, or Teflon. They may be laminates of metal and plastic designed to rupture along prescored lines to obtain full opening of the panel. They may be loose doors held in place by their own weight or by suction.

Latched or loose explosion doors on outdoor baghouses must have rain hoods built around them. Weighted doors, unattached by any other means, should have chains connected to them, which are fastened to an internal structural member, so that the doors do not become missiles when blown open.

Every explosion is rapidly followed by an implosion that can be as damaging to the structure as the explosion. Explosion doors that are pivoted or hinged should be designed to prevent automatic reclosing. Such doors should be reclosed manually.

Dust Hoppers and Dust Discharge

The material removed from the bags may be dropped directly into a bin or silo if the baghouse is used for venting that bin or silo. Otherwise the material is dropped into a collecting hopper that is shaped to direct the flow to the dust outlet usually fitted with an airlock-type discharge valve.

On multicompartment units, each compartment should be fitted with its own dust hopper to permit its isolation during cleaning or maintenance. To eliminate a multiplicity of airlock discharge valves, the bag compartments may discharge into a common V-shaped hopper equipped with a screw conveyor that delivers the dust to a single discharge valve. Such a screw conveyor must operate at the baghouse pressure. Therefore, the conveyor must be equipped with effective seals at the ends of the trough and at the bearings, which should be outboard of the hopper. The conveyor must also be able to handle the collected dust at the proper rate.

The flow properties of the collected dust should be known, since they may be quite different from the properties of the material from which they originated. The slope angles of the dust hopper may have to be steeper than those of the storage silos of the original material because (1) fine dusts tend to pack more readily than coarse materials, (2) the dust may contain moisture picked up in the conveying air or formed by condensation within the collector, (3) collected material should be removed as rapidly as it is deposited on the hopper walls. Steep slope angles increase the overall height of the baghouse and may limit the space available for installing a discharge valve. However, failure to provide the proper slope will create problems after installation.

A conical hopper attached to a cylindrical baghouse presents no problem other than the increase in height required to obtain a steep slope. Attachment of a square, rectangular, or conical hopper to a square or rectangular baghouse requires study—especially of the valley angles formed by the slope sheets of inverted pyramidal hoppers, or by transitions to conical hoppers. These valley angles should be at the minimum value required to obtain material flow. If the wall-slope angles to the horizontal are known, the valley angles may be obtained from Figure 11–1 or from the formula from which the chart is derived:

$$\cot^2 \theta = \cot^2 \alpha + \cot^2 \beta$$

where θ is the valley angle, and α and β are the slope angles between the horizontal and wall A and wall B of the hopper, respectively, as shown in the sketch in Figure 11–1.

On baghouses cleaned by reverse-air blast (using compressed air), the slope-sheet angles can be reduced by a few degrees, because the air blast helps to propel the collected powder down the slope.

Materials may still hang up even though the proper slope angles are provided on dust hoppers, because of process upsets, changes in moisture content, etc. To ensure flow, mounting pads should be installed on the slope sheets for attaching mechanical or pneumatic vibrators for inducing flow. The vibrators should be placed at the centroid of each surface. The flow inducers should be operated only when needed. Cylindrical hoppers require only one pad, located at the centroid of the projected surface.

The dust hopper and its supports should be designed on the basis that it could become full of material and may also have to support a surcharge. This situation may arise if the baghouse becomes full of material because of a broken bag, a malfunction of the dust-discharge valve, or stoppage of a downstream dust conveyor.

FILTER SELECTION

Dust filtration and collection is as much an art as pneumatic conveying itself. The sizing and selection of dust filters for use in pneumatic conveying systems is based on past experience and on actual tests using specific materials at specific material-to-air loadings. The many types of collectors and filter fabrics available, combined with the numerous material properties that affect filtration, indicate that selection of a filter should be made by actual test at the design load using the cheapest fabric available, consistent with service requirements. All reputable manufacturers of filter units maintain pilot plants for such tests and can assist the pneumatic-conveying-system purchaser in his selection by scale-up and application of experience gained on similar applications of dust filters.

The many types of dust filters make selection difficult unless basic criteria are established to limit the choice. Even after establishing such criteria, it would be difficult to find a specific type of filter that meets all of the set requirements. The final selection of a unit would be obtained from analysis of bids submitted by vendors whose equipment meets most of the criteria and whose estimated cost of a full-size unit is within budget limitations. This unit would then be tested, or the test may be waived, if past experience indicates that the filter is suitable for handling a specific material at the proposed dust loading.

The factors given below should be analyzed and decisions made as to the features considered particularly desirable in a dust filter for the type of material to be handled and the service conditions (continuous or intermittent).

Method of Cleaning

Factors to be considered include: (1) complexity of mechanism to effect cleaning; (2) amount of auxiliary equipment and number of motors required; (3) location of filter cake—inside or outside of tubes; (4) possibility of plugging up the tubes; (5) direction of dust flow in relation to normal airflow to obtain minimum redeposit of dust on clean fabric; (6) effect of cleaning on pressure drop; (7) type of fabric required—felt or woven—and its effect on filter cost, space requirements, and filter area; (8) utilities (air, electricity) required to effect cleaning and to operate the exhaust fan.

Filter Tube Replacement

This needs study of: (1) accessibility of tubes for removal—from inside or outside of unit; (2) cleanliness of operation if men must enter filter—work to be done on dirty side or clean air side; (3) auxiliary equipment—installation of temporary platforms or scaffolding within filter; (4) ease of tube removal and replacement—simple clamp adjustment of ferrules, deformation of retainer rings built into the tubes, or removal of grids from the unit in order to strip and replace the tubes or envelopes.

Mechanical Moving Parts

The following need consideration: (1) availability for servicing—unit must be shut down or may continue to operate; (2) accessibility for servicing—mechanism is inside or outside of unit—requires permanent or temporary platforms and ladders; (3) protection—covers are required for outdoor installation—may become loaded with dust if a bag breaks; (4) contamination—lubricants may enter dust hopper or dust may enter bearings or seals and foul the lubricants; (5) structural—effect of reciprocating or vibrating action on equipment supports and tightness of joints. (The large mass of the parts that are in motion during cleaning of filters having a tube-shaking mechanism imposes a heavy impact and vibrational load on the supporting structure. In time, bolted structural connections and panel joints and rolled tube ferrules work loose.)

Floor Space

The floor space available in a plant may determine the type of filter unit to use. Multicompartment, continuous-service filters with tube shaking cleaning methods generally have a very low air-to-cloth ratio and occupy more space than other types of filters. Units cleaned by reverse airflow using compressed-air nozzles require the least amount of floor space because of their high air-to-cloth ratio.

Housing Construction

Structural arrangements to be considered are: (1) modular design to permit installation as a complete, factory-assembled unit, alone or in a group of parallel-connected units; (2) knocked-down, panel construction for assembly at the jobsite; (3) type of assembly—bolted joints or all-welded; (4) materials of construction to resist action of the filtered material or of the elements; (5) exterior and interior coatings—galvanized, paint, bare, epoxy, dust-releasing coating, etc.; (6) form of unit—square, cylindrical, rectangular.

Baghouses may operate under positive or negative pressures. Baghouses that receive air from a fluidized-bed process or from a positive-pressure pneumatic conveying system generally operate under positive pressure. The pressure is limited by the pressure drop across the bags because discharge is directly to atmosphere.

Most baghouses operate under negative pressure, which imposes stringent design requirements for the bag enclosure walls.

Although the pressure drop across a baghouse may be 3 to 6 in. water column, the negative pressure imposed by the exhaust fan is determined by the duct-system requirements that may be five to ten times those of the pressure drop. If the air inlets to the baghouse are blocked by powder or closed inlet dampers, the suction may increase to values well above the normal operating level. The user must determine the maximum suction that may be imposed on the baghouse and seek a unit that is rated for this condition.

If the dust-air mixture can become explosive, the baghouse walls must also resist a positive pressure that is determined from dust-explosion tests. These fix the vent area in the walls so as to limit pressure buildup. The user must specify the allowable positive pressure for baghouse design.

Cylindrical baghouses are simpler to design for high suction or high pressure than are flat-sided enclosures. Standard suction ratings for cylindrical units range from 15 to 160 in. water column, depending on diameter. Rectangular or square baghouses have ratings that range from 15 to 30 in. water column. These units require external battens and cross-bracing to stiffen them. Multicompartmented units have interior walls that require special bracing to resist differential pressures and stress reversal during cleaning operations. Tubesheets are usually made of heavier plate than are the walls, and they are fitted or formed with stiffeners because the tubesheets not only serve to support the bags and retainers but also act as walkways during tube replacement. Distortion of tubesheets can cause top-suspended bag retainers to touch each other and retain dust. Some baghouses are only available as cylindrical units; others may be square, rectangular, or cylindrical, depending on size and application.

Receiver-Filter Units

In selecting a filter, particular attention should be given to analyzing combination receiver-filter units since these units attempt to combine in one piece of compact equipment the functions of a cyclone-type receiver and a dust filter. The features usually provided for separation of material are a tangential entrance nozzle into a hopper at the base of the unit, and internal skirts or baffles to deflect most of the dust from the tubes. These features are only partially effective since no true cyclone action can take place because of limited space available and interference with the air circulation or vortex during cleaning of the tubes.

FILTER TESTING

The testing of a dust filter is time consuming and there is a tendency to bypass this phase of equipment selection and rely on a vendor's application experience in sizing his particular type of filter. However, a reputable vendor would request that several cubic feet of the dust be sent to him for test in his pilot plant unit. The unit would be equipped with brand new filter tubes of his choice based on analysis of the dust characteristics by his laboratory. A typical test procedure would be as follows:

1. A volumetric feeder would be set to obtain a high discharge rate and the actual amount of test material discharged over a period of time would be weighed to determine the dust feed rate.
2. A blower would exhaust air from the filter and a calibrated venturi tube in the suction line would be used to meter the airflow.
3. The feeder would be set to discharge dust into the filter inlet connection at the calibrated rate. The discharged powder would be dispersed by an air jet to simulate plant operating conditions.
4. The cleaning cycle programmer would be set at the vendor's recommended rate and the necessary filter operating utilities turned on.
5. The filter would be operated for at least eight hours in order to build up a filter cake. However, if the differential across the bags builds up to the vendor's standard pressure drop prior to this time, the dust feeder is recalibrated to a lower rate and the airflow reduced accordingly.
6. The standard pressure differential is then maintained continuously at various rates of dust loading and airflow as determined by trial and error. Each combination of dust loading and airflow rate is maintained for several hours.

During this type of test, the data is plotted continuously as dust loading in grains per cubic foot of air entering the unit versus the filter ratio expressed as cubic feet of air per square foot of cloth area. Typical curves for various materials are shown in Figure 10–11. These are called loading curves. A loading curve is useful only for sizing the type of unit tested, when it is handling the material for which the curve was drawn.

TEMPERATURE CORRECTION

The dust loading test is run using ambient air temperatures. When the filter is intended to be operated at a higher air temperature, the filter rate must be corrected to account for temperature above stan-

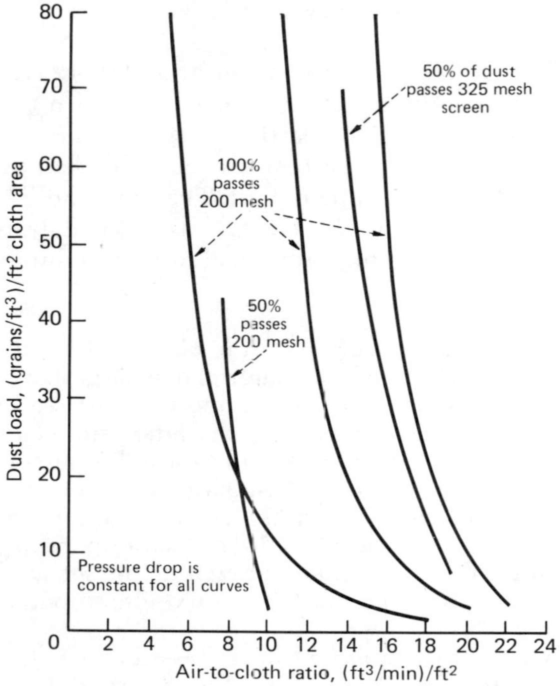

FIGURE 10–11. Characteristic dust loading curves for a filter cleaned by high-pressure reverse air blast. (MikroPul Corporation.)

dard. The correction factor is obtained from Figure 10–12 which is a curve for constant volume flow. This factor, in effect, increases the required cloth area. It compensates for the increased air viscosity at high temperatures (the viscosity of air increases almost in direct proportion to the increase in temperature). In general,

$$\frac{F_1}{F_2} = \frac{\phi_1}{\phi_2}$$

where F_1 = filter rate at a given temperature and 3.5 inches of water pressure drop

F_2 = corrected filter rate

ϕ_1 = filter-rate correction factor for ambient temperature on test

ϕ_2 = filter-rate correction factor for operating temperature.

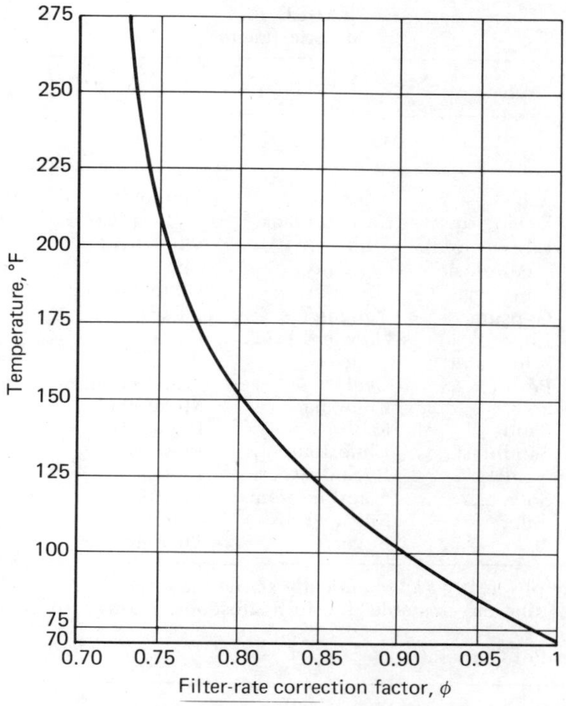

FIGURE 10–12. Correction factor to account for temperature above standard. (MikroPul Corporation.)

After the filter rate is corrected, a separate calculation must be made to allow for expansion of the air at the higher temperature. (See Gas Laws, page 35.)

ESTIMATING THE FILTER RATE

A preliminary estimate of the test-filter rate can be made by using a filter rate guide developed by one manufacturer for application to pulse-jet or blow-ring baghouses that have continuously cleaned felt-filter media. Such baghouses maintain a pressure differential of 3.5 in. water column at design flow and loading conditions. The guide consists of five factors, multiplied together to arrive at a filter rate. These are:

Material factor, F_m, from Table 10–2 allows for the properties and characteristics for a material that would affect its tendency to form a

TABLE 10–2
Material Factor

Factor, F_m				
15*	12	10	9	6†
Cake mix Cardboard dust Cocoa Feeds Flour Grain Leather dust Sawdust Tobacco	Asbestos Buffing dust Fibrous and cellulosic materials Gypsum Lime, hydrated Perlite Salt Sand Sandblast dust Soda ash Talc	Alumina Carbon black, fine Cement Ceramic pigments Clay and brick dusts Coal Fluorspar Kaolin Limestone Rock dust, ores and minerals Silica Sugar	Ammonium phosphate fertilizer Coke Diatomaceous earth Dyes Fly ash Metal powder Metal oxides Pigments Plastics Resins Silicates Starch Stearates	Activated carbon Carbon black, (molecular) Detergents Fumes and other dispersed products direct from reactions Powdered milk Soaps

* In general, any physically and chemically stable materials.
† Includes solids that are unstable due to hygroscopic nature, sublimation or polymerization.
SOURCE: MikroPul Corporation.

TABLE 10–3
Application Factor

Process	Factor, F_a
Nuisance venting Relief of transfer points, conveyors, packing stations, etc.	1.0
Product collection Air-conveying venting mills, flash dryers, classifiers, etc.	0.9
Process-gas filtration Spray dryers, kilns, reactors, etc.	0.8

Source: MicroPul Corporation

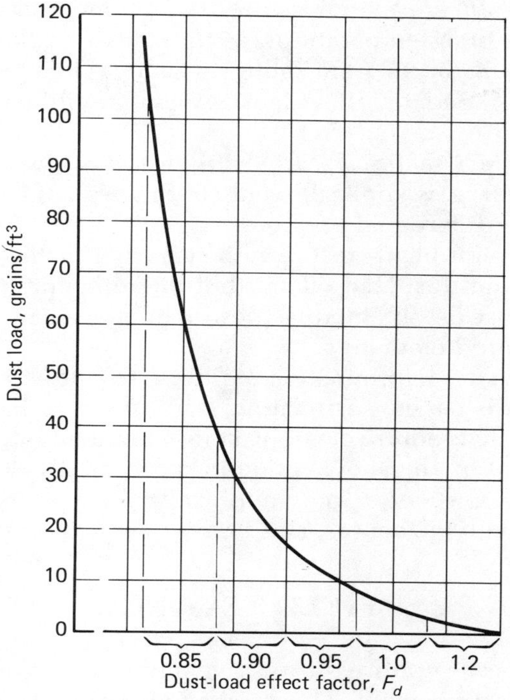

FIGURE 10–13. Correction factor to account for the effect of dust loading. (MikroPul Corporation.)

filter cake during normal airflow and release it during the cleaning cycle for the bags by reverse airflow.

Application factor, F_a, from Table 10–3 allows for the type of process delivering dust to the baghouse and for upsets that cause variations in the dust load.

TABLE 10–4
Particle-size factor

Size	Factor, F_f
Over 100 micron	1.2
50 to 100 micron	1.1
10 to 50 micron	1.0
3 to 10 micron	0.9
Under 3 micron	0.8

Source: MikroPul Corporation

Temperature effect, ϕ, from Figure 10–12 compensates for increased air viscosity at high temperatures, as previously explained.

Particle-size factor, F_f, from Table 10–4 allows for the fact that fumes and fine dusts tend to blind fabrics more rapidly than do coarse particles.

Dust-load effect factor, F_d, from Figure 10–13, which is a typical performance curve, is applicable only to this type of baghouse operating at constant differential pressure.

An actual pilot-plant test eliminates the factors for material, particle-size, and dust-load effects, but the temperature and application factors must be obtained, as described above, in order to determine the required cloth area.

The preliminary determination of filter rate can be used to roughly size a baghouse for cost-estimating purposes. For baghouses using woven fabrics that require cleaning at less-frequent intervals, the filter rate established by the above guidelines can be divided by five. The resulting rate, however, is only applicable for dust loads within the capability of such baghouses, and for cost-estimating purposes only.

FIELD TESTING

Some vendors can supply a prospective purchaser with a field test unit for installation at his plant. This method of evaluation should only be undertaken under one or more of the following circumstances: (1) when there is no precedent for filtering a specific material; (2) when a filter design is deemed novel and the purchaser has had no previous experience with it; (3) when the filter must be evaluated under a variety of plant operating conditions; (4) when the effect of weather conditions on filtration is desired for a proposed outdoor installation.

CONDITIONS OF FIELD TESTING

The test unit is generally supplied on a rental basis with an option to purchase it. The accrued rental charges would be applied to the purchase price of the test unit or of a larger unit to meet project requirements. The agreement with the vendor should provide for the following items: (1) freight, handling, and installation expense (usually borne by the purchaser); (2) services of an installation supervisor if one is desired; (3) allocation of reconditioning expenses for damage caused beyond normal wear and tear during a test; (4) responsibility for loss of the unit from any cause; (5) access to the test area by the vendor's representatives and furnishing of test data to them; (6) a schedule of charges and method of billing; (7) statement of credits allowed upon return of the test unit; (8) a list of auxiliary equipment furnished with the test unit.

Test Procedures

Installation of a field test filter in an operating plant usually requires that a small percentage of dust laden air be drawn from an existing exhaust system to match the estimated capacity of the test unit. This requires insertion of a special nozzle into a straight section of the existing duct—not just a tee or lateral connection. The open end of the nozzle should face upstream and be centered so as to receive a representative sample of the dust load. The air drawn through the test filter should enter the nozzle at the same velocity that normally exists at the cross-section from which the sample is drawn. This method is called isokinetic sampling. The exhaust fan for drawing this side-stream through the test unit must be sized for the airflow required to obtain the isokinetic velocity and a suction pressure low enough to overcome the suction in the main plus the estimated resistance of the test filter and its ductwork. Some form of shut-off gate should be placed in the suction line close to the test unit so that (1) it may be opened for inspection and modifications made without affecting flow in the suction main, and (2) the test unit may be shut down without causing reverse airflow through the exhaust fan which will cause it to rotate as if it were running, but in reverse rotation.

Collected dust should be arranged to discharge, through some form of airlock, into a drum for weighing. If no airlock is used, the dust may flow through an airtight drum hood and the drum kept under suction. However, every time the drum has to be emptied or weighed, the test unit must be shut down. Provision should be made for periodic disposal of the collected dust, otherwise, dust may back up and fill the filter thus voiding the test results for that period.

Measuring Performance

Means for measuring performance variables should be installed and data taken as follows:

1. Airflow rate. Install a calibrated venturi meter or orifice in the clean air discharge duct with the required lengths of straight duct before and after these devices, or, install two connections in a straight section of duct for making a Pitot tube traverse. These connections should be at right angles to each other and at the same cross section of the duct. When testing, obtain the flow rate before, during and after cleaning. A rough, fairly reliable, check of airflow may be made by measuring the pressure differential across the exhaust fan and referring to the characteristic fan curve of static pressure versus air volume, for the operating speed of the unit. When the characteristic curve rises and then drops so that it is uncertain which of two identical static pressures has been read,

the motor horsepower should be measured, using clamp-on instruments for quick readings. The horsepower-versus-volume curve should then be used to determine which side of the static pressure-capacity curve the static pressure reading belongs. (The horsepower curve is usually a straight line increasing proportionally with the capacity.) During a test, a flow measurement higher than that required to obtain isokinetic velocity indicates the presence of an air leak in the test equipment, usually at the dust discharge connection.

In a filter using reverse flow of atmospheric air for cleaning, the reverse airflow rate should be obtained by installing a calibrated orifice at the reverse air inlet and measuring the pressure drop across it.

2. Dust load. Install a platform scale close to the dust collecting drum. Weigh the load in the drum at regular intervals and use the weights of filled sample-sized containers of known volume to determine the average powder density.
3. Temperature. Install thermometer connections at the filter inlet and outlet. The thermometers can be of the indicating or recording type. A glass stem indicating thermometer inserted in a cork-stoppered hole in the duct may suffice for most tests.
4. Pressure. Install manometer connections on the inlet and outlet ducts close to the filter in order to measure pressure differential across the entire unit. Also install connections on the clean and dirty sides of the filter housing in order to obtain the pressure differential across the filter tubes. When testing, take pressure differential readings across the tubes and across the housing just before a filter cleaning period starts and just after the cleaning period ends.

   On filters using compressed air for cleaning, install an air pressure gauge on or close to the air supply manifold furnished as part of the filter.
5. Timing of cleaning cycles. Before starting the test, operate the program timer and check the time between closing and opening of air reversal dampers and starting and stopping of shakers in each compartment of multicompartment units with tube shaking and reverse-airflow cleaning. Check the time between the termination of the cleaning period in one compartment and the start of the cleaning period in the next successive compartment as set up in the program timer.

   On units with reverse-airflow cleaning, time the period between each successive blow and the duration of the airflow.

   The total elapsed time required to clean all the tubes in a filter once is called the cleaning cycle time.

11 Storage Bins

BIN SELECTION FACTORS

Before selecting a storage bin, consideration should be given to all storage factors and the characteristics of a material that may affect its properties and flowability during storage. Especially important are the storage factors that affect flowability such as: moisture absorption; surface changes of particles due to chemical action; compression of material in storage; effect of external vibrations; effect of changes in temperature and humidity. Information on these factors may be obtained from previous experience in handling the material in smaller containers or from other firms who have handled the material in bulk. The factors that affect the flowability of material as it is discharged from the bin are: the lateral pressure of the material on the bin walls; the coefficient of friction of the particles against adjacent particles; the coefficient of friction of the particles on the bin walls; the ratio of lateral to vertical pressure; and the shear strength of the compacted materials. Information on these factors must be obtained by experiment for the design of a mass-flow bin—one in which all material flows uniformly toward the bin outlet when the material is being discharged. Some of the properties of the material, along with the small size of the outlet required to match the inlet flange of a discharge feeder, will preclude consideration of designing the bin hopper for mass flow. However, sufficient information will have been accumulated to permit the use of a standard type of bin with necessary hopper modifications for the attachment of bin dischargers or flow inducers to ensure complete emptying.

The testing of the flow properties of a material for the design of a mass-flow bin is time-consuming and requires the use of expensive equipment such as a flow-factor tester and a consolidating bench that holds a minimum of six shear cells. However, many more shear cells are required when time consolidation tests are contemplated. Furthermore, the procedure for packing material into the cells requires the services of a trained technician. The type of shear cell to use and the proper procedure for packing them in order to ensure valid results is still controversial. Therefore such testing, if desired, should be done in a commercial laboratory that specializes in the flow testing of solids.

Even though the mass-flow mode is not required, the principles for the design of mass-flow bins should be applied to the design of any bin in order to ensure complete emptying of the bin and steady flow of material out of it.

Other considerations that may affect the choice of a bin design are as follows: mass-flow bins are normally designed to handle free-flowing, noncohesive materials with relatively constant flow properties; the source of material may change and the new material may have different flow properties; future process modifications may require storage of an entirely different material in the bin; extended periods of storage will cause the material to consolidate because of vibration, surface binding, moisture effects, etc.

Cylindrical Bins

Cylindrical storage bins with conical hopper bottoms and gravity discharge outlets are most commonly used for storing pneumatically conveyed materials, despite the tendency of some of these materials to arch, stick, or flood. This form of bin is low in cost; can be readily obtained in a range of standard diameters and heights to suit any location; can be made free-standing or set on columns to clear other equipment or interferences; can be fitted with flow inducers to insure continuous discharge; can have its hopper easily modified to suit any discharge condition; and can be obtained in a choice of materials or fitted with liners to suit any corrosion or flow condition. Special forms of bins should be considered only where space and headroom are limited; where the materials in storage have unusual flow characteristics, or where segregation of material must be avoided.

Insuring Free Flow

Regardless of the form of bin, the means of ensuring material flow toward the discharge connection must be analyzed and selected to suit the flow characteristics of the material stored. The sides of sloped sheets should be as steep as possible; never less than the angle of slide of the material.

Steep slope angles increase the overall height of a bin and may limit the space available for installing discharge feeders and conveyors. However, failure to provide the proper slope angle will create problems after installation.

Consideration should be given to the fact that on square or rectangular hoppers or transitions to conical hoppers, the valley angles formed by two inclined slope sheets meeting at a corner have less slope than the slope sheet angles. These valley angles should be at the minimum angle required to obtain material flow. If the wall-slope angles to the horizontal are known, the valley angles may be obtained from Figure 11–1, or from the formula from which the chart is derived:

$$\cot^2 \theta = \cot^2 \alpha + \cot^2 \beta$$

where θ is the valley angle, and α and β are the slope angles between

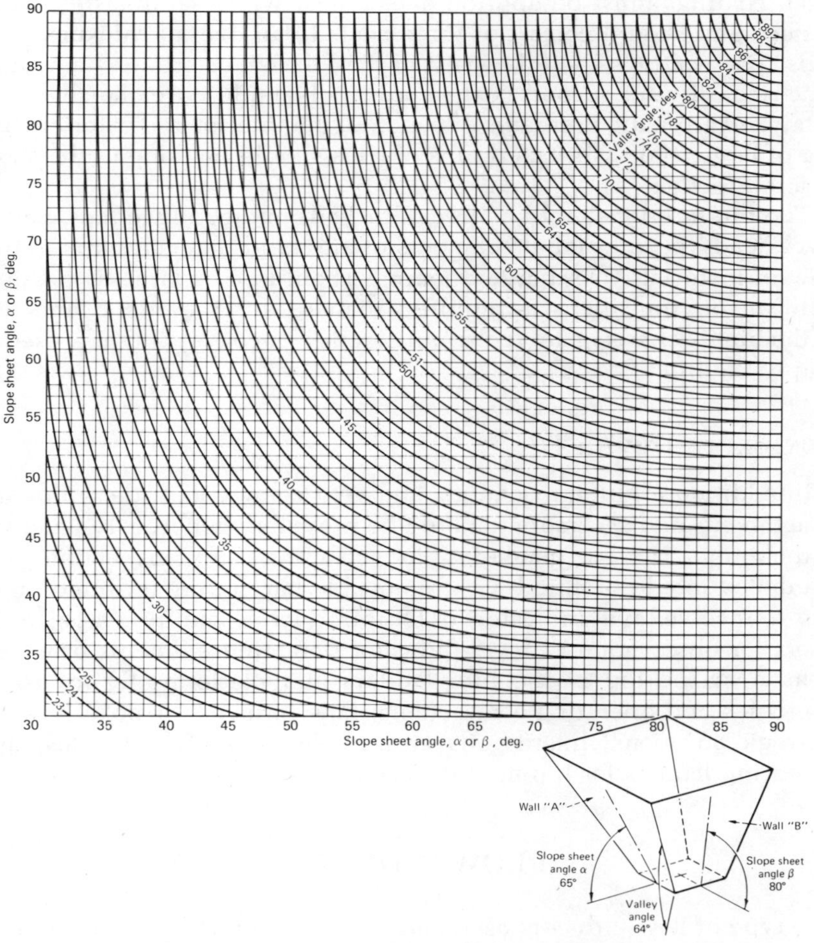

FIGURE 11–1. Valley angles for bins and hoppers having flat slope sheets.

the horizontal and wall *A* and wall *B* of the hopper, respectively, as shown in Figure 11–1.

Provision of the proper valley angle at an intersection does not ensure material flow, because materials can bridge across the intersection. To minimize bridging, all intersections should be formed to as large a radius as possible when joining the flat slope sheets internally to corners, to knuckle points at vertical walls, and to each other.

All surfaces in contact with stored material should be as smooth as possible and polished or lined with a self-releasing coating—one that will not retain product. For many tacky materials, an epoxy painted surface has been found to be smoother than polished stainless steel. Such coatings must be applied over a near-white sandblasted metal surface and should consist of three coats to an overall thickness of 6 mils. The coatings should consist of the following: primer—inorganic zinc-rich silicate; second coat—epoxy polyamide cure; finish coat—catalyzed aliphatic urethane. If possible, the interior and exterior coating of large, shop-fabricated bins or silos should be done in the shop, before erection.

The bin interior should be free of tie rods, projecting ledges, bolt or rivet heads, or internal attachments of any kind that can impede free flow of materials. Discharge openings should be as large as possible with smooth transitions, as required, to mate with the inlet connection of the bin discharge gate, reclaiming feeder, or conveyor that is used to empty the bin.

Auxiliary Devices

In addition to inlet, discharge, and vent connections, the following attachments and accessories should be provided: generously sized top and bottom access manholes for cleaning and inspection; top manholes should be fitted with removable bar grating for safety; nozzles to receive level indicators or for introducing sounding tape; small doors for observation of level-indicator detecting elements or conditions near the top of the bin; ladders and platforms for access to manholes and operating connections; flow inducers or poke holes at strategic locations; guard rails around the roof of high silos; and mounting flanges for bin-mounted vent dust filters.

FLOW INDUCERS

The type of flow inducers used and their location must be selected to suit the characteristics of the stored material and the configuration of the bin. A flow inducer is any means or device used to cause relative motion of particles of material within a bin. Electric or pneumatic vibrators may be used. They should be mounted on reinforcing plates attached to those slope sheets that are in contact with as large a surface of the stored material as possible and located at the estimated center of pressure of the surface. Vibrators should be sized to suit the rigidity and extent of the slope sheet. They should be arranged to operate only when dislodged material can flow freely out of the bins, since vibration without free flow will compact the materials and aggravate the bridging problem.

Aeration pads of woven fabric, porous metal, or porous ceramics may be installed with the active face flush with the interior of the bin slope sheets and connected to a low-pressure air-supply system. These pads may be isolated, grouped in strategic locations, or may line an entire slope sheet, if required. The air piping may be manifolded to activate any group of pads or all pads at the same time, depending on the available capacity of the air-supply system. Air may be admitted to the pads under manual control or by pushbutton-operated solenoid valves in the air-supply piping. As few aeration pads as necessary to induce flow should be activated, otherwise, the material may flood uncontrollably.

Air nozzles in lieu of aeration pads are useful for some coarse materials, but only tend to core holes without inducing flow in packed, finely powdered materials. Air-inflatable rubber pads may be installed inside the bin, connected to an air-supply system in such a manner that they flex alternately on a timed cycle. However, these are used only when dislodged material can flow freely out of the bin; otherwise, material may be compacted by expansion of the pads.

Air Supplies

The airflow and pressure required for air-operated flow inducers on a bin should be studied since an auxiliary air supply may be required. Air-operated vibrators of the piston type require a considerable airflow at high pressure and an air line lubricator. The exhaust should be directed to an area where the oil mist will not create a problem. Air-operated vibrators of the rolling ball type also use a considerable amount of air, but can operate at lower air pressure than the piston type. These also require lubrication and the dispersion of oil mist can present a problem. These two types of vibrators are noisy. Both can use air from a plant-air-supply main.

Aeration pads can have a positive-displacement blower as an air source since they use air at pressures less than 5 psi. The airflow required for aeration pads varies with the pressure, but those depending on diffusion through porous ceramics, porous metal or very thick woven canvas require only 3 to 12 cfm/ft². The actual airflow required per square foot of aeration surface at a given pressure should be evaluated before selecting a particular type of aeration unit. In designing the piping for a system of aeration pads, provision should be made for the following: Provide an inlet strainer on the blower; provide drains at all low points; provide a shut-off valve at each aeration unit or group of units; provide controls so that aeration pads are used only when necessary.

The air supply of air-inflated rubber pads may be obtained from plant air mains. Operating pressures vary from 5 to 15 psi when supplied through a control panel that receives air at 50 to 150 psi.

Live Bin Bottoms

When material with very unusual characteristics is encountered and none of the flow inducers described above would be effective, the bin may have to be fitted with a live bottom. This may be in the form of one or more screw conveyors arranged to withdraw the material, or patented devices, such as a rotating, mechanical arch breaker; a sweep unloading arm for a flat-bottom bin; or a vibrating bin hopper.

Vibrating Bin Bottoms

A form of vibrating bin bottom that has proved successful in handling many problem materials consists of a cylindrical hopper fitted with a baffle plate centered over the outlet so that material must pass through the annulus formed between it and the shell. The unit is supported by suspension bars, fitted with shear rubber bushings, hung from isolating channels welded to the main shell. A flexible joint consisting of rubber strips strapped to the main shell and hopper and protected by an internal metal skirt isolates the hopper from the shell. The discharge nozzle is also fitted with a flexible sleeve connection to isolate the hopper from downstream equipment. A motor driven gyrator is mounted vertically on a baseplate welded to the hopper and stiffened by several gusset plates (Figures 11–2 to 11–4).

The gyrator itself consists of two pairs of eccentric weights driven by an eccentric pin attached to one of the weights and an arm attached to the motor shaft. By shifting a set of movable eccentric weights with respect to a set of fixed eccentric weights, the unbalanced centrifugal force can be varied to suit the characteristics of the material in the bin.

The diameter of the hopper is usually one-half to one-third of the diameter of the bin to which it is attached, depending upon the material characteristics. Very slow moving material may require a larger

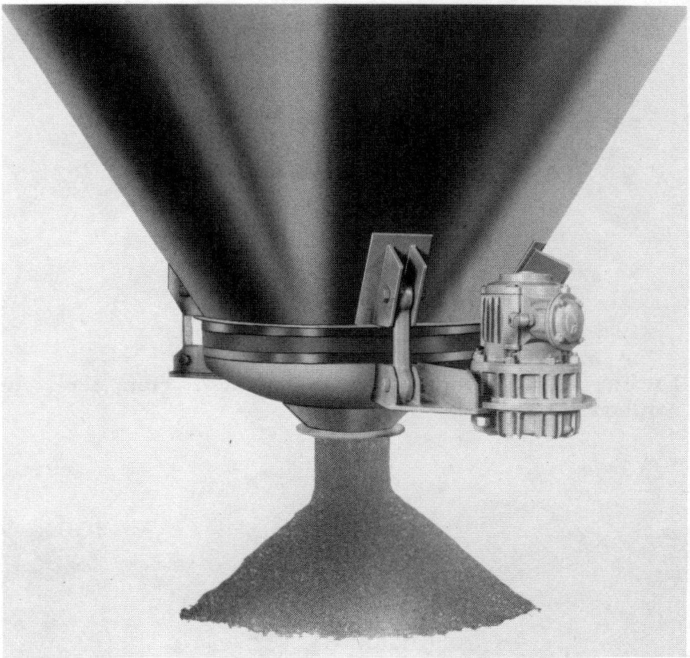

FIGURE 11–2. Elevation of bin activator. (Vibra Screw Incorporated.)

diameter ratio—almost equal to the bin diameter. The motor is sized for the eccentric force required to move the material as determined by test.

The material discharge through the annular space between the hopper and center baffle tends to remix material flowing into it from the sides and core of a cylindrical bin. This reduces segregation and ratholing and thus promotes mass flow. For problem materials, the dished baffle can be replaced by a conical baffle or can even be eliminated. The internal parts of the hopper should have the same coating or finish as the walls of the bin.

Certain precautions are necessary when using this type of hopper. They are as follows: (1) Bond the bin and hopper electrically using

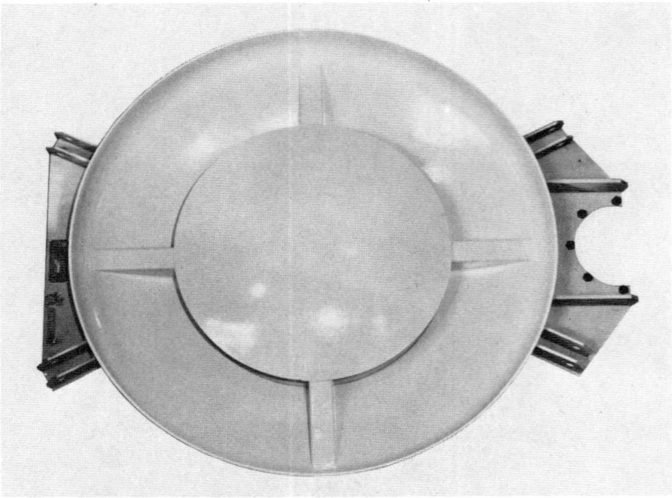

FIGURE 11–3. Interior of a bin activator. (Vibra Screw Incorporated.)

FIGURE 11–4. Installation of a bin activator. (Vibra Screw Incorporated.)

extra flexible No. 2 welding cable. This will prevent motor burnout if a welder inadvertently attaches his ground clamp to the hopper. (2) Isolate detector units from capacitance type probes mounted on small bins to prevent damage to them caused by vibration transmitted to the bin when starting and stopping the gyrator. (3) Tighten all hinge bolts and latching dogs on access doors to prevent them from loosening when vibrating through the critical speed of the gyrator. (4) Isolate access ladders on small bins—make them free-standing. (5) Isolate lighting fixtures from bin sight ports on small bins to lengthen the life of the bulbs.

HORIZONTAL INDOOR BINS

Horizontal indoor bins with low headroom that can hold a carload of material must be custom-designed and built to receive pneumatically conveyed materials. For convenience in fabrication and erection, such bins are of modular design (Figure 11–5). The modules can be assem-

FIGURE 11–5. Exterior of a low-head bin, showing modular design.

bled to form a long, rectangular bin with single or multiple V-bottoms (depending on the width of bin), designed for screw conveyor (Figure 11–6) or air-activated gravity-conveyor discharge. Bins with the air-activated gravity-conveyor discharge have a design essentially similar to the bodies of the covered hopper railroad cars with air-activated

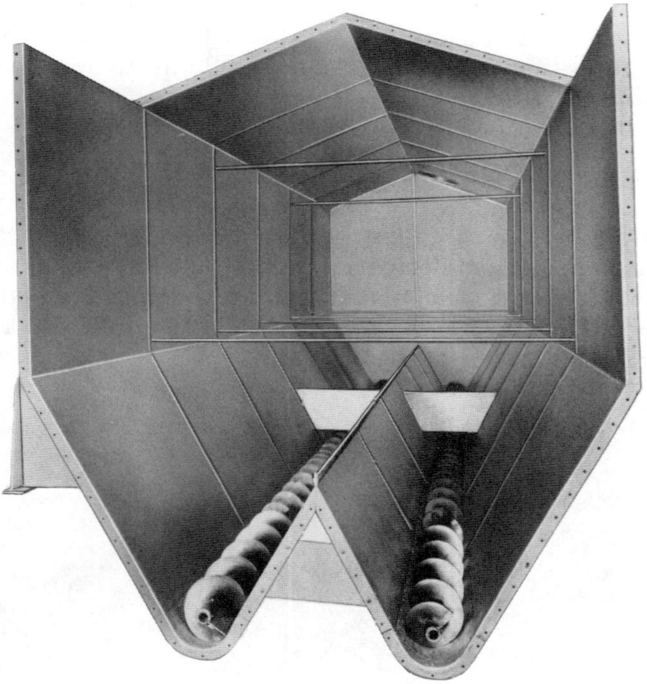

FIGURE 11–6. Interior of a low-head bin showing modular design.

gravity conveyors. (See Figure 6–3, page 106). The bottom slope sheets have the same angles and form as the prototype hopper car, with the long axis of the conveyors pitching toward two discharge connections on the centerline, normal to the length of the bin. The air-supply piping to the conveyors is arranged as on the railroad car, but air-pressure and capacity requirements are determined specifically for the material to be stored.

SCREW CONVEYORS

Bins with screw-conveyor discharge have each V-bottom fitted with a continuous screw conveyor having variable-pitch flights so that material can be withdrawn evenly throughout the length of the bin. An inverted V-baffle is placed over each screw to take the weight of stored material off the screw shaft and to obtain side feeding of material into the screw. The screws may be arranged for center discharge, off-center discharge, single-end or double-end discharge (Figure 11–7). Single-end discharge reduces the discharge capacity of each screw by one-half for a given size screw, so center, off-center or double-end discharge is preferred. The flights adjacent to each discharge opening must be shrouded and closely spaced in order to form a material seal and prevent material from flushing through the screw when it is not in operation. The screw conveyors on a bin with multiple V-bottoms may be motor-driven individually or linked together by a chain drive.

Structural Supports

Bins are generally made up in 4-ft modules with bolted and gasketed stiffener flanges and feet at the junction of each module. Since these feet may not be in alignment with existing structural supports in a building, the purchaser will have to design and furnish steel grillage to transfer the bin loading to the structural members of the building housing the bin (Figure 11–8)

Long beams under the legs, such as the ones shown in Figure 11–7, are satisfactory for support on basement or grade-level slabs, but not on slabs supported by framed structures. On framed structures, a 1-in.-thick steel bearing block must be placed between the load-distributing beam and the floor slab at a point directly above the framing beam or girder that supports the slab.

Distributing Material in Bin

The discharge end of the pneumatic conveying line enters one end of the bin and may distribute material in several ways depending on the terminal velocity of the conveyed stream and the material characteristics. A short entrance nozzle may be sufficient to fill the entire bin. If not, the conveying line may be extended along the longitudinal center line of the bin in the form of a distributing pipe with distributing slots facing downward. As each slot becomes blocked by material, the conveying stream is automatically extended to the following slot, and so on, in succession, until the bin is filled. If there is doubt as to the probable distribution of material within the bin, a distributing or leveling screw conveyor may be installed near the top of the bin to aid distribution of material. This conveyor may be reversed, if necessary, to suit operating conditions.

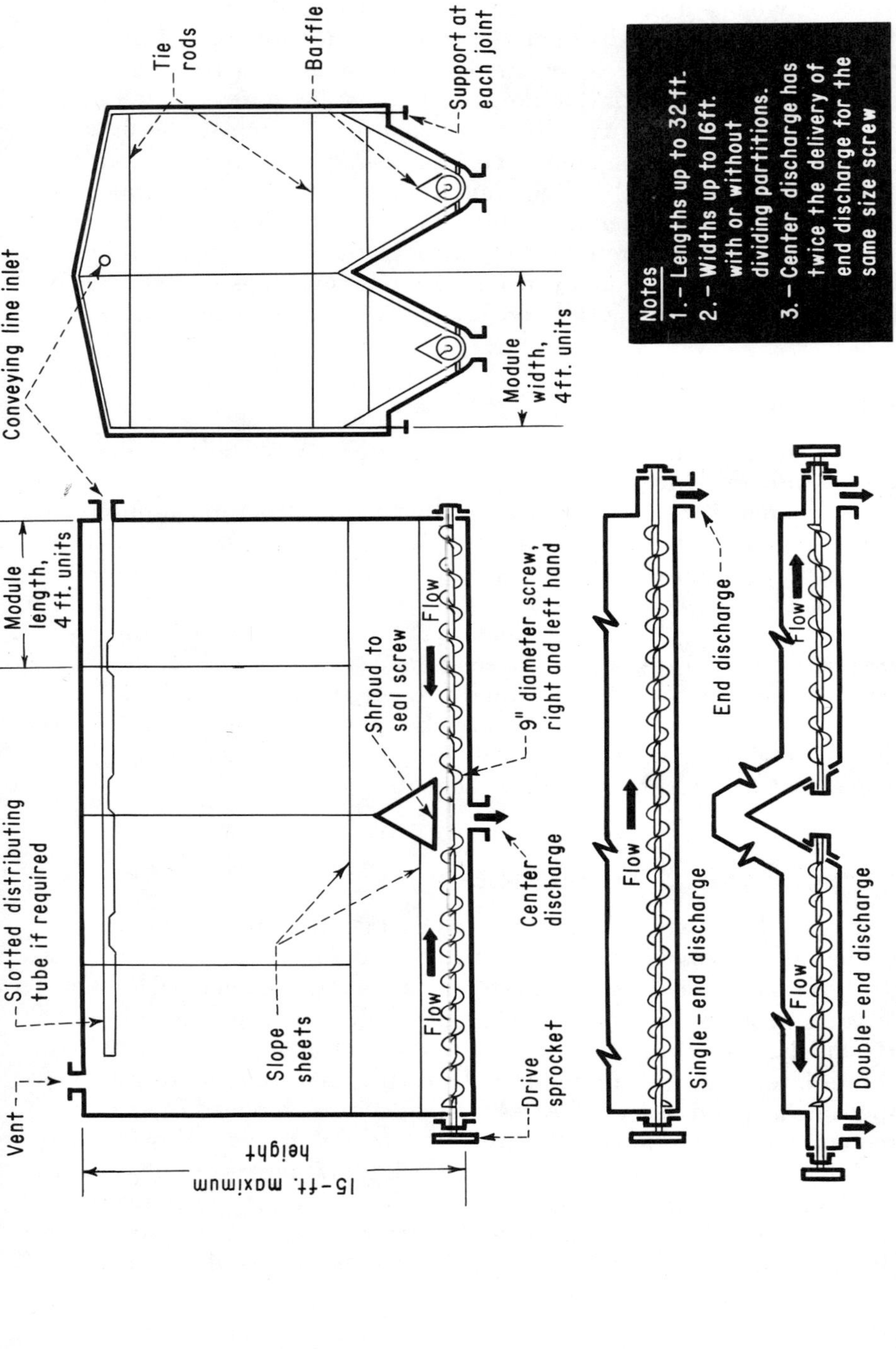
Vent
15-ft. maximum height
Slotted distributing tube if required
Module length, 4 ft. units
Conveying line inlet
Tie rods
Baffle
Support at each joint
Module width, 4 ft. units
Slope sheets
Flow
Shroud to seal screw
Flow
Center discharge
9" diameter screw, right and left hand
Drive sprocket
Flow
Single-end discharge
End discharge
Flow
Flow
Double-end discharge
Notes
1. - Lengths up to 32 ft.
2. - Widths up to 16 ft. with or without dividing partitions.
3. - Center discharge has twice the delivery of end discharge for the same size screw

FIGURE 11-8. Installation of a low-head bin, showing supporting grillage.

The proper location of vent connections and high-level indicators on long horizontal bins presents a real problem. If possible, flexible or alternate locations of these connections should be provided depending on the probable locations of high material levels, and anticipated quiescent areas suitable for vent takeoff. The best method of venting a low-headroom bin is to mount a bin vent dust filter on the roof of the bin, at the end farthest from the material inlet connection, even if this requires an opening through the ceiling or roof to accommodate the filter.

DESIGN CONSIDERATIONS

The following should be considered when laying out bins:

Venting. (1) Where long horizontal lines are required to vent two or more bins to a dust filter, it is preferable to mount a bin-venting type of dust filter on each bin, equipped with its own exhauster. Long horizontal lines can plug-up if a bin is overfilled. (2) Where a common dust filter is used to vent several bins, the fine dust should be discharged into a live bin. If fines are discharged into a dead storage bin, they will accumulate and cause bridging or conveying problems when the dead bin is finally activated.

Inventory. Horizontal indoor bins with low headroom should be weigh-scale or load-cell mounted, if close inventory control is required. Inventory on such bins without scales can only be estimated by sounding the sides for presence of material, or by inspecting the interior and making due allowance for hills and valleys.

Future Use. (1) In laying out filling and discharging conveyors for groups of bins handling various materials, consider the requirements for conversion of one or more bins to handle a different material, in case the original bin containing that material cannot be used, or, if extra storage for that material is temporarily required. (2) In designing supports for a bin holding lightweight material, consider the possibility of future conversion of the bin for the storage of heavier material that may be required by a process formula change.

12 Piping

CONVEYING PIPING

Piping Materials

Conveying piping can be fabricated from any one of several materials such as carbon steel, stainless steel, monel metal, aluminum and plastic. The material and wall thickness are selected to suit the characteristics and properties of the material to be conveyed and, occasionally, on the basis of cost. Since the conveyed materials usually scour the internal surface, many inert materials may be conveyed in carbon steel piping without harming the piping or the materials handled. When a chemically active material is to be conveyed, corrosion resisting materials such as stainless steel or monel metal are used. When sanitation is imperative, stainless steel piping is selected. When freedom from discoloration is important, but cost is also a consideration, aluminum would be selected. When abrasion resistance is required, stainless steel may be selected, or a combination of materials may be used, carbon steel for straight piping and another material for pipe bends.

Some vendors use cast sand-spun chrome-iron alloy pipe with matching bends for ash handling, but these can be adapted to handle other abrasive materials. This type of piping may be rotated on its axis in order to obtain uniform distribution of wear.

Plastics, resins, and similar soft materials that tend to smear on smooth walls are best conveyed through aluminum or stainless steel piping that is roughened in such a manner that an abruptly descending velocity at the walls prevents the formation of skins and ribbons that can foul processing of these materials. The methods of roughening the interior surface are patented and take several forms: closely spaced longitudinal serrations, helical grooves, annular rings, or coarse indentations produced by sandblasting.

Pipe Sizes

Standard sizes of piping are available in several wall thicknesses for aluminum and stainless steel. Standard Schedule 40 wall thickness carbon steel piping is used in all pipe sizes, but thinner piping such as Schedule 20 or Schedule 10 wall thickness may be selected for large pipe sizes in order to reduce cost. In many cases, lightweight tubing may be used in lieu of standard piping sizes, but the choice will depend on the availability of matching tube bends of equal wall thickness.

The size of the conveying line should be as recommended by the vendor after tests of the material in his pilot plant.

Bends and Couplings

All changes in direction are made with factory-formed bends, having a radius of 6 to 12 times the nominal diameter of the pipeline depending upon the abrasiveness or compaction characteristics of the conveyed material. The longer the bend radius, the greater the resistance to abrasion, the lower the possibility of compaction in the bend, and the lower the pressure drop in the bend. Some highly abrasive materials may require special reinforced wearback bends. The reinforcing is usually on the outside radius of the bend where all the wear is concentrated. The wearback may be a metal box welded to the bend and filled with concrete; or it may be half of a larger size pipe that is split longitudinally and bent to fit the back of the bend. The principle of the hollow design is that a hole worn through the pipe wall would fill the hollow wearback with the material conveyed so that it forms its own wearing surface. The bend may be a special 4 or 5 ply, abrasion resistant rubber hose, bent to the proper radius and held in position by attaching it to a formed structural steel channel. Instead of hose or piping, the bend may be a square or rectangular sheet steel duct, formed to proper radius having a bolted-on backplate made of special abrasion resistant steel or rubber-faced steel. Sometimes an impact box lined with renewable rubber, epoxy and silicon carbide blocks or carborundum blocks may be used instead of a pipe bend. A pipe tee may be used as a target or impact box at the end of a horizontal line, before entering a bin; the material enters one end of the run and is deflected downward through the branch by material packed into the opposite end of the run during conveying.

All bends at strategic points where dimensional errors may occur should have long tangents so that field cuts may be made to correct these errors during erection. When the bends are made, a minimum length of tangent is required to form them and it may be desirable to retain this length rather than have the tangent cut to a smaller dimension. This avoids a cutting charge and will aid in field erection.

The usual type of piping joint is the heavy-duty, compression-type sleeve coupling (e.g., the "Dresser" Coupling) with gaskets made of material to suit the service conditions. This type of coupling can be quickly loosened and slid over the pipe to break the joint for cleaning out plugged lines. Lightweight designs of compression couplings are available for joining small sizes of tubing. These resemble the Dresser type, but are not as rugged and have three bolts instead of four to simplify installation. Also available, for small sizes of tubing, are sanitary couplings that have counterbored flanges clamped to each pipe and a rubber seal ring seated inside the counterbore so that, when the flanges are made up, the pipe ends butt against the seal ring and eliminate the crevice that can harbor bacteria. Another form of compression coupling is built like a repair clamp. A gasket is clamped around the joint by a split, sheet metal band that is bolted and drawn together at the split section. Flanged joints are used at connections to equipment and on trussed pipe lines where stiffness is required at the joint. The flanges should be one of the standard types that match the tubing or piping selected for the conveying line, but care must be taken to make a true butt joint. Welding-neck and true Van Stone flanges are best. Slip-on welding flanges are sometimes used welded only on the outside, so that the pipe ends are flush with the flange face. However, the full strength of this type of flange cannot be developed unless the pipe is recessed and the internal weld at the end of the pipe is made.

Piping should be made up by butting the ends together to eliminate crevices or potential points of material buildup. Burrs must be removed and crushed ends straightened before butting ends of pipe.

Sight Tubes

Sight tubes for installation in conveying lines are available from vendors of pneumatic conveying system accessories. These tubes are made from clear acrylic plastic tubing with metal tubing ends cemented to it for insertion in a line using compression couplings. These are easily removable for cleaning. A section of piping or tubing may be hung near the tube to replace it in case of breakage. Flanged sight tubes are available with toggle clamps on the gasket compression rings so that the tubes may be easily replaced. These are used at the entrance or end of the conveying line where the flange may be readily attached to equipment.

When handling materials that may build up a static electrical charge on plastic, a built-in ground wire is required to maintain a continuous electrical path through the conveying line. When handling materials that may smear the plastic tube as it flows by, a sight glass that is normally used for liquids may be used, since the sight ports are recessed out of the path of the conveyed material. The glass should be replaced with a nonbreakable plastic such as Lexan.

Piping Runs

In general, piping should be laid out to run directly, in a straight line, from the source of supply to the point of delivery, with as few changes in direction as possible, consistent with the location of known interferences. Inclined piping of any appreciable length should be avoided regardless of the type of material conveyed. Material in an inclined line tends to roll or float down the slope as air passes over it, creating congestion at the bottom of the slope. This can cause pulsating or sluggish flow and excessive pressure drop.

The open ends of conveying lines on negative-pressure systems should be designed as a bell mouth or A.S.M.E. nozzle and provided with a heavy expanded metal screen to prevent the entrance of foreign material. If outdoors, the nozzle should be protected from the direct entrance of rain, snow, sleet, and chemical fumes that may come from nearby truck unloading operations or process centers.

Often, where scheduling and material characteristics permit, one pipeline may be used to convey several materials. However, if contamination must be avoided, a separate pipeline should be provided for each material. Separate lines for each provides flexibility for future increases in the demand for the material handled by the line.

Static Electricity

When the materials conveyed tend to build up an electrostatic charge on the piping or on plastic sight tubes, these charges may be dissipated by grounding the pipeline. The extent of grounding depends on the possible hazard involved. Where flammable materials are handled, it may be necessary to install bonding straps and cable across each compression coupling or flanged joint that is insulated from contact with the pipe by its gaskets. The piping should then be solidly grounded to a water distribution system, to building steel or to an earthed grounding rod.

Structural Systems for Pipes

Where a pipeline spans large open areas, it is usually made self-supporting to resist swaying or sagging by designing it as a truss as in Figure 12–1 using the pipeline itself as the compression member. Joints in these lines are usually flanged for stiffness and located at points of zero moment. Tension members may be cables or rods that bear on steel struts that form the bearing points to support the piping. A short, trussed line may be supported as a king truss (one strut in span) or as a queen truss (one containing two struts in span). Longer trusses having more than three span panels become indeterminate structures but may be made determinate by designing all connections to axially loaded members as pin-connected joints. This manner of computing stresses is preferred by municipal building departments, when applying for a building permit to cross a street. Pinned joint members should have electric welded and forged eyes and clevises at their ends. Struts set normal to the piping in order to take the bearing loads should be all-welded structural-steel frames. Care should be taken to avoid skewed truss members when laying out trussed lines for a piping run that is at an angle in each of three planes. Calculations should be made on the basis that the lines may be plugged with powder. Snow and wind loads should also be taken into account in the design.

Provision for Blowout of Line

Some materials may tend to plug up a pipeline and the line may have to be blown clear with compressed air. High-pressure air blow connections may be installed at several strategic locations in a pipeline. These connections should be made with a pipe union and a check valve located as close to the conveying line as possible to prevent powder from plugging up the air connections. To prevent the high-pressure air from escaping through rotary feeders or open-ended lines, a line blind may be installed in the conveying line close to the feeder or opening into the line.

Line blinds are essentially pivoted spectacle flanges located between fixed flanges that can be forced apart sufficiently to permit the spectacle flange to be swung to its alternate position. This type of

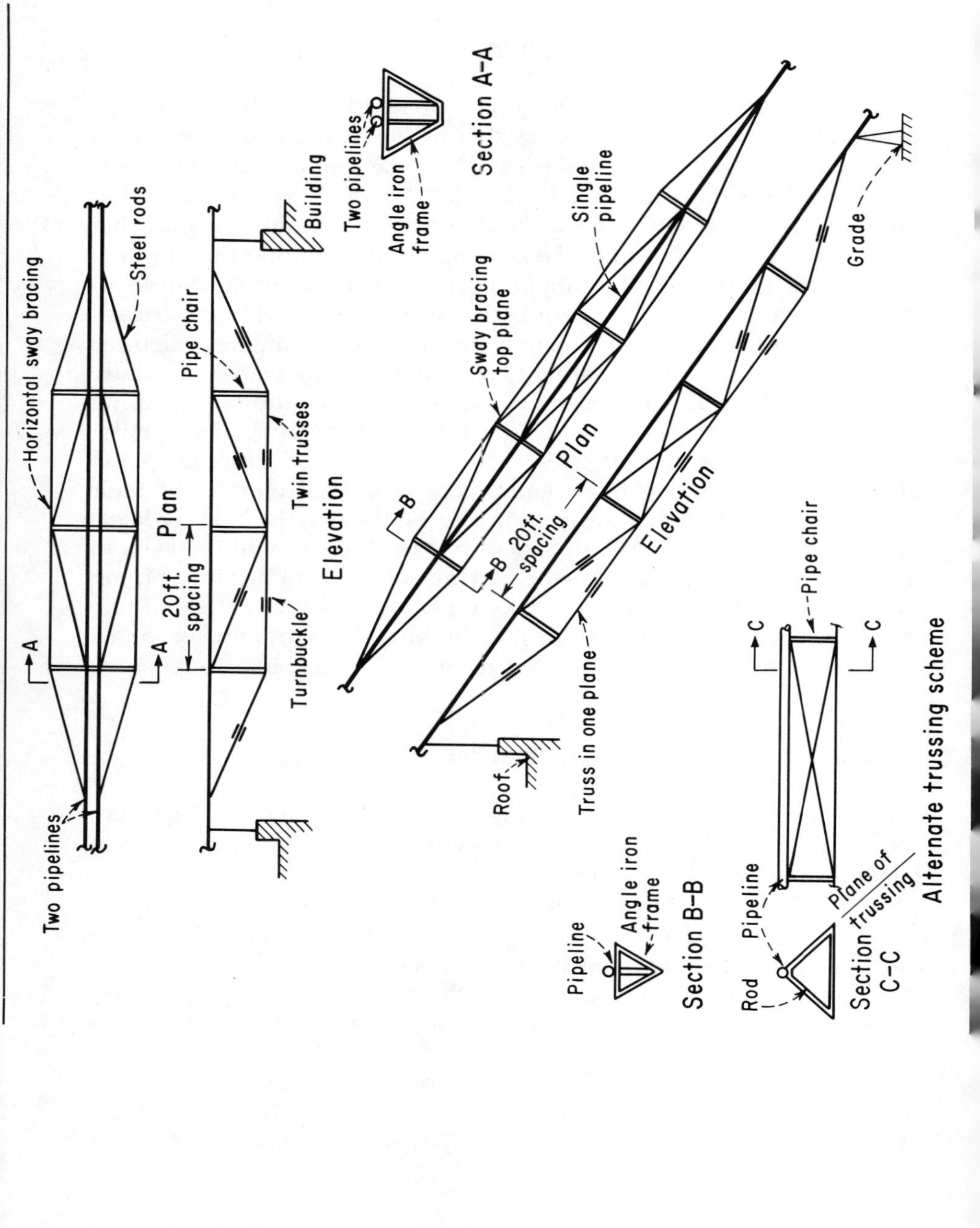
Horizontal sway bracing
Steel rods
Two pipelines
A
A
Plan
20 ft. spacing
Pipe chair
Building
Twin trusses
Turnbuckle
Elevation
Two pipelines
Angle iron frame
Section A-A
Single pipeline
Sway bracing top plane
B
B
20 ft. spacing
Plan
Elevation
Grade
Roof
Truss in one plane
Pipe chair
C
C
Pipeline
Angle iron frame
Section B-B
Pipeline
Rod
Section C-C
Plane of trussing
Alternate trussing scheme

closure can be operated even when the line is plugged. In service, the open end of the spectacle flange is aligned with the piping, then the bolts on the fixed flanges are tightened to seal the O-rings that are built into the flanges. When used to blow out a line, the bolts spread the fixed flanges so that the blind side of the spectacle flange can be pivoted to block the conveying line. The open end of the spectacle flange, including its plug of powder, is thus swung out of the way. The bolts are then tightened so that the joints are sealed. The bolts are made with a hexagon head to receive a wrench and holes through the head so that they can also be turned by inserting a bar through the holes (Figure 12–2).

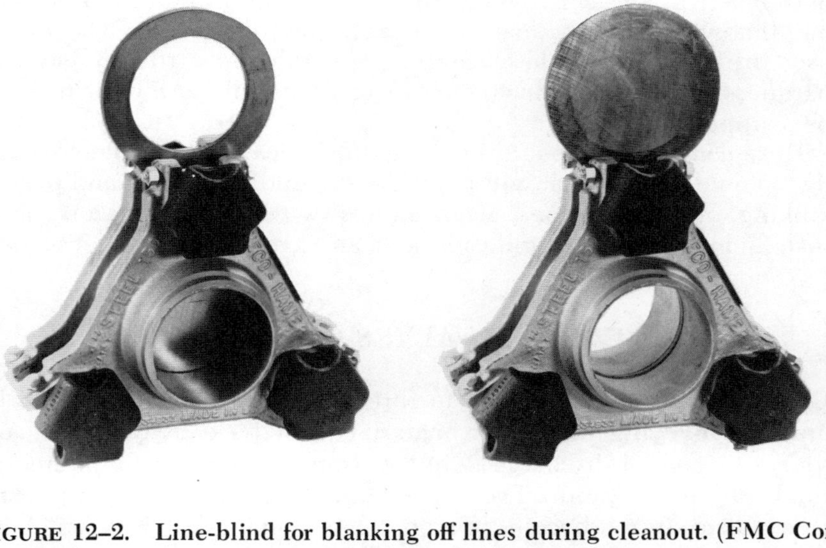

FIGURE 12–2. Line-blind for blanking off lines during cleanout. (FMC Corporation.)

HOSE CONNECTIONS

Hose connections and hoses are required at railroad car or truck unloading stations. The standard hose used for truck unloading is 4 inch size rubber hose. Railroad cars may be equipped with larger diameter hose connections, but rubber hose is too unwieldy and intractable in sizes larger than 4 inches. Hose larger than 4-in. size should be made of rolled metal stripping with interlocked joints, except when handling corrosive material that may pack in the crevices. Metal hoses are available with special hardened liners, where required. Since metal hose is not watertight, special precautions must be taken to prevent entrance of water when unloading under negative pressure in rainy weather. Provide V-blocks to keep hose off the ground and plastic covers or plastic tape wrapping to seal the joints. The pipe terminals and hose at unloading stations should be provided with the type of toggle-clamp, quick-coupling connections selected as the plant standard. The pipe terminal should be fitted with an airtight cap or cover when not in service. Covers should be chained to the piping to prevent loss. When a group of hose connections are installed to serve a battery of bins, each connection should be properly identified to show the product handled and its destination. When materials must not be mixed, the hose connections should be fitted with limit switches that permit activating a conveying system only when certain conditions are met that will direct the material to its proper destination. The gaskets used in toggle-clamp hose couplings should be a rubber having a hardness of not more than 40 Durometer. This will permit the toggle to be snapped tight easily.

Hose connections should be made to vertical sections of conveying line at an elevation that will permit easy bending of hose and prevent kinking. In tight locations, a short-radius swivel elbow of metal may be installed to permit swinging a hose in an area where it cannot be bent.

DIVERTER VALVES OR "GATES"

Diverter valves are required to direct the flow of material into the proper conveying line when material must be conveyed from one source to several discharge points or from several supply points to a single discharge point. The valve tends to be a rather complicated mechanism since it becomes part of the conveying line into which it is directed and must prevent leakage of material from the live line to the dead line. Also, in most applications, it is remotely operated. There are many valve designs available, each of which should be analyzed before application to see if the above requirements have been met completely.

Two types of conveying line diverters are available commercially:

1. Line diverters, which switch the flow of material into the proper conveying line or bin. These are usually activated when no material is flowing through the conveyor.

2. Double-acting diverters, which divert flowing material from one receiving hopper to another during sequential filling. (The receivers are usually scale-mounted.)

The numerous types makes it difficult to classify these valves. Figure 12–3 shows some of the types available and indicates the princi-

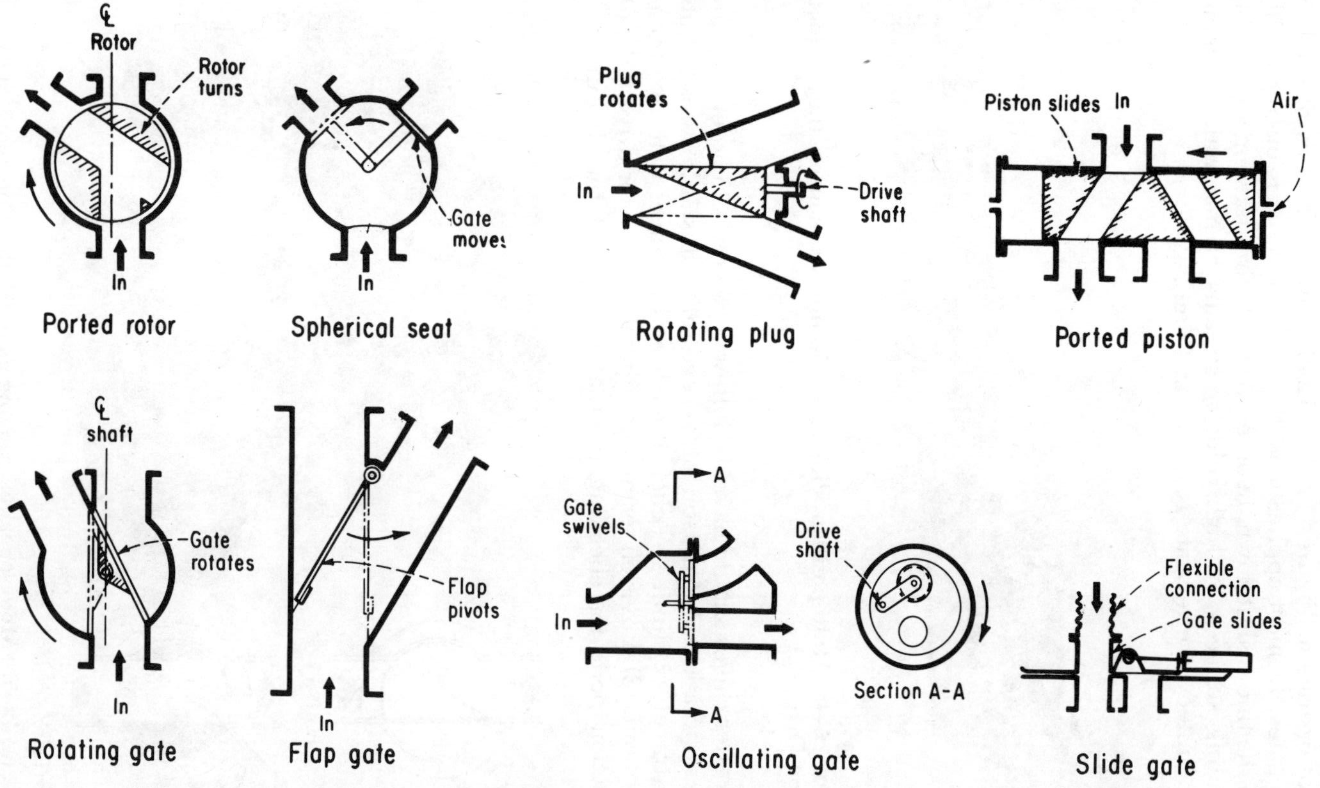

FIGURE 12–3. Diverter valves for switching conveying stream from one pipeline to another.

ples of operation. Figures 12–4 to 12–6 show assemblies of some of these gates. In most applications, the diverter valve is operated only when the line is shut down, in order to avoid wire-drawing and cutting of sealing surfaces by high-velocity streams of material. A special bin-mounted diverter gate is shown in Figure 12–7.

FIGURE 12–4. Multi-position slide gate. (Koppers Company Inc., Sprout-Waldron Division.)

Diverter valves should have the following features in order to insure proper operation and adjustment: (1) external adjustable stops to limit valve travel; (2) limit switches at each end of travel with overtravel adjustment; (3) provision for manual operation if motor drive fails; (4) inspection ports for checking internal position of the diverter and

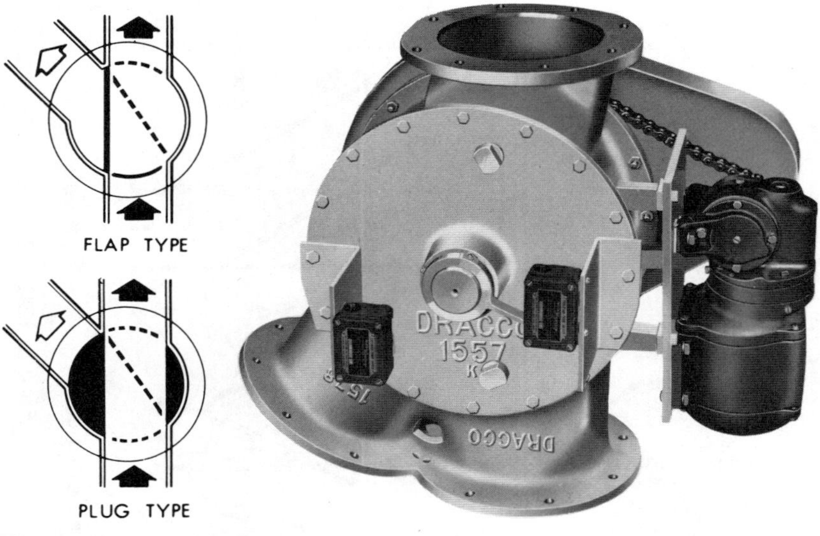

FIGURE 12–5. Motor-operated rotating gate. (Fuller Company/GATX.)

FIGURE 12–6. Ported-piston-type gate. (Shick Tube-Veyor Corporation.)

clearances; (5) horizontal chain-drive sprockets guarded from slipping downward by set collars; (6) an external indicator showing diverter position. In the case of a high-speed valve, a brakemotor should be used to prevent damage to limit stops.

The term "high-speed" valve is relative since too high a speed may impose a duty on the brake that it cannot perform. An example of too high a speed is as follows: A gate with a valve rotation of 55 degrees from one line to another tended to push the limit switches off their

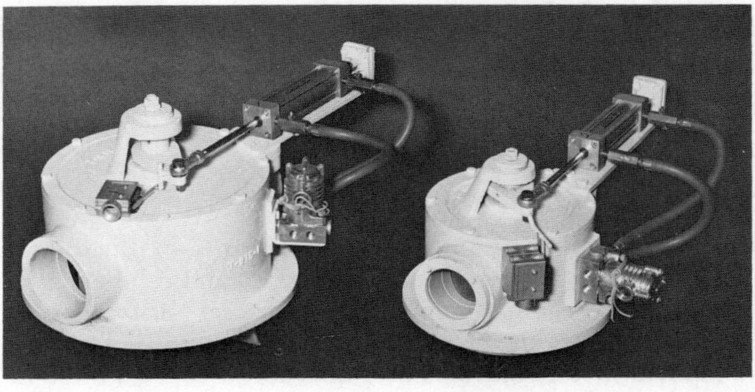

FIGURE 12–7. Bottom diverter gate for on-stream diversion of flow from one bin into another. (Shick Tube-Veyor Corporation.)

mounting clips. The gearmotor drive output was 24 rpm, the chain drive sprocket ratio was 1/3.

$$\frac{\text{Gearmotor output (24 rpm)} \times \text{chain sprocket ratio (1/3)} \times 360 \text{ degrees}}{60 \text{ seconds}}$$
$$= 48 \text{ degrees of valve movement per second}$$

$$\text{Time for valve travel} = \frac{55}{48} = 1.14 \text{ seconds}$$

This timing required very high-speed operation of the brake since the motor had to stop in milliseconds in order to eliminate overtravel of the gate. The gearmotor speed and chain sprocket ratio had to be changed to obtain a reasonable operating speed for application of the brake.

A speed check of a successful diverter gate drive was made. This gate required a valve travel of 152 degrees between outlets:

$$\frac{\text{Gearmotor output (2.4 rpm)} \times \text{chain sprocket ratio (8/60)} \times 360 \text{ degrees}}{60 \text{ seconds}}$$
$$= 2 \text{ degrees of valve movement per second}$$

$$\text{Time for valve travel} = \frac{152}{2} = 76 \text{ seconds}$$

When high-speed operation of a diverter gate is required, air cylinder operation is sometimes preferred despite the possibility of cold weather and lubrication problems.

It is advisable to provide jog buttons close to remotely operated valves to permit checking valve operation and positioning while the system is not operating. These may also be used to move the gates periodically when handling corrosive materials that may freeze them.

Where connections must be made from one source of material supply to several destinations, a single hose may be manually connected from that source to any one of several lines leading to those destinations. This solves what may at first appear to be a complicated diverter-control problem.

Electric Control Circuit

A schematic wiring diagram shown in Figure 12–8 illustrates a control circuit for a two-way valve, and the functional relationship of the contacts to the conveying system equipment. The wiring for each contact must be run from the control device in which it is located to the control panel or motor-control center in order to be integrated into the control circuit. Figure 12–8 shows how extensive the wiring must be even for a relatively simple piece of equipment.

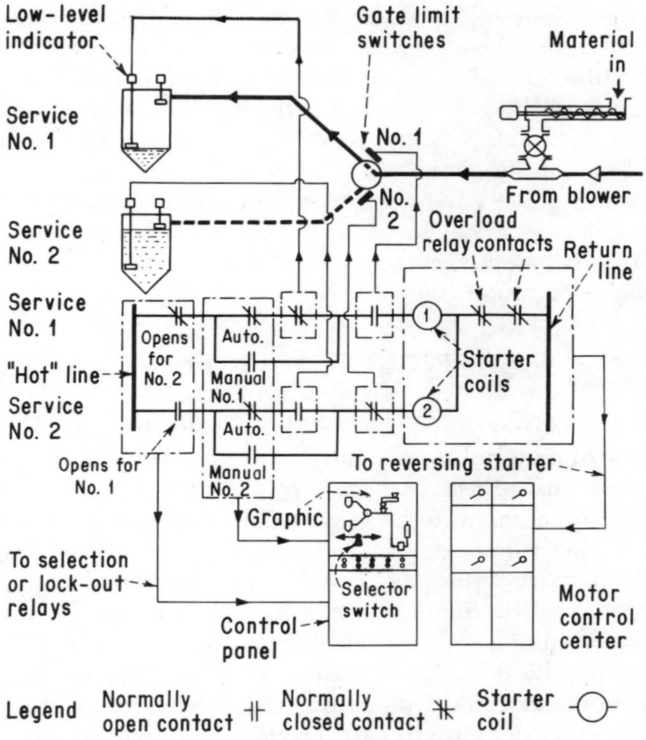

FIGURE 12–8. Two-way diverter valve control circuit.

Here we have a single-phase a-c control circuit that energizes the starter coils to actuate the contacts that start a three-phase motor. These coils are energized alternately when their respective control contacts are closed, so as to complete the circuit from the "hot" line to the return line.

Operation is as follows: The selector switch is set to close the contacts for one of three modes of operation: automatic positioning of the gate on demand from the low-level bin indicators; manual switching to service No. 1; or manual switching to service No. 2. For the service No. 1 position shown in Figure 12–8 the selector switch is on "automatic" and the gate has completed its travel to open the contacts in limit switch No. 1 and stop the motor. The operation of the conveying-air blower has actuated a selection or lockout relay to open a contact in the circuit to starter coil No. 2. This prevents operation of the gate in the reverse direction during the conveying period.

When the low-level bin indicator for service No. 2 is activated, its contacts close, but the gate cannot move toward service No. 2 until the conveying blower is shut down and the lockout relay in starter coil No. 2 circuit is closed. When this occurs, the gate proceeds to move toward service No. 2 and stops when it actuates limit switch No. 2.

Manual switching connections from a common pipeline to several pipelines may solve what appears to be a complicated diverter valve problem. Limit switches at each line are installed in order to set up control circuits for activating a particular line only when properly connected to the common line.

PICKUP DEVICES

Pickup hoppers, nozzles, valves and car-unloading pans are used to induce flow of material into negative-pressure systems. Car-unloading nozzles may be built-in and form part of a railroad car, or may be portable for attachment to the gravity discharge connection on the car. In general, these devices are designed to permit the entrance of air into the conveying line so as to sweep the material along with it. The air may flow past the main entrance spout, or it may be given a 180-degree sweep to entrain the material as it flows by gravity toward the nozzle. Figure 12–9 shows several types of pickup devices.

Pickup hoppers should be fitted with easily removable, gasketed access doors for checking the air nozzle adjustment and for removing lumps that may block the entrance to the conveying line. The hopper should be fitted with high- and low-level indicators to control the operation of the material feeder and prevent flooding or plugging up of the hopper.

Car-unloading pans should be single, shallow and rugged, but light in weight so that it can be fitted under a loaded car by one man. The suction nozzle should have an adjustable air-inlet gate and an adapter to permit quick attachment of car-unloading hose.

SIMPLE METHOD OF ATTACHING UNLOADING PAN TO CARS

Most attachments to the outlet connections of air-activated discharge-hopper cars are fastened by clamping the outlet and attachment flanges together, using C-clamps, locking-plier wrenches or toggle clamps. The handling of the attachment and its clamping is a difficult operation for one man, unassisted. In addition, the clamping devices may quite frequently become lost or stolen.

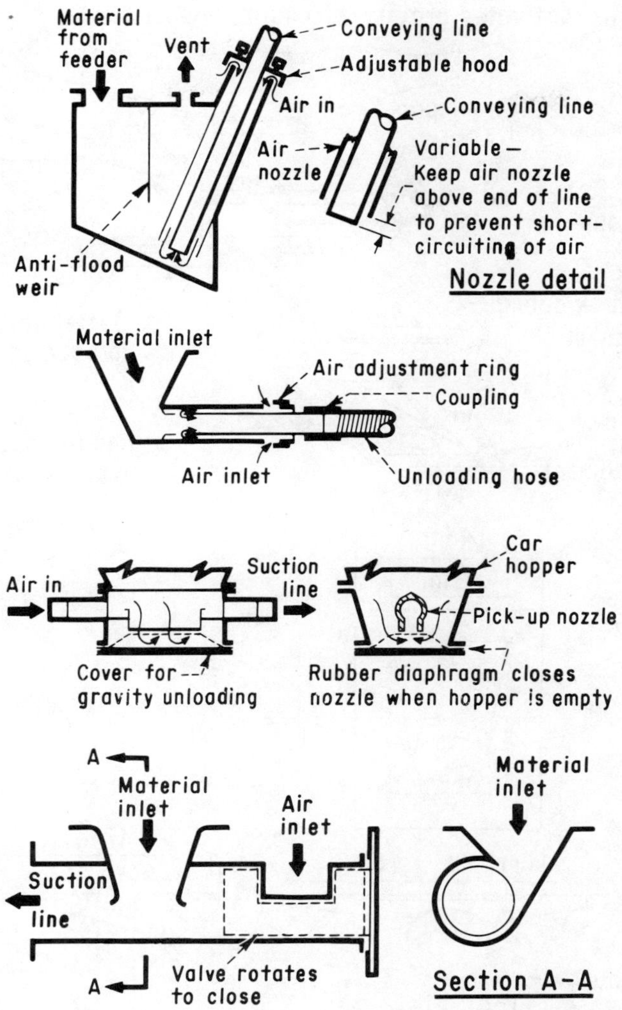

FIGURE 12–9. Several pickup arrangements for vacuum systems.

Figure 12–10 shows how the car unloading-pan attachment can be modified to make use of the car's existing bolts, which retain the outlet cover. The following modifications must also be made: the thread on each bolt should be extended upward another inch, the outlet-cover pivot-bearing should be made flush with the bar joining it to the cover plate, and a heavy washer should be welded to the upper end of the cover to prevent loss of material, and to have fewer parts to handle.

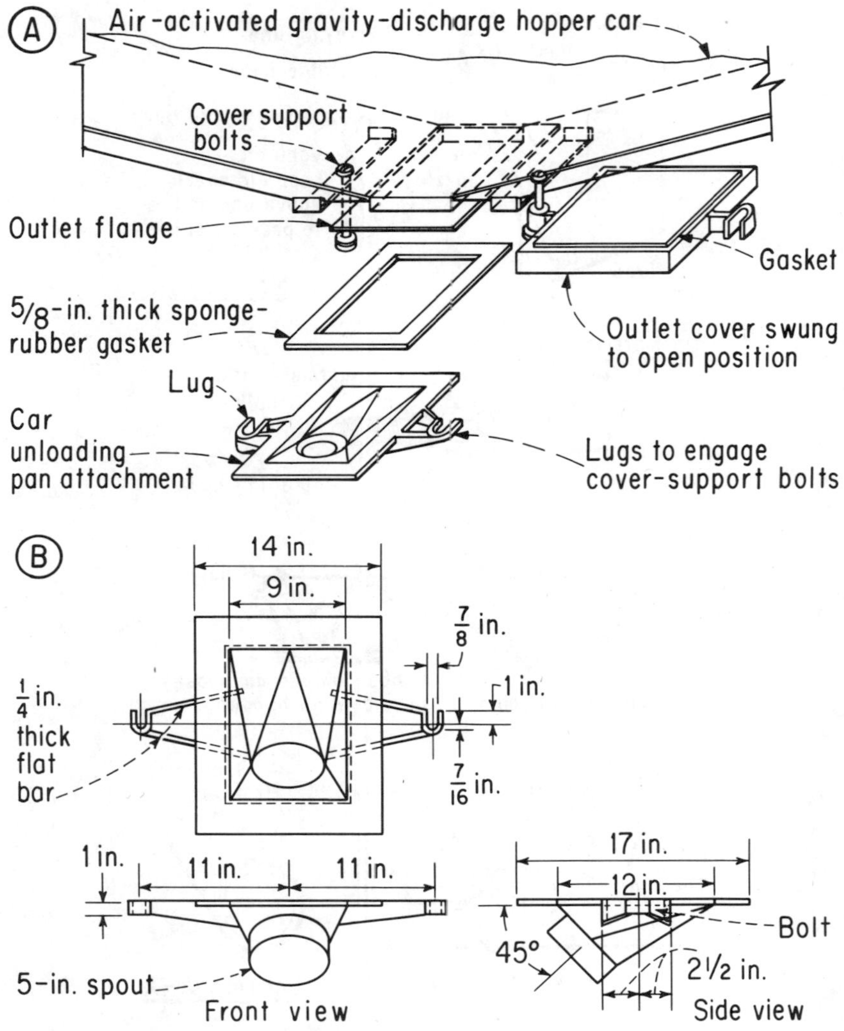

FIGURE 12–10. Simple car-unloading pan arrangement.

One man can make the attachment by loosening the outlet-cover nuts and lowering them to obtain proper clearance. He then swings the outlet cover to the open position, inserts the unloading pan and gasket, and finally tightens the nuts to clamp the pan, all without help.

The detail of the car unloading-pan, Figure 12–10, shows additional features that are intended to promote sanitation and ease of handling such as shallow design (to retain a minimum amount of material and cut down weight) and a hose-connection spout off to one side to clear the rail. The spout can be faced to make connections outside of the track or between the rails, but the outside location is preferred for easy access. The gently sloping sides of the pan present no problem because all material leaving the car has a low angle of repose due to aeration.

EXHAUST PIPING

All ductwork for venting bins and hoppers, including all their connections to dust filters and exhausters, is considered exhaust piping. Since these ducts and their associated equipment operate under slight negative pressure and are essentially air pollution control systems, their design is regulated by local or state laws. The various rules and regulations governing exhaust systems contain much useful information and copies of them for a particular community should be procured before purchasing a pneumatic conveying system.

In general, the construction and installation section of these rules and regulations calls for the following: (1) Use of light gauge galvanized sheet steel ducts, the gauge increasing with the size of duct. Other materials are permitted where required to resist corrosion or any element that will attack galvanized steel. (2) Ducts handling abrasive dust are required to be two gauges heavier than ducts handling non-abrasive dusts. (3) All elbows, bends and fittings are required to be two gauges heavier than the runs of corresponding diameter. (4) The throat radius of elbows must be not less than twice the diameter of the elbow. (5) All joints and seams must be airtight. (6) Pipes must enter main pipes at an angle not exceeding 45 degrees and only one branch pipe may enter a main pipe at the same point. (7) All piping must be braced and supported every 10 feet. (8) Clean-out openings must be provided in the lower side of the pipe at 20 ft intervals, they must be large enough for access to the interior. (9) The length of transition pieces must be 6 times the diameter of the inlet pipe and maintain a straight run on the bottom of all horizontal ducts. (10) Where piping connects to the suction side of an exhaust fan, a detachable sleeve must be used at least 18 inches or one pipe diameter in length. (11) Main ducts must be blanked off at the tail end with removable caps and the last branch connection should be not more than one diameter from the capped end.

The design section of these rules and regulations requires the following: (1) The capacity of exhaust systems shall be calculated on the basis that all dampers are wide open unless locked in a throttled position, except when the system is interlocked so that only a portion of it can be operated at a given time. (2) Vents from bins carrying different dusts that might combine into explosive or flammable products shall not be connected to the same exhaust system. (3) All parts of the system shall be as free as possible from air leakage into or out of the system except at points where air is taken in or out by design. (4) Ducts may be sized for constant resistance, that is, equal pressure drop per foot of pipe. (5) Transport velocities for conveying dust shall be adequate to prevent settlement of material. (6) All reinforcing of ducts shall be done on the outside of the duct. (7) Dampers, gates or orifices for balancing airflow in an exhaust system shall be permanently fastened in position. (8) Where an exhaust duct passes through a fire wall, an automatic fire door or shutter shall be provided on each side of the wall. (9) Fans shall be located on the clean air side of separator or filter equipment.

All of the above are considered minimum requirements and are excellent rules to follow even if they may not be legally required in a community.

One method of reducing the amount of exhaust piping in a plant is to use bin vent filters wherever possible. These discharge dust back into the bin and eliminate the problem of disposing of the collected dust.

When a group of bins filled by positive pressure is vented to a common dust filter, the vent piping from each bin should enter a manifold at the dust filter and be isolated from adjacent bin vents by a butterfly valve or damper. On automatic systems, each damper may be air-motor or electric power unit operated and interlocked so that it opens only when material is blown into the bin that it vents.

AIR SUPPLY PIPING

Pipe Sizes and Types

Air supply piping to a conveying system can vary in wall thickness and type depending on air pressure requirements of the system. The detailed requirements for air piping operating at 30 psi and over are thoroughly covered in the Code for Pressure Piping of the American Society of Mechanical Engineers. Below 30-psi pressure there are no standards other than those arbitrarily set by the manufacturers of lightweight, longitudinal and spiral-welded piping. Therefore, low-pressure-air piping materials may vary considerably. Sheet metal ductwork may be used on systems with centrifugal blowers; lightweight spiral-welded piping may be used on systems with positive-displacement blowers and standard piping may be used on systems with sliding-vane or reciprocating types of air compressors. In many cases where relatively small pipe sizes are required, it is more economical to install standard piping than to fabricate special sheet metal piping.

The airflow requirements on many material-into-air-stream systems are so high that large sizes of piping are required in order to reduce the pressure drop between the blower and the injector tee for positive-pressure conveying, and between the receiver and exhauster for negative-pressure conveying. This, plus the low pressure requirements, makes the use of lightweight pipe for these systems economically attractive.

Lightweight piping is made in two ways: (1) by rolling a plate into a cylinder with a longitudinally welded joint and (2) by winding narrow strips of plate helically into a cylinder and welding the spiral joint.

Spiral-welded piping is made to conform to A.S.T.M. Specification A 211. This standard indicates that spiral-welded pipe can be obtained in sizes from 4 to 48 inches in diameter and in wall thicknesses ranging from $^{1}/_{16}$ to $^{11}/_{64}$ inch (No. 16 to No. 8 U.S. Standard Gage). This piping may be obtained in 20-, 30-, or 40-foot lengths.

The lightweight pipe manufacturers list several wall thicknesses for each size of pipe, except for sizes below 6 inches—these are only available in 14 gage. Sizes of 6 inches and over may be obtained in thicknesses ranging from 14 gage to 8 gage (U.S. Standard Gage for Plate).

Pipe Joints and Fittings

Spiral-welded piping may be butt welded or lap welded; it may also be lock seamed and then welded depending on the manufacturers standard. The ends may be made plain to receive a slip-on welding flange, or they may have special thick-walled ends welded to them. The inside diameter of these heavy ends is made flush with the inside diameter of the piping, but the outside diameter is made to match the outside diameter of standard piping so that standard "Dresser" type couplings may be used. These ends can also be grooved to receive any one of several patented couplings such as the Naylor "Wedge-Lock" coupling and the "Victaulic" type coupling. The heavy ends may also be beveled for butt welding.

Lightweight flanges are made to standards originally prepared by the riveted pipe manufacturers. These flanges are thinner than the American Standard 125-lb flanges used on standard pressure piping and have different drilling dimensions. However, most manufacturers can furnish these flanges drilled to the 125-lb standard, if specified. The 125-lb drilling must be specified on many systems because the flanged piping connections on positive-displacement blowers, silencers, gates, diverter valves and other piping accessories are made to the 125-lb standard dimensions. The lightweight flanges are made to slip on the pipe for welding and are made in three types: (1) A forged steel type with a short neck, (2) a plate type, and, for sizes over 10 inches in diameter, (3) a rolled type that has a thickness almost half that of the plate type.

Lightweight fittings are also made to the standards originally set by the riveted pipe manufacturers. The center-to-face dimensions of these fittings conform to the 125 lb pressure piping standard, whether plain-ended or flanged. However, flanges are of lightweight design and wall thicknesses of the fittings are selected to match the adjacent piping. All lightweight fittings are fabricated by butt welding sections of piping that are mitered to suit the shape of the fitting. Elbows may be made from two, three or five mitered sections depending upon the required radius and the bend angle.

Piping Accessories

The accessories required in an air supply line are as follows: (1) A check valve, located close to the inlet of the injector tee on a pressure system, or close to the rotary feeder inlet on a mixing system, to prevent backflow of material from the conveying line into the air supply line in case the conveying line becomes plugged. (2) A vacuum or pressure relief valve to protect the air mover, as required to suit the type of conveying system. This should be located close to the air mover. (3) An injector tee, on material-into-airstream, positive-pressure systems, to inject air into the conveying line at the point of material entry. (4) Shut-off gates or valves to isolate unused piping when several air movers are interconnected for emergency service. (5) Diverter gates to interconnect air movers for emergency service or to supply air to any one of several services. (6) Dampers for volume control.

All of these accessories must be specially fabricated, or standard equipment adapted to fit them into a lightweight piping system. Some vendors of conveying systems have developed their own designs of piping accessories for lightweight service and can modify their designs to suit customer requirements, since they are not stock items, but must be fabricated to order.

Where two air movers must be interconnected for emergency service only, it is possible to eliminate shut-off or diverter valves by using offset sections of piping arranged so that they could be swivelled to connect any one of two pipelines. The offset must be unbolted when swivelled and bolted into its new position. This method avoids the hazard of air leakage between systems that exists when valves are used to isolate the lines.

Accessories for standard weight piping are commercially available as off-the-shelf items and, in some cases, may be adapted to lightweight piping. A good example is the use of flangeless wafer-type butterfly valves for shut-off or throttling service. This type can be bolted between flanges using long bolts that fit closely in grooves that center the body and align it with the pipe bore.

CHUTES AND SPOUTING

Chutes and spouting are used for gravity discharge of materials from hoppers, bins and receivers into a conveying system or into other types of equipment for processing or storage. There are no standard forms of chutes and spouts. These items are usually fabricated in shapes and configurations to suit the flow requirements and with a wall thickness to suit the structural requirements. The materials of construction are selected to suit the characteristics and properties of the materials handled. In many cases, lightweight or standard pipes and flanges are adapted to fabricate cylindrical chutes and spouts. Some vendors of accessory equipment for pneumatic conveying systems fabricate various spouting shapes, but their catalogs do not show wall thickness. These are not off-the-shelf items and are fabricated to order when the wall thickness and material of construction are specified. Since most fabrications shown in catalogs require some modification, installation time can be gained by having these made by a local sheet metal shop.

Joints may be flanged or welded, but flanged joints are preferred where chutes may become plugged and require removal. An alternative would be to provide cleanout doors where blockage may be anticipated. All flanged joints should be gasketed for dust tightness. Rubber gaskets are widely specified, but compressed asbestos gasketing can also be made tight and has the advantage of permitting welding of attachments to flanges during installation without burning the gasket or breaking the joint to remove it.

Cut-Off Gates

Chutes and spouts discharging from bins are preceded by cut-off gates for positive control of material discharge. Most gates are opened wide or shut tightly. Occasionally, they may require an adjustable opening as when feeding material into a vacuum system. There are several types of gates available and design and construction varies with the manufacturer. Some are lightweight fabricated weldments, others are heavy, rugged castings intended for power plant soot or coal piping. Almost all have rectangular inlet and outlet flanges that require the use of a transition piece to discharge into cylindrical spouting (Figure 12–11).

The sliding gate cut-off is the most abused of all gates since many lightweight varieties are used for shut-off and control on the common dust exhaust system. These lightweight types use channel guides to align the flat plate gate. The channels fill with material that cannot flow out and eventually the gate jams. Some gates are designed with a

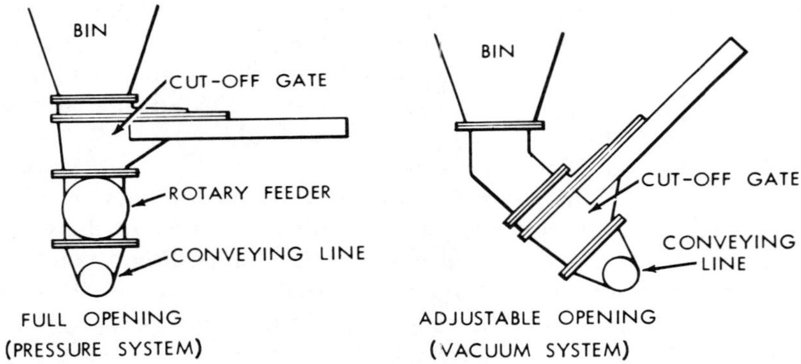

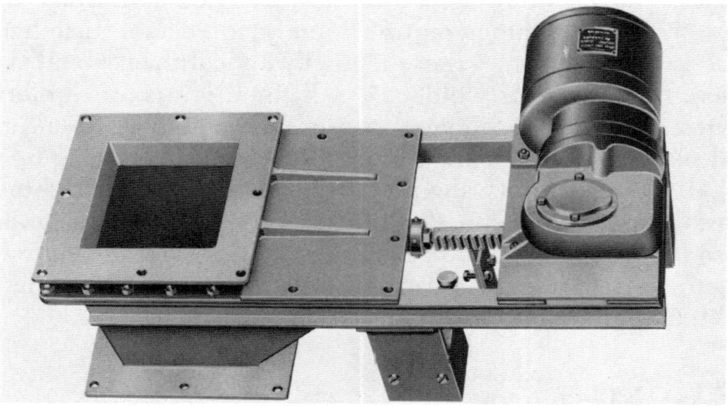

FIGURE 12–11. Motor-operated sliding cut-off gate. (Fuller Company/GATX.)

clean-out pocket at the bottom of the housing to receive material pushed out of the guides by the gate when it is closed. Some gates on negative-pressure service never jam, but are automatically cleaned out by the operating vacuum.

Bin shut-off gates should be installed with the gate horizontal, or as close thereto as possible when a head of material lies above the gate, the gate guides must be made self-cleaning and powder-sealed: powder sealing is simply the sealing of all crevices between the inlet section of the housing and the gate proper by fine material resting on top of the gate. The guides below the gate are made self-cleaning by using an open ladder type design that permits material to drop down into the body and not lodge on top of the guide to interfere with movement of the gate. The lower guides may be a series of inverted V-shaped teeth, or rollers built into the sides of the housing to support the gate. These supports are spaced so that powder can be pushed off the supports as the gate moves and can drop down into the gate outlet. These supports are generally located outside the path of materials flowing through the inlet. The end of the gate is made with a knife-edge to cut through built-up material, to seal the end of the gate, and to cut through lumps that may interfere with closing the gate. Some gates have bolted covers over access openings at the end of the gate and on both sides of the housing to permit cleanout and inspection of the guides.

Sliding cut-off gates are operated by direct application of force to a cylindrical stem attached to the plate and extending outside of the body through a packing gland. The force may be applied manually, by an air-operated cylinder or by a reversible electric motor that drives a pinion engaging a rack attached to the stem. Where leverage is required for manual operation, a handwheel is attached to a yoke threaded onto a screwed stem, or to a pinion engaging a rack. Where external dirt may lodge on the stem, the screwed stem is preferred. Any dirt caught between the teeth of a rack and pinion drive will prevent operation of the gate and tend to spring the pinion shaft. A motor-operated bin discharge gate should always be preceded by a manually operated isolating gate so that repairs could be made to the motorized gate without emptying the bin. The isolating gate may also serve as a metering gate on some systems and it may be set at a predetermined position (Figure 12–12).

Another type of cut-off valve that can be opened under a head of material and permits unrestricted flow is the rotary valve (Figure 12–13).

This valve is essentially a hollow drum rotating on a horizontal axis within a cylindrical housing. The ends of the drum have stub shafts

FIGURE 12–12. Arrangement of sliding cut-off gates and diverter gate for controlling gravity flow from a storage bin. (Fuller Company/GATX.)

that rotate in trunnions mounted in the end plates of the housing. The drum has two arcs cut out of its periphery on a common centerline so that a 90-degree rotation in either direction would open the valve by aligning the cut-out sections with the inlet and outlet openings. A 90-degree rotation in the opposite direction would align the drum surfaces with these openings and close the valve. The gate is thus powdersealed. The ends of the drum on better valves are built with slots that permit material that leaks into the ends to drop through, thus avoiding sticking due to packed material. The gate may be manually operated by a lever mounted at one end of the shaft, or motor operated

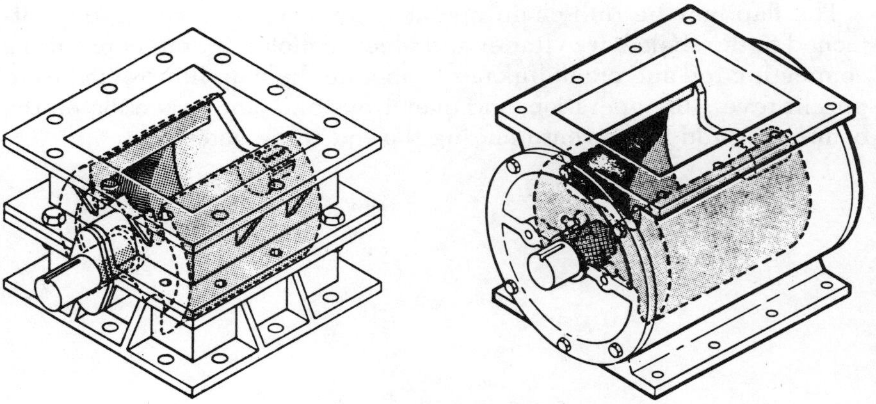

FIGURE 12–13. Rotary valves for bin discharge service (Fuller Company/ GATX.)

by a connecting rod and crank mechanism. The motor must be reversible and limit switches mounted on the valve body are required to obtain the reversing action. Limit switches may be mounted on manually operated valves in order to indicate the valve position or to interlock control circuits.

Diverter valves for gravity discharge are installed below bins, hoppers and receivers, or inserted into downspouts, in order to direct material into either of two, or even three, discharge points. These gates are generally square or rectangular in cross section thus requiring a transition piece for connecting them to cylindrical spouts. The diverter is basically a steel plate flap mounted on a shaft that can be pivoted in either direction within a steel plate housing. The two outlets are sloped away from each other at an angle that suits the flow characteristics of the material handled. The bottom-pivoted flap forms a slope sheet that directs material from the inlet to one of the outlets and, at the same time, blocks the entrance to the other outlet. When the flap is pivoted in the opposite direction, it directs the flow to the other outlet and blocks the original outlet.

Clearance is required between the flap and housing in order to permit the flap to swing. This clearance permits some material to dribble into the unused outlet. When this leakage is objectionable, the gate is modified by adding side walls to the flaps, forming what is known as a shrouded gate, and pivoting the flap at its center. The inlet port is fitted with a spout that discharges material between the shrouds so that it can flow downward without leaking into the unused outlet port.

The flap may be pull-chain operated by a center-pivoted lever attached to the shaft. Air cylinder and electric motor operators require a connecting rod and crank linkage to operate the flap, limit switches to permit reversible operation, and overstroke and jam protection on the connecting rod to prevent bending the rod if the gate jams.

13 Air Movers

THE POWER PACKAGE

The type of air mover selected for a pneumatic conveying system is primarily determined by the airflow and pressure required to convey a specific material through a specific system of piping at a given rate. This requires proper allowance for leakage into or out of the system, and a factor for the vendor's lack of knowledge and experience.

The air mover and its accessories are referred to by many vendors as a power package, since it is this package that determines the power required for transporting material through a conveying line. The power packages available, in order of increasing pressure (and cost) should include the following equipment.

1. Centrifugal blower; drive motor; V-belt transmission and guard, if required; and a suction strainer.
2. Positive-displacement blower; drive motor; V-belt transmission and guard, if required; inlet filter; intake and discharge silencers; pressure or vacuum relief valve; check valve; pressure indicators; mounting base for entire unit.
3. Reciprocating air compressor; drive motor; V-belt or flat belt transmission and guard; inlet strainer; air receiver with start-stop control switch mounted on it; intercooler for a multistage unit; after-cooler; pressure relief valve; pressure indicators; moisture drain trap; mechanical lubricator.

The motor horsepower supplied in a power package is a reflection of the vendors knowledge of pneumatic conveying and his conservatism. In general, motors are oversized in order to provide (1) for speeding up the air mover if conveying guarantees cannot be met, (2) for reserve power to blow out a plugged conveying line, and (3) to provide increased power to meet the no-load condition on a centrifugal blower installation.

Systems where material enters a stream of air require large volumes of air at relatively low pressure, due to the high air-to-material ratio used. These systems, whether operating under positive or negative pressure, are generally furnished with high-speed, positive-displacement blowers, or centrifugal fans, depending on the vendor's preference. Such units can be obtained in a wide range of sizes and can be belt-driven to operate at any speed required to meet operating conditions. The positive-displacement blower is favored for the majority of systems because it can develop high pressure (up to 18 psig) and sufficient capacity to move material when a conveying line plugs up. A centrifugal blower's characteristic is such that airflow decreases

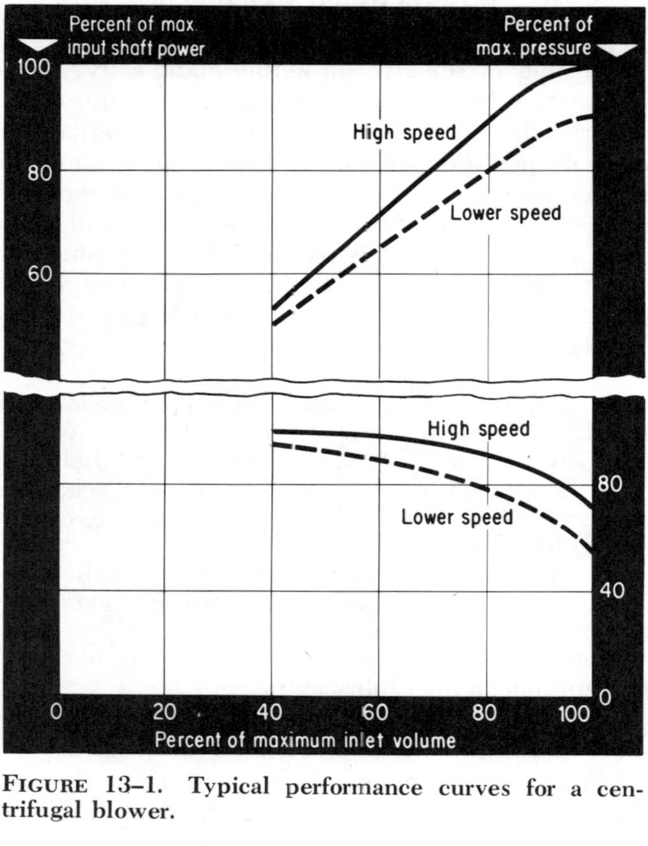

FIGURE 13–1. Typical performance curves for a centrifugal blower.

rapidly with increase in system resistance. See the curves of Figures 13–1 and 13–2.

Systems in which air enters a mass of material are generally furnished with positive-displacement blowers to meet the relatively low-airflow, medium-pressure requirements of pressurized tank-

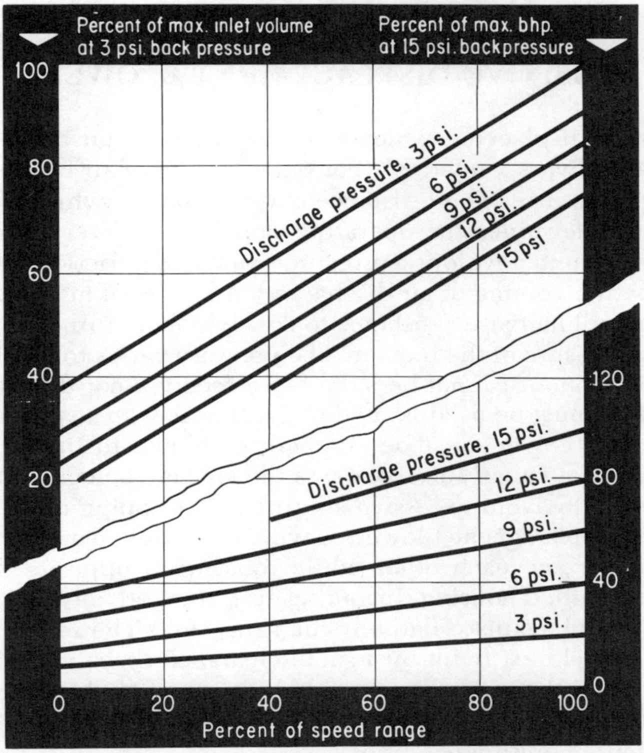

FIGURE 13–2. Typical performance curves for a positive-displacement blower.

trailer service and low-pressure blow-tank systems. However, for the low-airflow, high-pressure requirements of a high-pressure blow-tank system using very low air-to-material ratios, the choice is usually a reciprocating air compressor or a radial sliding-vane compressor, both of which require air and oil filters, and a supply of cooling water for the compressor cylinder.

Systems in which air and material are simultaneously intermixed and air-to-material ratios are low are generally furnished with either two positive-pressure blowers operating in tandem from a common drive for low-airflow, medium-pressure service (up to 20 psig), or with a radial sliding-vane compressor for low-airflow, high-pressure service (up to 40 psig).

The feed screw into air-nozzle mixing system may use single-stage reciprocating compressors or sliding-vane compressors.

POSITIVE-DISPLACEMENT BLOWERS

The positive-displacement blower is essentially a constant-speed machine that delivers a relatively constant volume of air over a range of discharge pressures. The resistance of the system to which the blower is applied determines the operating pressure. Delivery of air into a conveying system develops a pressure equal to the frictional resistance created by the volume of air discharged. Addition of material into the air stream will increase resistance to flow which, in turn, increases the operating pressure of the blower. The pressure varies to meet the load conditions. The air cannot be stored in a receiver nor throttled at the discharge; it must be used immediately. If airflow to a system must be stopped, the blower must be stopped; or blown to the atmosphere through a relief valve. In some cases the air may be cooled and recycled in order to avoid excessive starting and stopping of the blower.

Positive-displacement blowers have two timing-gear-driven lobed rotors that engage each other while rotating in opposite directions within a machined housing. Incoming air is trapped between the lobes and discharged against relatively high pressure. The rotors are separated at their closest point by a small clearance space and never contact each other. A single motor or V-belt-driven shaft drives the timing gears. V-belt drive is preferred to permit speed changes, if required to meet system operating conditions.

There are two basic types of rotor design, the well-known rotary piston type with two identical "figure-8" lobes (see Figures 13–3 and

Incoming air (right) is trapped by impellers. Simultaneously, pressurized air (left) is being discharged.

As lower impeller passes wrap-around flange, Whispair jet, equalizes pressure between trapped air and discharge area, aiding impeller movement and reducing power.

Impellers move air into discharge area (left). Backflow is controlled, resulting in reduction of noise relative to conventional blowers.

FIGURE 13–3. Operating principle of Whispair Roots blower. (Roots Blower Operation, Division of Dresser Industries, Inc.)

FIGURE 13–4. Exploded view of Roots blower features. (Roots Blower Operation, Division of Dresser Industries, Inc.)

13–4); and the more-recently developed axial-flow type that has a spirally lobed main rotor meshing with a spirally grooved gate rotor to cause air to flow axially toward the discharge connection (Figure 13–5).

Blower casings are generally of ribbed cast-iron or steel construction, designed to resist a limited pressure. Some manufacturers feature three pressure ranges designated as low (to 5 psig), medium (to 10 psig), and high (to 15 or 18 psig). However, since no industry standards

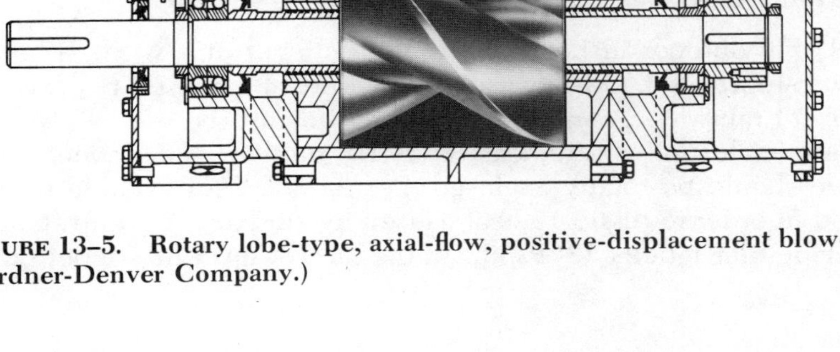

FIGURE 13–5. Rotary lobe-type, axial-flow, positive-displacement blower. (Gardner-Denver Company.)

exist, pressure and volume ratings vary considerably. The maximum rotor operating speed and allowable pressure for which the casing is designed should be determined in advance if future plans contemplate speeding up a blower in order to increase the capacity of a system. The radial and axial clearance spaces between the rotors and the casing allow some of the compressed air to flow back into the casing inlet. This back leakage is called "slip," since the blower delivers less than the quantity of air corresponding to the displacement of the rotors. The slip increases with pressure and is accounted for in capacity tables by specifying net displacement at a given speed or by stating the loss in capacity as slip rpm at a given pressure.

Large blowers are made with the casings horizontal, that is, with the impellers in the same horizontal plane. Thus inlet and outlet connections are vertical and the casing would require a raised foundation or a trench for connecting piping to the lower flange. Small- and medium-sized blowers can be obtained with the casings vertical or horizontal, thus providing a choice to suit any piping arrangement.

The positive-displacement blower generates considerable noise and pulsation energy that requires noise snubbers on the inlet and outlet of blowers and on the outlet of a blower used as a vacuum pump. The standard silencers are usually expansion chambers connected by perforated conduits. The noise generated by a blower becomes critical at pitch line velocities of the timing gears above 2,700 feet per minute. Above this velocity an additional silencer of the absorptive pack type is required at the blower to attenuate the higher frequencies that cause "pipe-ring" in downstream piping. The effect of noise snubber pressure drop on the inlet conditions must be considered when sizing a blower. Discharge snubbers are sometimes omitted on blowers for positive-pressure systems since downstream equipment tends to attenuate the noise.

Since positive-displacement blowers are fine pieces of high-speed equipment, machined to close clearances, the following details must be considered to insure successful operation.

INSTALLATION

(1) For outdoor installation, select a blower unit with all bearings grease-lubricated. Oil-lubricated units must be housed to prevent the entry of rainwater, and the oil changed to suit the season, especially where enclosures are unheated. (2) For outdoor installation, a V-belt drive should be totally enclosed, but the enclosure should be ventilated in order to dissipate heat caused by slippage. Takeup should be provided for V-belts. (3) Piping at the blower must be supported so as

to relieve strains on the casing. Or, one can install flexible, heat-resisting piping connections between the blower and the air piping. (4) Where two or more blowers can discharge into a common delivery line, install a tight-sealing isolating valve in each discharge line to prevent reverse rotation of idle blowers. (5) Do not locate silencers directly above an inlet or an outlet connection on a casing; place them to one side. Built-up rust or material on the silencer walls can drop between the rotors and prevent rotation. This will require dismantling the equipment in order to clean the rust out.

PROTECTION

(1) Use pressure- or vacuum-relief valves to protect the blower casing and limit the power drawn by the drive motor (Figures 13–6 and

FIGURE 13–6. Weight-loaded pressure relief valve. (Roots Blower Operation, Division of Dresser Industries, Inc.)

FIGURE 13–7. Spring-loaded vacuum relief valve. (Roots Blower Operation, Division of Dresser Industries, Inc.)

13–7). (2) Install a check valve in the discharge line on all positive-pressure systems and locate it as close to the material-entrance point as possible. (3) Install a low-resistance, renewable, dry-type air filter at the suction (or as close to the suction as possible), on all positive-pressure-system blowers. This should be in an area free from dust and odors. The filter cartridge should be of a type that will prevent the passage of dust of such size as would contaminate the conveyed material. (4) Install a low-resistance protective filter at the suction of blowers on negative-pressure systems so as to prevent material from entering the blower casing if a filter tube should break in the receiver-filter. Install a pressure-differential switch across this filter so as to shut down the system if the filter becomes blocked by powder. (5) Make certain that the casing enclosing oil-lubricated timing gears is isolated from the blower casing by an air space, with oil seals on the gear case and air seals on the blower case.

NOISE REDUCTION

(1) Provide desonated-chamber-type silencers at the discharge of blowers on negative-pressure systems and at the suction of blowers on positive-pressure systems. The silencers should be capable of attenuating sound corresponding to the frequency of impeller explosions. (2) Provide silencers on pressure- and vacuum-relief-valve discharge ports. (3) Use flat-bottom V-belts; notched belts generate noise.

(4) Pipe air leakage from casing seals back into the suction line on high-pressure blowers (those over 10 psi). (5) Operate blowers at the lowest possible gear tip speeds. (6) Apply sound insulation to silencers and to piping to and from the blower. (7) Use vibration isolating pads between the blower and baseplate. (8) Apply an acoustic lining to belt guards.

The blower casing should not be insulated, since its ribs serve to cool the casing and prevent binding of the rotors.

Sound-deadening enclosures should be ventilated through openings that are equipped with sound traps. Windows or sight ports should be provided so that oil-flow sight glasses and oil-pressure gauges can be observed. Hinged and latched doors should be furnished with access to grease fittings and oil fill caps on gear casings. If necessary, a separate ventilating fan should be supplied in order to prevent heat buildup within the enclosure.

Some of the details involved in selecting a positive-displacement blower are also applicable to the selection of a centrifugal blower. This is particularly true regarding protection and installation.

POSITIVE-DISPLACEMENT BLOWER CONTROL

When the rotary piston type of positive-displacement blower is used to supply air to a blow tank, some provision must be made to relieve the air pressure between blow-tank discharge cycles since no air storage is used. This can be done by blowing off the compressed air in several ways: (1) A weight- or spring-loaded relief valve may be set to open and discharge the air directly into the room. This method is noisy, even when the valves are equipped with noise snubbers. (2) The discharge may be directed into the suction side of the blower through a pilot operated solenoid valve that permits full-ported flow (up to 3-in. I.P.S.) using a pressure switch for control. (3) When the discharge must be maintained at a high pressure to supply air to aeration pads, the recycling of air may build up the temperature of the blower. To prevent seizing of the rotors, the discharged air may be directed outside of the area through a full-ported pilot operated solenoid valve. (4) When the blow cycles occur at widely spaced intervals, start-stop operation under pressure switch control may be used.

CENTRIFUGAL BLOWERS

Centrifugal blowers were originally used for the pneumatic conveying of wood chips, raw cotton, paper scrap, and many other light or fluffy materials. These materials actually passed through the fan into a conveying duct for delivery to a cyclone receiver. Three types of wheels were designed to handle these materials. The common feature being the straight radial blades evenly spaced around the inlet eye of the wheel. The standard material wheel had side plates on each side of the blades with one side plate forming a support for the hub of the wheel. A lint wheel omitted the side plate on the entrance side of the wheel, but the back plate was contoured to form a smooth transition from end of the hub to the blades and the blades themselves were contoured to prevent catching lint. A long-shaving wheel had no shroud plates, the radial blades were simply attached to arms radiating from the hub of the wheel. These types of blowers and the seemingly antiquated conveying systems associated with them are still in use today and are available to solve similar conveying problems.

The centrifugal blower with a standard material wheel is still used today for pneumatically conveying powdered or granular materials. However, whenever possible, the materials are handled so that they do not pass through the fan. The rotating wheel imparts velocity or kinetic energy to the air passing between the revolving blades, this energy is converted to static or potential energy, required to overcome the resistance of the duct system, in the stationary scroll case or fan housing. The pressure developed at the fan, and the volume of air handled at a given speed varies with the system resistance. The curve showing this relationship is called the fan performance characteristic. Associated with this curve, for the same speed, is another curve showing the variation in horsepower with the volume of air handled. Each fan speed would require a different set of curves to show the changes in fan characteristics peculiar to that speed. These characteristic curves apply only to airflow through the fan, not to flow of an air and material mixture.

The principal application of centrifugal blowers in pneumatic conveying work is on the small material-into-airstream systems operating under positive or negative pressure with high ratios of air to material. In these systems, the possibility of plugging a conveying line is usually remote and the blower can operate safely on its head curve. In most negative-pressure systems the blower is mounted atop a cyclone, with its shaft vertical, and is driven by a direct-connected motor. The inlet flange is bolted to the outlet connection of a high-efficiency cyclone. In operation, the suction conveying line may receive material from several locations and deliver them to the cyclone from where they are discharged through a rotary airlock feeder. The air on dustless materials passes through the blower to atmosphere. On loop systems, the blower discharges, through a pipeline filter, back into the suction line; or the dust may be recycled continuously. When handling very dusty materials, a dust filter must be installed between the cyclone and blower thus requiring mounting of the blower on its own foundation. At startup, the blower motor will draw its maximum horsepower, since the system resistance of the empty pipeline, ducts and attached equipment is so low that the blower will tend to draw its maximum volume of air. As material is drawn into the conveying line, the effect is exactly the same as closing a damper in the line, the system resistance increases. The blower will then tend to operate at another point on its head curve, namely, one with a higher head and a lower volume of air that corresponds to the system resistance. The horsepower will correspondingly decrease (Figure 13–8). If too much material is drawn into the conveying line, the volume of air drawn by the fan will decrease to the point where material will drop out in the line because of low conveying velocity. This characteristic requires that material feed to the conveying line be closely controlled or arranged to be delivered into the system at a constant rate. Many industrial exhausters have a slightly drooping performance characteristic and are not suitable for conveying service. Those exhausters with steep characteristics, where volume changes are relatively small for a given change in pressure, are preferred for conveying service.

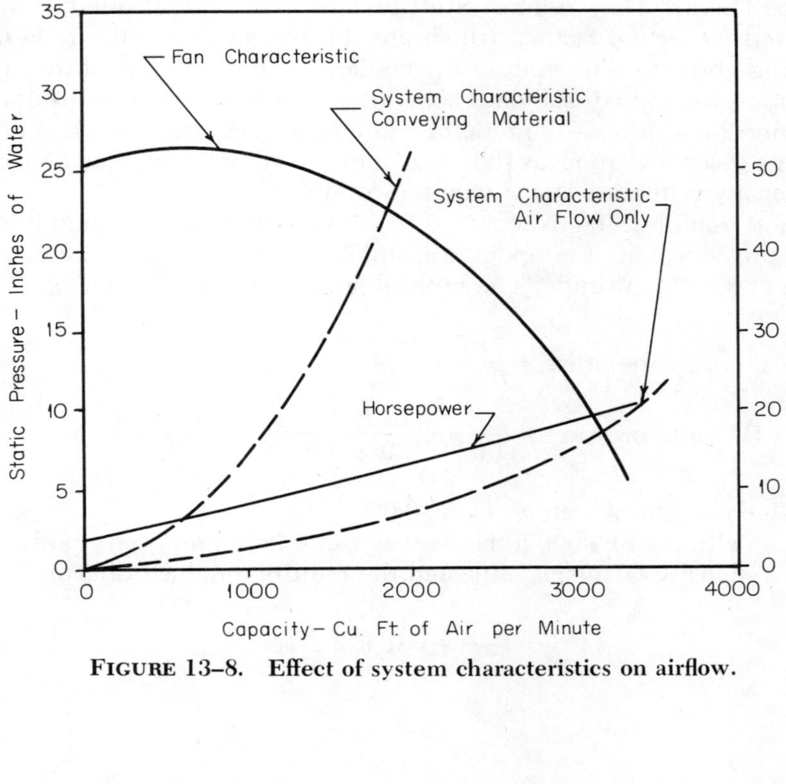

FIGURE 13–8. Effect of system characteristics on airflow.

SELECTION AND RATING OF FANS

Most fans used for pneumatic conveying or used as exhausters for venting are of the radial, flat-bladed wheel type. At a given speed, these fans will draw the same volume of air at a given pressure measured in feet of air regardless of air density. The horsepower, the pressure measured in inches of water and the weight of air handled will vary with the air density. Fan sizes are usually selected from performance tables, for each size of wheel, listing volumes in even increments of outlet velocity with speed and horsepower tabulated for various static pressures. These tables are based on air at standard conditions. Volumes must be increased, in the same manner as described below for blowers under the same intake conditions, to compensate for reduced inlet pressure and increased inlet air temperatures. However, when a definite airflow is specified for given conditions of pressure, temperature or elevation, other than standard, the fan must be selected for the given rate of flow, but for an equivalent static pressure. The equivalent static pressure is the working static pressure corrected for temperature, pressure and altitude to permit the selection of a fan from standard tables. The working static pressure must be multiplied by the required correction factors, which are: for temperature—the ratio of working absolute temperature to standard absolute temperature; for pressure—the ratio of standard absolute pressure to working absolute pressure; for altitude—the ratio of barometric pressure at sea level to that at a given elevation, as shown in Table 13–1, or the inverse ratio of the density reduction factor shown in some tables.

As an example, a fan is required to deliver 10,640 cfm of air at 6-in. static pressure, 250°F temperature, and a suction pressure of 20 in. of water at 3,000-ft altitude. The multiplying factors are as follows:

$$\text{For temperature} = \frac{710}{530} = 1.34$$

$$\text{For inlet pressure} = \frac{14.7}{14.7 - (20 \times 0.036)} = \frac{14.7}{13.98} = 1.05$$

(One-inch height of water weighs 0.036 psi.)

For an altitude of 3.000 ft the density factor, as found from a table of altitude-density factors, is .891, and the multiplying factor becomes

$$\frac{1}{0.891} = 1.122$$

TABLE 13-1
Barometric Pressures When Above Sea Level

Elevation, ft	Barometer, In.-Hg	Elevation, ft	Barometer, In-Hg	Elevation, ft	Barometer, In.-Hg
0	29.92	4000	25.84	8000	22.22
100	29.81	4100	25.74	8100	22.14
200	29.71	4200	25.65	8200	22.05
300	29.60	4300	25.55	8300	21.97
400	29.49	4400	25.46	8400	21.89
500	29.38	4500	25.36	8500	21.80
600	29.28	4600	25.27	8600	21.72
700	29.17	4700	25.17	8700	21.64
800	29.07	4800	25.08	8800	21.55
900	28.96	4900	24.99	8900	21.47
1000	28.86	5000	24.89	9000	21.38
1100	28.75	5100	24.80	9100	21.30
1200	28.65	5200	24.71	9200	21.22
1300	28.54	5300	24.61	9300	21.14
1400	28.44	5400	24.52	9400	21.06
1500	28.33	5500	24.43	9500	20.98
1600	28.23	5600	24.34	9600	20.90
1700	28.13	5700	24.25	9700	20.82
1800	28.02	5800	24.16	9800	20.74
1900	27.92	5900	24.07	9900	20.66
2000	27.82	6000	23.98	10000	20.58
2100	27.72	6100	23.89	10100	20.50
2200	27.62	6200	23.80	10200	20.42
2300	27.52	6300	23.71	10300	20.34
2400	27.41	6400	23.62	10400	20.26
2500	27.31	6500	23.53	10500	20.18
2600	27.21	6600	23.44	10600	20.10
2700	27.11	6700	23.35	10700	20.02
2800	27.01	6800	23.26	10800	19.95
2900	26.91	6900	23.17	10900	19.87
3000	26.81	7000	23.09	11000	19.79
3100	26.72	7100	23.00	11100	19.71
3200	26.62	7200	22.91	11200	19.64
3300	26.52	7300	22.82	11300	19.56
3400	26.42	7400	22.74	11400	19.48
3500	26.32	7500	22.65	11500	19.40
3600	26.23	7600	22.56	11600	19.33
3700	26.13	7700	22.48	11700	19.25
3800	26.03	7800	22.39	11800	19.18
3900	25.94	7900	22.31	11900	19.10

The equivalent static pressure becomes

$$6 \text{ inches} \times 1.34 \times 1.05 \times 1.122 = 9.475 \text{ inches}$$

Entering the fan tables of a typical catalog, a 43½-in. diameter wheel shows the following data:

Capacity (cfm)	Inlet Velocity (ft/min)	9-in. S.P.		10-in. S.P.	
		rpm	hp	rpm	hp
10,640	3,000	941	21.8	985	24.4

Interpolating, performance for 10,640 cfm at 9.475 inches equivalent static is 961.9 rpm and 23.07 hp. The correct horsepower at the working static pressure would be

$$\frac{23.07}{1.34 \times 1.05 \times 1.122} = 14.6 \text{ hp required at the fan shaft}$$

The fan would be purchased to handle 10,640 cfm at 9.475 static pressure, 962 rpm, 70° F temperature. A 15-hp motor would be purchased having a speed that would provide a reasonable speed ratio for a V-belt drive. In operation, the fan static pressure would become 6 inches and the motor would draw about 15 hp. However, if the fan were operated with cold air, the horsepower drawn would exceed the capacity of the motor and the airflow to the fan would have to be throttled until the equipment warms up.

Fan Performance Curves

The fan performance for a given size and at constant speed and gas or air density is shown on characteristic curves furnished by the vendor. The standard curves show total static pressure in inches of water, horsepower input and total or static efficiency as ordinates versus airflow in cfm as abscissa. These curves are useful when duct or system changes are made since they show the effect on airflow of resistance changes. For each system of ductwork and interconnected equipment there is only one resistance curve. The intersection of this resistance curve with the fan characteristic curve of total static pressure is the operating point of the fan for that system. Any change in the system, even if it is only a slightly throttled damper, or increased dust filter resistance is equivalent to forming a new resistance curve that will intersect the fan characteristic curve at a different operating point. When the same fan is used on a revised duct system and the air density does not change, the three basic fan laws may be used to determine the fan characteristics at another speed. These laws are as follows: (1) Capacity is proportional to speed. (2) Static pressure is proportional to the square of the speed. (3) Horsepower is proportional to the cube of the speed. These three proportionality factors apply for the specific condition represented by operation at the same point on the fan characteristic. The effect of changes in density on horsepower, static pressure and the weight of air handled can be combined with the fan laws by multiplying these terms by the proper density ratios found as described below for blower intake density correction factors. See page 280.

RECIPROCATING AIR COMPRESSOR

The reciprocating air compressor is a well-known piece of equipment the construction of which is common knowledge, but certain precautions must be taken when applying one to a blow-tank type of pneumatic conveying system. The unit must be of the oil-free type with carbon or Teflon piston rings. It must be equipped with a suitably sized air receiver matched to the compressor displacement and to the required service. An aftercooler followed by a chemical air dryer is necessary in order to ensure moisture-free air for conveying. The chemical dryer may require periodic regeneration or renewal of the drying agent.

Blow-Tank Pressure Characteristics

Air requirements are generally determined on test by constructing a curve of blow-tank pressure vs. discharge time, by accounting for air used by the booster jets discharging into the conveying line and by recording the air-receiver pressure during a complete cycle. An analysis of these data can reveal several interesting facts. A succession of sharply rising and falling pressure cycles during discharge shows that the material is free-flowing and that the air supply is ample. A large, sudden drop in pressure between the receiver and the blow tank may seem to indicate wasted energy, but this energy is actually used to activate the material in the tank. A long, flat discharge pressure curve shows effective use of air and a fairly constant air-to-material ratio in the conveying line. A gradually rising discharge-pressure curve, followed by a sudden drop, shows that material flow is sluggish and that a booster jet may be required to transport the material.

Variation in the air-receiver pressure is the best indicator of whether or not the compressor is properly sized. Full pressure must be recovered in the time it takes to perform a complete filling and discharge cycle on the blow tank. Otherwise, time will be wasted waiting for full pressure to be restored. The compressor delivers air continuously during the discharge cycle, but pressure recovery starts near the end of the discharge cycle and should be completed by the time the blow tank is refilled with material. See Figure 7–14, page 145. The compressor should be 20 to 30 percent oversized to allow for leaks, future requirements and auxiliary uses.

A useful formula relating these variables is:

$$Q = V\left(1 - \frac{P_0}{P_f}\right)\frac{1}{\theta}$$

where Q = compressor capacity, ft³/min
V = receiver volume, ft³
P_0 = lowest absolute pressure in receiver
P_f = highest absolute pressure in receiver
θ = time available to recover full pressure, min.

The recovery time should equal, or be less than, the blow-tank filling time in order to obtain the greatest system capacity. The effect of total cycle time on blow-tank-conveying capacity is shown by the equation:

$$C = 3\rho V_2/100t$$

where C = capacity, tons/hr
ρ = loose density, lb/ft³
V_2 = blow-tank volume, ft³
t = total cycle time, min.

This computation should be made for each specific material and for each blow-tank size considered.

It should be noted that some blow-tank units make use of the expansive force of the air by cutting off airflow to the tank prior to the end of the discharge cycle so that residual air pressure empties the tank.

Airflow into Blow Tanks

During a field test, the free air capacity at the intake of a blower, or compressor, and the time during which air enters the blow tank when discharging is sometimes used as a convenient estimate of the airflow rate. This is based on the premise that the compressor will start operating when air is first admitted to the tank and continue to operate until the air is cut-off. This is true on systems equipped with low-pressure blowers discharging directly into the blow tank. However, on systems having air compressors and air receivers, the pressure in the receiver must drop to the compressor cut-in point before it starts. Nevertheless, the method is used as a good approximation of air usage. For example, on a high-pressure system having a 250-cfm compressor and a receiver pressure control set to start the compressor at 80 psi and to stop it at 100 psi the time interval between air inlet and cut-off on a blow tank discharging 2,500 lb of a particular material was 4.5 minutes. The air compressor started 30 seconds after air was admitted to the blow tank; therefore air was supplied by the receiver during this time. The air usage during the cycle is estimated to be

$$250 \times 4.5 = 1125 \text{ cfm}$$

giving an air-to-material ratio of

$$\frac{1125}{2500} = 0.45 \text{ ft}^3/\text{lb material}$$

or a material-to-air ratio of

$$\frac{0.45 \times 2500}{13.35} = 84.25 \text{ lb material/lb air}$$

Methods of Measuring Compressor Capacity

The quantity of air discharged by a high-pressure compressor may be measured by one of the following methods given in order of greatest accuracy: (1) Gaging tank—air flows through an orifice or nozzle attached to a gaging tank in which pressure is maintained constant. The tank is installed on the discharge side of an air receiver with a throttling valve between them. The orifice is sized to permit as much throttling as possible between the receiver and tank. The reduced pressure caused by throttling attenuates pressure fluctuations and permits use of a low-pressure nozzle or orifice—one in which the pressure drop is small—so that flow may be calculated by the standard formula. This method may be used to test a vacuum pump by measuring flow into a gaging tank on the intake side of the vacuum pump. (2) Pipeline flow—air flows under reduced pressure through a standard, primary metering element installed in the pipeline as described in Chapter 4. This method is used when installation of a gaging tank is not practicable. (3) Receiver capacity—air is pumped into a receiver of known volume and the time required to increase the pressure between two predetermined intervals is noted. Temperature readings of the air in the receiver are taken simultaneously with pressure readings. The weight of air pumped during the pressure interval, $W_2 - W_1$, is determined by use of the perfect gas equations using the volume of the receiver as V thus

$$\text{Weight added in } \theta \text{ seconds, pounds} = W_2 - W_1 = \frac{V}{R}\left(\frac{P_2}{T_2} - \frac{P_1}{T_1}\right)$$

This weight is then converted to volume at free-air conditions. This is a rough method of determining capacity due to the error in measuring the temperature of the air in the receiver.

Example. Air is discharged from a compressor into a receiver having a volume of 500 cubic feet. The pressure increases from 50 psig to 70 psig in 236 seconds. The temperature reading corresponding to 50 psig was 110° F and that corresponding to 70 psig was 125° F. What is the capacity of the compressor in cfm of free air?

Solution.

$$W_2 - W_1 = \frac{500}{53.3}\left(\frac{144 \times 84.7}{585} - \frac{144 \times 64.7}{570}\right)$$
$$= 9.39(20.85 - 16.35)$$
$$= 42.20$$

$$\text{Capacity lb/min} = 42.20 \times \frac{60}{236} = 10.75$$

$$\text{Average absolute pressure} = \frac{(144 \times 84.7) + (144 \times 64.7)}{2}$$
$$= \frac{12{,}200 + 9{,}310}{2}$$
$$= 10{,}755 \text{ lb/ft}^2$$

$$\text{Average absolute temp., °R} = \frac{585 + 570}{2} = 577.5$$

$$\text{Capacity, cfm} = 10.75 \times \frac{10{,}755}{2{,}116} \times \frac{530}{577.5}$$
$$= 502 \text{ cfm of free air}$$

A more accurate variation of this method measures the time and entering air temperature for equal increments of pressure increase. The density of the air is determined for each interval by the pressure and by the temperature of the air in the receiver, assuming adiabatic compression. The weight of air delivered during each interval is calculated

$$W_1 = \frac{\text{Receiver volume} \times \text{pressure increment}}{k \times R \times \text{Mean temperature } T_1 \text{ during interval}}$$

$$W_T = W_1 + W_2 + W_3 \ldots$$

then the pumping rate is

$$\frac{W_T \times 60}{\text{Total time, minutes}} = \text{lb/min}$$

from which the volume at free-air conditions may be calculated.

Air Compressor Control

Since blow-tank operation is intermittent, some form of control is required to start and stop the compressor action. The receiver pressure is used to initiate the control action. Three forms of control are available. (1) Automatic start-stop control—here a pressure switch stops the compressor when the receiver pressure is at its maximum and restarts the compressor when the receiver pressure drops below a predetermined low-pressure setting—usually 20 psi below the high setting. (2) Constant-speed, suction valve unloading—here the high pressure in the receiver actuates an unloader that admits air to diaphragm operators, built into the suction valve cages, to hold the valves open so that the air cannot be compressed, but is simply drawn in and pushed out of the cylinder by the piston without compression. (3) Dual control—this is a combination of the two controls described above with a selector switch that permits the choice of either method of control at the operators option. Start-stop control is used where the blow cycles are widely spaced so that air demand is intermittent; constant speed suction valve unloading control is used where the blow cycles are so closely spaced that the demand for air is almost continuous.

ESTIMATING COMPRESSOR HORSEPOWER

The simplest method of estimating the horsepower of any type of air compressor is to refer to Table 13–2, which shows the theoretical horsepower required to compress 100 cfm of free air to various pressures from atmospheric pressure. For conditions other than those shown in the table, or for high altitudes, the air horsepower formulas in Chapters 2 and 8 may be used to obtain the theoretical horsepower. The brake horsepower may then be determined as described in Chapter 8.

TABLE 13-2
Theoretical Horsepowers Required to Compress Air from Atmospheric Pressure

Discharge Pressure		Isothermal Compression, Single or Multistage		Adiabatic Compression* Single-Stage		Adiabatic Compression* Two-Stage	
Psig	Psia	Mep†	Theoretical hp/100 ft.³	Mep	Theoretical hp/100 ft.³	Mep, psi Referred to Low-Pressure Air Cylinder	Theoretical hp/100 ft.³
5	19.7	4.13	1.8	4.48	1.96		
10	24.7	7.57	3.3	8.21	3.58		
15	29.7	10.31	4.5	11.4	5.0		
20	34.7	12.62	5.5	14.3	6.2		
25	39.7	14.68	6.4	16.9	7.4		
30	44.7	16.30	7.1	19.2	8.4		
35	49.7	17.90	7.8	21.4	9.3		
40	54.7	19.28	8.4	23.4	10.2		
45	59.7	20.65	9.0	25.2	11.0		
50	64.7	21.80	9.5	27.0	11.8		
55	69.7	22.95	10.0	28.7	12.6		
60	74.7	23.90	10.4	30.3	13.3		
65	79.7	24.80	10.8	31.9	13.9		
70	84.7	25.70	11.2	33.3	14.6	29.2	12.8
75	89.7	26.62	11.6	34.7	15.2	30.3	13.3
80	94.7	27.52	12.0	36.0	15.7	31.3	13.7
85	99.7	28.21	12.3	37.3	16.3	32.3	14.1
90	104.7	28.93	12.6	38.6	16.9	33.2	14.5
95	109.7	29.60	12.9	39.8	17.4	34.2	14.9
100	114.7	30.30	13.2	40.9	17.9	35.0	15.3
110	124.7	31.42	13.7	43.2	18.9	36.7	16.1
120	134.7	32.60	14.2	45.2	19.8	38.3	16.8
130	144.7	33.75	14.7	47.2	20.7	39.6	17.3
140	154.7	34.67	15.1	49.2	21.5	40.8	17.9
150	164.7	35.59	15.5	51.0	22.3	42.3	18.5

* Based on a c_p/c_v ratio of 1.3947.
† Mean effective pressure.
Reprinted from Compressed Air and Gas Handbook, 4th ed., 1973. Copyright, Compressed Air and Gas Institute.

Figure 13–9 is an idealized record of simultaneous pressure and volume in one end of the cylinder during one complete cycle of an air compressor. This corresponds to an indicator diagram that would be taken on test, except that the ordinate axis is drawn at zero volume and the abscissa axis is drawn at absolute zero pressure. The pressures P_1

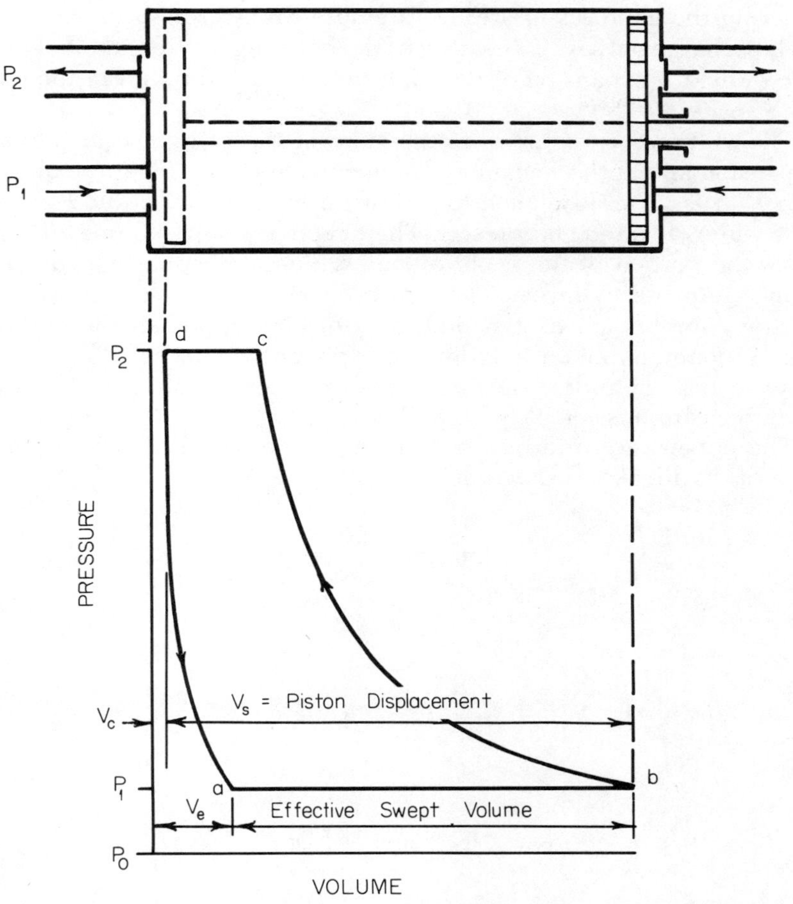

FIGURE 13–9. Conventional indicator card for estimating power requirements for a reciprocating air compressor.

and P_2 are the absolute intake and discharge pressures respectively, which are set by operating conditions external to the compressor. The volume remaining at the end of the discharge stroke is known as the clearance volume, V_c. Its relative amount is evaluated by the proportion it bears to the piston displacement, V_s. Thus volumetric clearance, or clearance ratio is V_c/V_s—it may vary from 2 percent in high-grade compressors to 12 percent or more in poorly designed units.

The direct effect of clearance is to reduce the volume of air drawn into the cylinder at each stroke so that the free air delivered is less than the piston displacement. The factors that reduce the amount of free air entering the compressor are: (1) pressure drop through the air intake and discharge valves; (2) heating of the entering air from the hot cylinder walls; (3) expansion of the high-pressure air trapped in the clearance space.

The volume, V_e, represents the state of the clearance air after re-expansion and at the instant the intake valve opens. The re-expansion curve *da* is generally taken to follow an exponential curve $PV^{1.2} = C$ for a water jacketed compressor. The effective swept volume *ab* represents the portion of the cycle during which atmospheric air is being drawn into the cylinder. The curve *bc* represents the state change during compression of the total mass of air trapped in the cylinder. This is generally taken to follow an exponential curve $PV^{1.35} = C$ for a water-cooled cylinder. Line *cd* is the cylinder pressure during delivery after compression.

The power required per cycle may be computed directly from the area of the idealized diagram

$$\int_b^c V\,dP - \int_a^d V\,dP$$

where

$$\int_b^c V\,dP = \frac{n}{n-1} P_b V_b \left[\left(\frac{P_c}{P_b} \right)^{\frac{n-1}{n}} - 1 \right] \qquad (n = 1.35)$$

and

$$\int_a^d V\,dP = \frac{n'}{n'-1} P_a V_a \left[\left(\frac{P_d}{P_a} \right)^{\frac{n'-1}{n'}} - 1 \right] \qquad (n' = 1.2)$$

If the exponent n' is taken as equal to n, these equations may be combined to give the following approximate equation for indicated work per cycle

$$\text{Indicated work per cycle} = \frac{n}{n-1} P_b(V_b - V_d) \left[1 + C - C \left(\frac{P_c}{P_b} \right)^{\frac{1}{n}} \right] \times \left[\left(\frac{P_c}{P_b} \right)^{\frac{n-1}{n}} - 1 \right]$$

where

$V_b - V_d$ = the piston displacement per stroke

$\left[1 + C - C \left(\frac{P_c}{P_b} \right)^{\frac{1}{n}} \right]$ = the volumetric efficiency of the compressor

C = the clearance ratio.

The area of the indicator diagram is equivalent to the terms P_mLA in the familiar formula for indicated horsepower

$$\text{Indicated hp} = \frac{P_mLAN}{33{,}000}$$

where P_m = mean effective pressure per cycle, psi
L = length of stroke of the piston, ft
A = area of the piston, in.^{2}
N = number of working strokes per minute
N = rpm for a single-acting compressor
$N = 2 \times$ rpm for a double-acting compressor.

SLIDING-VANE COMPRESSOR

The sliding-vane type of air compressor is generally used only on systems in which air and material are intermixed in a chamber at the entrance to the conveying line. The compressor supplies air to a series of nozzles, and airflow is relatively constant.

This type of compressor requires lubrication and special attention must be given to installation of oil eliminators and oil traps in the connections to the transport pump. Air-inlet filters must be of the oil-bath type and must be properly maintained in order to insure removal of all dirt prior to entrance into the compressor. An accumulation of dirt below the sliding vanes can wreck the compressor by preventing complete retraction of the vanes.

The amount of oil in the discharge air stream makes it impractical to use absorptive pack types of silencers on the discharge of those units operating above the critical pitch line or peripheral velocity of 2,700 ft per min. Where critical silencing is required, an externally lagged silencer should be used.

A cooling-water supply is a must, and a flow switch should be installed in the cooling-water-inlet line, electrically connected to prevent operation of the compressor if water is not flowing to the compressor jacket.

CAPACITY RATINGS OF AIR MOVERS

The capacity of compressors, blowers, and fans is usually rated in cubic feet of gas or air under intake conditions of temperature and pressure. The distribution of pressure between suction and discharge must be specified in order to make the proper correction for density of the gas at the inlet. Failure to take cognizance of these facts may result in the selection of an undersized air mover. Unless otherwise specified, an air mover manufacturer will assume that intake conditions for air are for a standard density of 0.075 lb/ft³ at standard conditions of 70°F temperature and a barometric pressure of 29.92 inches of mercury.

A common error is to specify a blower to deliver a given airflow at a given differential pressure across the blower. If the blower was intended to operate on a vacuum-pressure system it would be undersized. Suppose that a combination vacuum-pressure system required an inlet airflow of 2,000 cfm at a suction pressure of 10.2 inches of mercury (equivalent to 5 psig) and at a discharge pressure of 5 psig and the blower was specified to deliver 2,000 cfm of free air at a differential pressure of 10 psig, the blower would have a capacity determined as follows:

The rise in temperature due to adiabatic compression from standard conditions to 10 psig would be

$$\frac{T_2}{T_1} = \left[\frac{P_2}{P_1}\right]^{\frac{k-1}{k}}$$

$$\frac{T_2}{530} = \left[\frac{144 \times 24.7}{144 \times 14.7}\right]^{\frac{1.4-1}{1.4}} = 1.68^{0.286} = 1.16$$

$$T_2 = 530 \times 1.16 = 615°\ \text{R} = 155°\ \text{F}$$

The exponential factor may be worked out by use of a pocket calculator, a table of logarithms, or reference to a handbook table of Adiabatic and Polytropic Expansion Factors for perfect gases.

The airflow at 10 psig would then become

$$2{,}000 \times \left[\frac{14.7 \times 144}{24.7 \times 144}\right] \times \left[\frac{615}{530}\right] = 1.380\ \text{ft}^3$$

If this blower had been applied to the vacuum-pressure system, the suction would draw only

$$2{,}000 \times \frac{9.7 \times 144}{14.7 \times 144} = 1{,}320\ \text{ft}^3$$

When compressed to 5 psig, the temperature rise would be

$$\frac{T_2}{530} = \left[\frac{14.7 \times 144}{9.7 \times 144}\right]^{0.286} = 1.517^{0.286} = 1.1267$$

$$T_2 = 596°\ \text{R} = 136°\ \text{F}$$

The airflow at 5 psig would then become

$$1{,}320 \times \left[\frac{9.7 \times 144}{14.7 \times 144}\right] \times \left[\frac{596}{530}\right] = 980 \text{ cfm}$$

The superficial air velocity would decrease considerably and the material may not be conveyed.

Increasing Air Volume

When air is drawn from the atmosphere or from equipment, and is subsequently increased in temperature or reduced in pressure because of friction losses in piping and equipment, or because of increased elevation above sea level, the volume drawn by the air mover must be increased in order to maintain a given weight rate of flow. The following formula may be used to determine the flow

$$\text{cfm} = \frac{\text{weight rate, lb/min}}{\text{air density at inlet}}$$

Correct the air density for temperature by multiplying the density at the source of air by the ratio of the absolute temperatures at the air source and at the air mover inlet. Correct the air density for pressure by multiplying the density at the source of air by the inverse ratio of the absolute pressures at the source and at the air mover inlet. Correct air density for altitude by multiplying the density of air at sea level by the ratio of barometric pressure at altitude (from Table 13–1) to the barometric pressure at sea level or by a density reduction factor from a table of altitude-density factors which can be found in many engineering handbooks.

INERT GAS GENERATORS

The application of inert gas generators to loop systems was described in Chapter 7. These systems use centrifugal blowers and reciprocating compressors for supplying inert gas to a pneumatic conveying system or to storage bins. The flow diagram for a typical low-pressure system is shown in Figure 13–10. Operation of the system is as follows:

Filtered air is compressed by a blower and measured by a fixed orifice. The amount of air is determined by downstream demand on the generator. Fuel gas—propane or natural gas—enters the regulators and passes through the limiting orifice. The amount supplied is determined by combustion air cross-loading the fuel-gas regulator. Air and

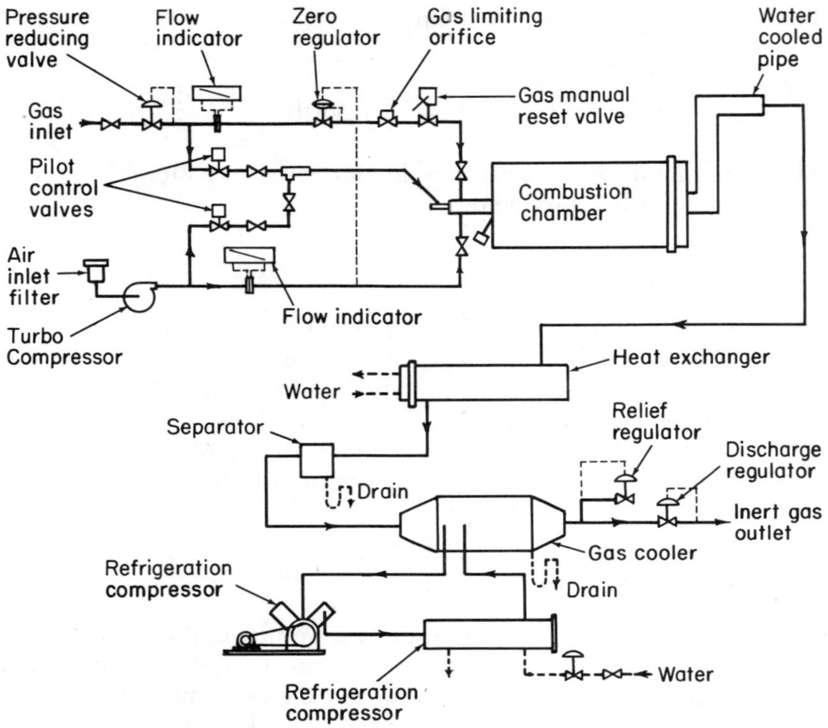

FIGURE 13–10. Flow diagram of a low-pressure inert-gas generating system. (The C. M. Kemp Manufacturing Company.)

fuel gas are piped separately to the burner, mixed and burned in the closed combustion chamber. After combustion, product gases are cooled in a direct water spray tower, or indirectly in a shell-and-tube heat exchanger.

Condensed water vapor is separated from the gas by a stainless steel demister and removed.

The packaged generator ends with the separator, directly downstream from the heat exchanger. A second packaged unit consisting of a refrigeration gas cooler is required to cool the discharged gas to a dew point of plus 40°F. If desired, the cooling unit may be mounted on the same base as the gas generating unit. Inert gas from this system is usually admitted to the suction side of the centrifugal compressor or blower in the conveying system. Where a single generator is used with a single conveying system, the airflow meter on the generator provides an indication of the amount of gas entering the system.

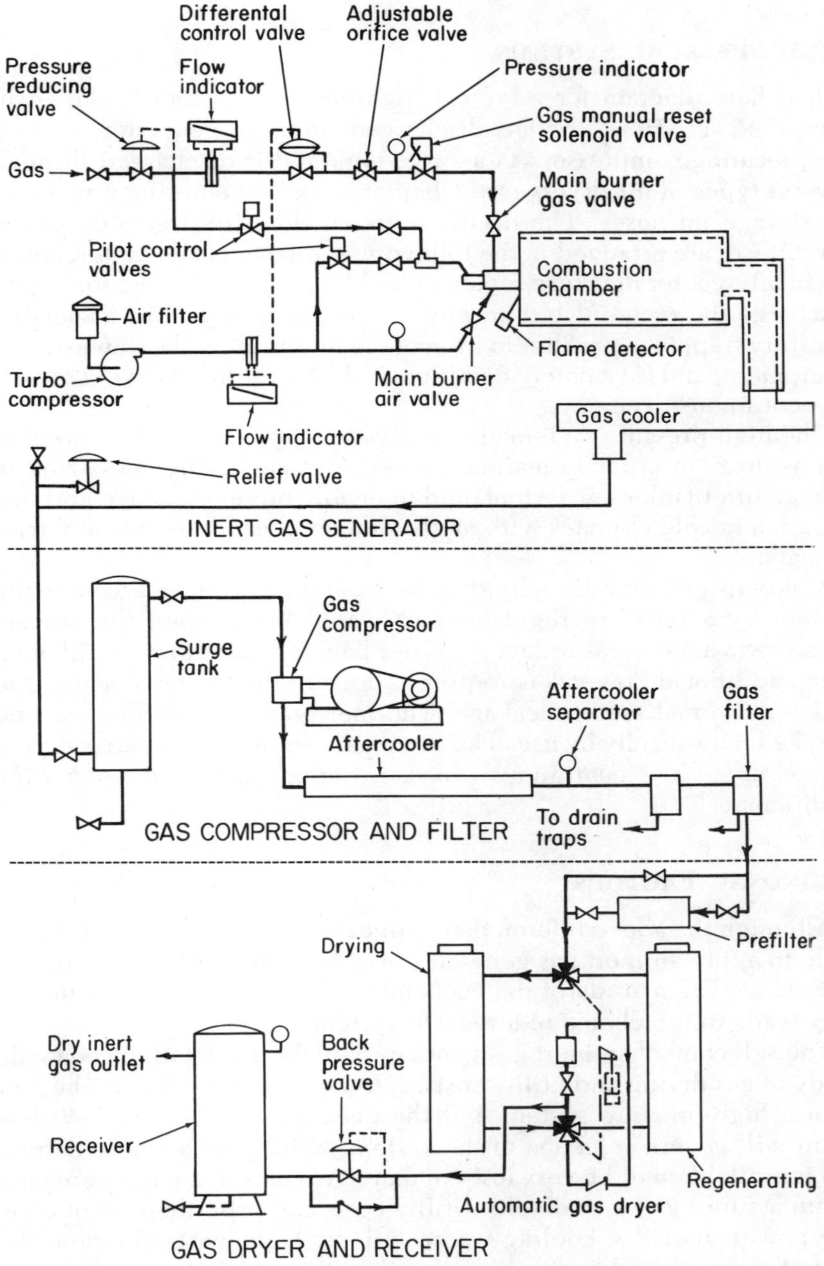

FIGURE 13–11. Flow diagram of a high-pressure inert-gas generating system. (The C. M. Kemp Manufacturing Company.)

High Pressure Systems

The flow diagram for a typical high-pressure system is shown in Figure 13–11. Here, the gas discharged from the generator enters a reciprocating compressor via a surge tank and is discharged through several types of air dryers (see Chapter 3) before entering a receiver for storage purposes. The drying units on the discharge side of the compressor are arranged in the following sequence: (1) An aftercooler; (2) an aftercooler moisture separator and its associated moisture trap to discharge the accumulated moisture; (3) a gas filter and its associated moisture trap; (4) a prefilter to protect the desiccant in the downstream equipment; and (5) a pair of continuous, desiccant-filled drying towers for continuous drying.

The high-pressure system almost always dries gas to a dew point of minus 40°F, in order to maintain a very dry gas in the conveying or storage bin blanketing system, and to enable outdoor storage and distribution in cold climates without the danger of condensation and frost formation.

Make-up gas may be admitted to a blanketing or to a conveying system by a pressure regulator that would open when the suction pressure to a blower is reduced, or by a flow regulator that would open when additional flow gas is required, beyond the quantity needed to make-up normal system leakage. The inert gas composition must be checked periodically by use of an Orsat gas sampler, or, where safety is paramount, by continuous oxygen analyzing and recording equipment.

Economic Factors

Although the above information applies to recycle systems, it is possible to apply an inert gas generator to a once-through conveying system. However, a study of the economics of these two systems invariably leads to the choice of a recycle system.

The selection of an inert gas generator involves a detailed economic study of gas drying and utility costs. A low-pressure system is cheaper than a high-pressure system, but the need for obtaining a low dew point will govern selection of the system. Utility costs where natural gas is available will be very low compared to costs in areas where only manufactured gas is available. Utility costs comprise the cost of electric power, fuel gas, cooling water and, for nitrogen generation, the cost of monoethanolamine (Mea) for absorption of CO_2.

14 Accessory Conveying Equipment

Pneumatic conveying systems are not always completely pneumatic. In many cases, they are dependent on other means of conveying for delivering material—from supply bins to rotary feeders; from receiving bins to process; from railroad cars or bulk trailers to inlet hoppers; or from receivers to storage bins. Almost any type of conveyor that can be enclosed in a dusttight housing can be adapted to convey material between these points, with only a few limitations that depend on the properties and characteristics of the material handled and conveying distance. The conveyors that are in this category are as follows: (1) The screw feeder and screw conveyor; (2) the air-activated gravity conveyor; (3) the natural-frequency vibrating conveyor.

THE SCREW FEEDER AND SCREW CONVEYOR

The most common type of conveyor or feeder used in conjunction with pneumatic conveying systems is the screw conveyor. Although the screw conveyor and feeder are well-known pieces of equipment, their usual design must be modified when used either to withdraw material from aerated bins or to deliver to the rotary feeder of a pneumatic conveying system.

A screw conveyor is used to transport material over relatively long distances and it usually operates with material well below the level of the shaft. The screw feeder normally is used to withdraw material at a definite rate from a bin or silo, and operates with its trough completely filled with material. Wherever possible, screw feeders should be installed below bin discharge connections so that they are completely out of the bin and not subject to pressure from the material within. This method of installation also permits a gate to be installed between the bin and the screw to permit servicing the feeder in the event of a breakdown.

Since the feeder runs full, its length should be no more than that required to meter the material properly. Beyond the metering point, the feeder should discharge into a conveying section having flights with a larger diameter or greater pitch, so that the material falls below the level of the drive shaft. The feeder screw must have a shroud at the point where the feeder meets the conveying section or it must be completely enclosed by a barrel-type housing.

Where bins are equipped with aeration pads, special care should be taken to reduce the clearance between the screw and housing to a minimum so that aerated material cannot flush into the conveying section and overload the screw-drive motor. When the material is of such nature that it would tend to pack in the bin or silo and must be kept "alive," it may be advisable to run the screw feeder through the bottom section of the bin or silo, rather than outside of it. This method is satisfactory for shallow bins, or tall bins having lightweight material (less than 70 lb/ft^3), but is not recommended for bins holding dense material.

When "live" screws are required, it is advisable to build a pressure-break baffle across the top of the screw and reduce the screw length to a minimum. It is also advisable to taper the screw so that material will be drawn uniformly from across the length of the bin. Otherwise, material at the rear of the screw will be withdrawn first and core its way through the remaining material. The preferred method of obtaining material from a tall bin requiring a "live" screw is to have the material discharge into the side of the screw feeder rather than have the screw located directly under a head of material.

When selecting a "live" screw-feeder design, it should be realized that any breakdown may require emptying the bin by some alternate method in making repairs.

On additions to existing equipment where more than one feeder must draw material from an existing bin or silo, it is possible to place one or more additional feeders through the bottom cone of the bin above an existing feeder so that they will act as "live" screw feeders.

To prevent flushing of aerated material through a feeder so that the discharge can be controlled, the shrouded section should enclose at least six pitches of flighting to impede the free flow of material. In order to do this in a limited distance, two schemes are used: (1) Fractional or short-pitch flighting within the shrouded section. This requires higher rotating speeds in order to equal the capacity of a standard-pitch screw. (2) Multiple-pitch flighting within the shrouded section. This comprises two standard-pitch flights wrapped around the same shaft, but having the same effect as fractional pitch insofar as resisting free flow of material is concerned, without the necessity of increasing the rotating speed of the shaft to obtain a given capacity.

Material should be fed into conveyors at a controlled rate so that the level of the material is always kept below the level of the shaft. Manual control by means of slide gates is not satisfactory since the operator has no method of determining the level of material in the screw. The conveyor should be designed for the process discharge rate, and the feed to the conveyor should be controlled so that its maximum conveying rate will not be exceeded.

Where the conveyor discharges into a series of bins through spouts not equipped with cut-off gates, the last bin should be equipped with a level indicator, or the conveyor trough fitted with a diaphragm type pressure switch arranged to shut down the conveyor when this bin is filled with material or the conveyor is jammed full.

When conveyors discharge into aerated bins, or feed pneumatic conveying systems, vent connections may be required on the troughs. For feeding pneumatic conveying systems, an automatically cleaned dust collector may have to be mounted on the screw conveyor housing.

Conveyor Bearings

Bearings recommended for powdered materials are as follows: (1) For internal hanger bearings, use open-type hangers with large, hard-cast-iron, nonlubricated bearings supporting hardened steel couplings. Hanger bearings generally are located clear of trough joints and inlet and discharge openings. (2) For exernal shaft bearings, use Dodge-Timken Type "C" pillow blocks or equivalent, mounted on a bracket welded to a steel-plate trough end, with a tell-tale space between the trough-end dust seal and the bearing.

Flanged cartridge-type bearings bolted to the trough ends may be used for conveyors to obtain a less costly installation, but these must be backed up by standard seal glands to prevent entry of dust. However, it should be realized that there would be no way of knowing when dust or powder is entering the bearing until a bearing failure occurs. There are flanged cartridge type bearings available that provide a tell-tale space between the dust seal and the bearing and make a more acceptable installation. Normally, the discharge end of a screw feeder or conveyor is not subject to material pressure, unless the down stream equipment is blocked up. The flanged bearing is safer to use at this end than at the opposite end that is always under pressure.

A screw conveyor or screw feeder is normally driven from the discharge end so that the shaft is always in tension. This requires that the drive-end bearing be a fixed bearing to take the shaft thrust and that the end-shaft bearing be a floating bearing so as to permit elongation of the shafting due to thermal expansion and thrust. Where possible, the bearings should bear against a shouldered collar or be keyed to the shaft so that the shaft cannot be pulled through the bearing-adapter mounting.

Outboard pedestal bearings are preferred and should be designed in such a manner as to permit the free discharge of powder that may leak through the dust seal at the trough end. Where the screw housing may be subjected to air pressure from aeration pads of bins or from back leakage of air from pneumatic conveying systems, a shaft slinger should be mounted on the shaft between the bearing and the dust seal so that the air leakage from the dust seal will not blow powder directly into the bearing.

End shaft bearings should be provided with end covers, except where shafts are extended for power takeoff to other conveyors. Trough ends with rigid, welded-steel outboard pedestals should be used for supporting pillow-block-type bearings. They should be fitted with stuffing-box-type seal glands, with gland followers having no more than two bolts. The stuffing box should hold at least three packing rings.

Other Design Considerations

Additional construction modifications that are sometimes made to adapt the screw conveyor and feeder for application to a pneumatic conveying system are as follows: (1) The angle iron flanges used at trough joints and along the top to receive a dust-tight cover should be continuously welded to the trough. Standard practice is to use intermittent welds. (2) The flights on feeders or conveyors that may jam full of material should be continuously welded to the shaft. Here again, standard practice is to use intermittent welds. (3) Where close clearances are required for the full length of a section of conveyor (10 to 12 feet), a section of heavy piping may be split and adapted to form the screw housing, otherwise, a special trough may be required, shrouded for its full length. Standard clearance between the screw flight and trough is as much as ⅜ to ½ inch. This is too much for handling aerated materials. (4) Mounting plates for flanged bearings should be continuously welded to trough ends. The plate should be drilled and tapped to receive cap screws or studs in order to obtain a dust-tight bearing mount.

THE AIR-ACTIVATED GRAVITY CONVEYOR

This type of conveyor is essentially a sloping rectangular duct with an air-permeable porous material forming its bottom. Below the porous medium, for its entire length, is a channel-shaped plenum chamber to which air is admitted (Figure 14–1). In operation, air enters the

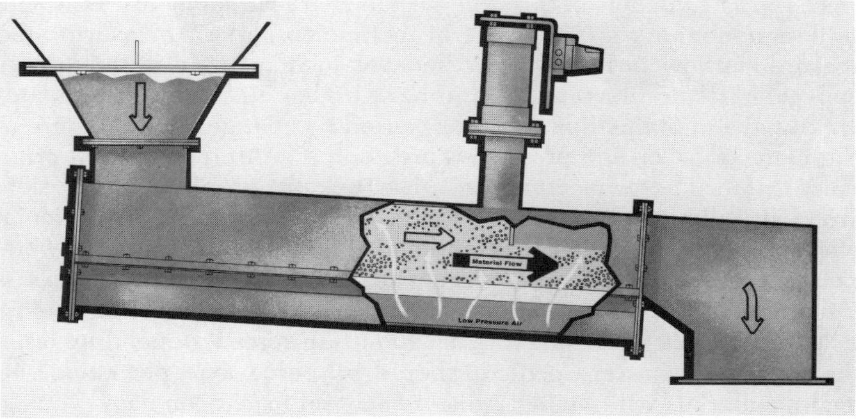

FIGURE 14–1. Section through an air-activated gravity conveyor. (Kennedy Van Saun Corporation.)

plenum chamber and is distributed uniformly through the porous membrane; material entering the duct, above the membrane, is aerated and the material starts to flow through the duct. The principle of operation is, basically, the fact that certain aerated powdered materials have a lower angle of repose than the deaerated material and try to seek that angle, however, since the slope of the conveying duct is always greater than this angle of repose, the material will continue to flow. The addition of air to powdered materials to create this condition is sometimes referred to as fluidization, since the powder displays characteristics similar to those of fluids, such as (1) exerting a pressure on the bottom and sides of a vessel equal to the head of powder above the point of measurement and thus assuming a horizontal surface, (2) finding the same horizontal level for a given head regardless of the shape of the container, and (3) flowing freely through a conduit and issuing in the form of a trajectory. These properties are commonly observed on some materials when collecting them from the discharge hoppers of dust collectors. The material will appear to be boiling, as bubbles of air break the surface, and the dust will hold the air for a considerable period of time.

LIMITATIONS

The use of the air-activated gravity conveyor is limited to the so-called fluidizable materials. The only way of determining the conveyability of any material on a particular type of porous membrane is to test it. The vendors of this type of conveyor maintain a test unit with provisions for varying the angle of inclination and for measuring the required airflow and pressure. Whenever such an air-activated gravity conveyor fails to function, or requires a higher air pressure or greater air flow than obtained on test, it is caused by changes in the nature of the material when the properties are compared to the material originally tested. The two factors that cause the most problems are particle size distribution and moisture content. The particle size distribution of material to be conveyed by this method must vary from coarse, say 100 percent through a 20-mesh screen, to fine. The percent of fines is important and vendors require that the fines be a minimum of from 20 to as much as 60 percent through a 200-mesh screen, depending upon the material. Moisture content should be below one per cent, but, again, materials with higher moisture content have been conveyed. In a few cases, non-conveyable material has been made conveyable by mixing it with fine material, if too coarse; or by mixing it with dry material, if too wet.

Since one of the purposes of using this type of conveyor is to obtain near-horizontal conveying, the angle of inclination is made as small as possible, consistent with low associated airflow and pressure requirements. The rate of material flow for any given inclination tends to be reduced to a minimum regardless of increase in air flow for a given size of unit. Variations in air pressure and airflow tend to have a greater effect on conveying than changes in angle of inclination for a given size of unit. Material flow can be stopped readily when air flow is cut off and re-started when air flow is resumed, even where large sections of the porous membrane are not covered with material.

Air Distribution

The airflow required to convey a particular material is determined primarily by the fluidization characteristics of that material. The nature of the porous medium determines the air pressure required to produce the required airflow across it. The airflow across any air-permeable membrane varies directly as the pressure drop across it. Although all available porous media can pass air, the uniformity of distribution of air flow across the surface is the most important feature of the commercial air-activated gravity conveyors, and their manufacturers try to preserve the secret of obtaining this uniformity. The pressure drop for a given airflow should be determined for each type of porous membrane so that the penalty, in terms of extra horsepower requirements, can be determined. The usual air-activating surfaces that are commercially available are cast ceramics, such as aluminum oxide and silicon carbide; urethane rubber; woven fabrics such as cotton and silicone treated cotton; and, for hot materials, laminated asbestos with stainless steel wire reinforcement and aluminum oxide or silicon carbide depending upon the temperature of the material handled. Other porous media are available and are used as aeration pads for inducing flow in bins and hoppers, but these may be adapted to form an air-activated gravity conveyor for short runs of chutes. These materials may be woven canvas belting; porous teflon, sintered bronze or sintered stainless steel. Their selection may be based more on their resistance to chemical attack rather than to their uniformity of air distribution. The permeability of any porous membrane can vary with time due to build-up of atmospheric dust on the air side of the membrane; or smeared material and moisture effects on the conveying side of the material. Cotton fabric can become mildewed and hole through, or lose porosity when wet thus blocking airflow. Cotton or other fabric type membranes may be pretreated to prevent mildew and prevent moisture absorption. Such pretreatment will extend the life of the fabric and permit washing it down with water.

The activating air may be supplied at any point along the conveyor plenum, even for very long conveyors. The air may be supplied from a centrifugal fan or from a positive-displacement blower, depending upon pressure and flow requirements. The use of a central air supply unit for a number of air-activated gravity conveyors requires special precautions to secure uniform air distribution to each unit. The pressure and flow requirement of each unit must be as determined by test at the vendor's laboratory. An orifice may be installed at the air inlet to each plenum chamber, sized to obtain the airflow required for the number of square feet of membrane in the unit. In some cases, butterfly type control valves are used at the inlets in order to regulate the airflow, but exact conditions within each plenum on an extensive system cannot be determined accurately. The best method of air distribution is to provide an air supply unit for each related group of conveyors that require air simultaneously. An alternate scheme is to use one air supply unit for several groups of conveyors and to install a properly sized orifice at the inlet to each plenum, but with a by-pass arrangement around each orifice, containing an adjustable butterfly valve. This method will provide for manual adjustment to suit any eventuality. Excessive airflow may classify the material and result in the fines being picked up in the vent line.

The prevention of blockage of the porous medium due to the accumulation of atmospheric dirt on the underside is so important that special attention should be given to selection of the suction filter on the air supply unit. A wire mesh type of filter is inadequate for this service, since large particles of dust can be pulled through it. The best type is an oversized, washable, dry-type filter having wool felt media with deep accordion pleats, armored with heavy galvanized steel cloth. A spare filter element should be available so that a dirty element may be removed and cleaned without the necessity of shutting down the conveying system. This type of filter can be cleaned by back-blowing the element with an air hose, by shaking off the dust, or by washing in a commercial solvent.

The activating air on the conveying side of the porous medium must be vented. This can be done anywhere along the length of the conveyor, but venting downstream is preferred, since the receiving hopper, bin or other equipment is generally vented and no special connections on the conveyor would be required. However, on long conveyors and conveyors where material may pack at the outlet, the vent connections should be at the inlet end of the conveyor.

Figure 14–2 shows a special application of an air-activated gravity conveyor. In this case, a near-horizontal offset in a vent line accumulated material and plugged the line, creating a dust nuisance. Replacing the duct with a section of an air-activated gravity conveyor (sloped

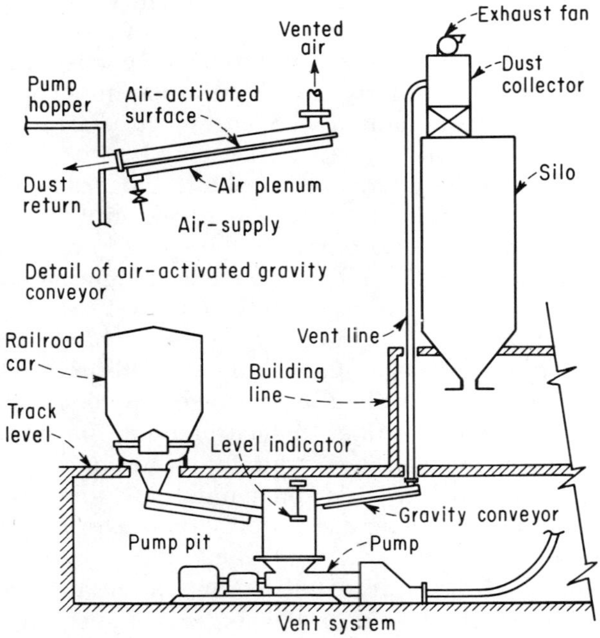

FIGURE 14–2. Special application of an air-activated gravity conveyor.

about 8 degrees to suit the material handled) eliminated plugged vent lines permanently. The air-supply valve was normally throttled to the minimum required to keep the vent line clear, but was opened wide to clear out the conveyor when an overfilled hopper backed material into it.

Flow Gates

Material flow into the conveyor from fully loaded hoppers, bins or transports must be controlled by flow gates. These are dusttight side gates, operated like an outside screw and yoke gate valve for fluids, but seating against the porous medium and enclosed in a dusttight housing (Figures 14–3 and 14–4). In some cases, dead material at the bottom of a bin may require a flow feeder and a weir box to start the flow. The feeder section is air-activated to fluidize the dead material and start flow; excessive flow is prevented by forcing the material to flow under a sluice gate and then over the weir. When overfilling of the conveyor must be avoided as when receiving material from gravity discharge transports, a flap gate is installed at the inlet and held in position by a quadrant and thumbscrews on each side of the conveyor duct.

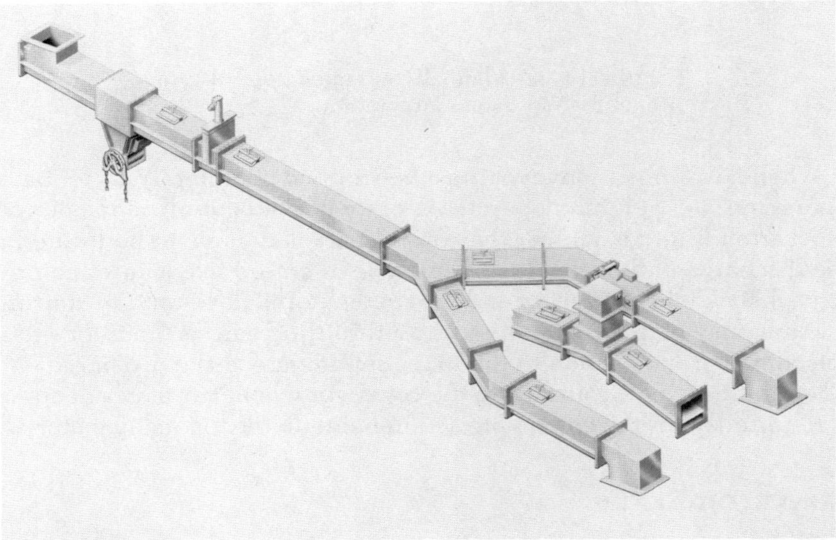

Figure 14–3. Arrangement of an air-activated gravity conveyor, showing the application of accessory equipment. (Kennedy Van Saun Corporation.)

FIGURE 14–4. **Manually operated control gate. (Kennedy Van Saun Corporation.)**

When flow from a conveyor must be stopped completely, as for batch operations, for weighing operations, or for overfill cut-off on the hopper of a carton filling machine, the control gate may have to be located at the discharge end of the duct. This type of cut-off may cause material to pack inside the conveyor and prevent material flow upon re-starting. Packing may be prevented by using a throttling gate at the inlet end of the conveyor in addition to the main cut-off gate at the discharge end. Also, shutting off the airflow to the conveyor whenever material cut-off is required stops the conveying action immediately on many materials.

Conveyor Bends

The material flow may be guided in any direction by use of horizontal bends or curves. These, too, must maintain the same angle of inclination as the straight sections of conveyor (Figures 14–3, 14–5 and

FIGURE 14–5. Corner piece. (Kennedy Van Saun Corporation.)

14–6). Thus curves and bends must be specified as left hand—curving downward counterclockwise, or right hand—curving downward clockwise when viewed from the center of curvature. Flow can be directed in two, near-horizontal, directions from one conveyor by use of a Y-piece, or can be directed to either of two, near-horizontal, directions by a stream diverter.

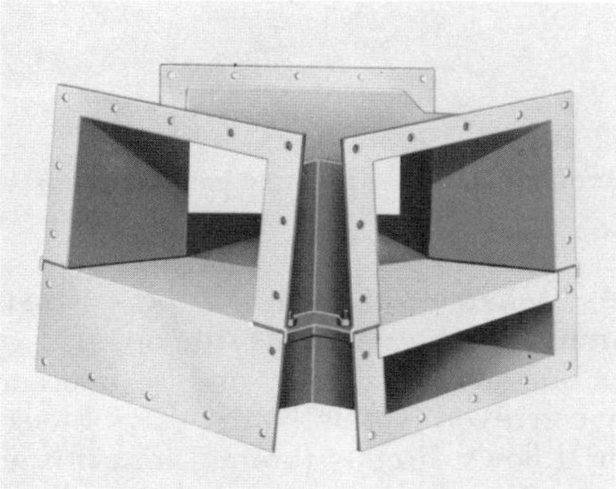

FIGURE 14–6. Y-piece. (Kennedy Van Saun Corporation.)

Spouts and Diverters

Inlet and outlet spouts are made to suit the angle of inclination of the conveyor. When spouting leads from a conveyor facing in one direction to a conveyor leading in another direction, the spout is pitched downward at an angle in the direction of flow of the downstream conveyor, rather than chuted vertically into it. Material may be discharged from one conveyor to two or more discharge points by use of side discharge valves or boxes. See Figure 14–7. When

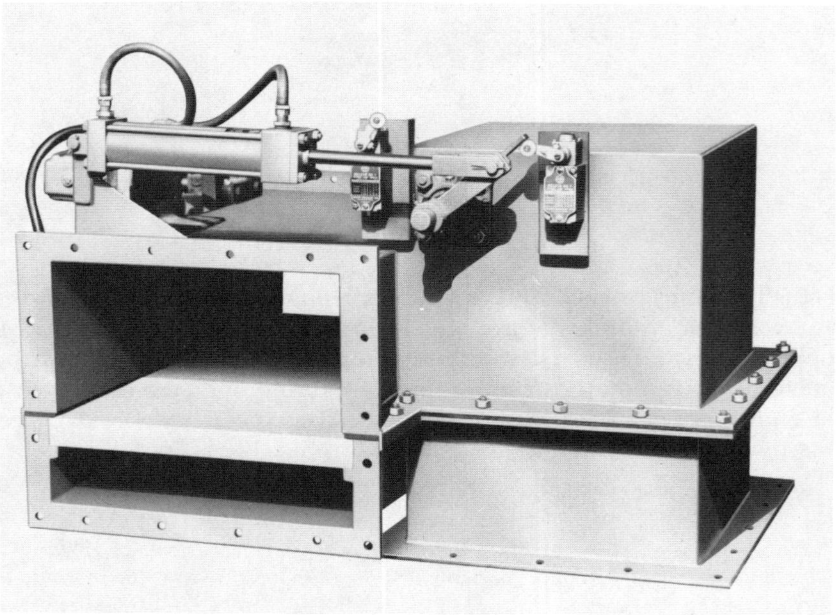

Figure 14–7. Double-gate side-discharge box. (Kennedy Van Saun Corporation.)

operated, a side discharge valve blocks the downstream section of conveyor and shunts the material to the side opening where it can drop vertically into a bin, a flow-splitter, or into another section of air-activated gravity conveyor. Normally, the valve blocks the side opening so that material flows directly through it as if it were just another section of conveyor. Two types of side discharge boxes are available. One uses a single flap gate pivoted on the downstream side of the side opening. In normal flow, the flap gate blocks the side opening; in diverted flow the flap gate is pivoted across the flow path and seals against the opposite side of the duct at such an angle as to shunt the material into the side opening. Another design uses a double set of flaps at right angles to each other and pivoted at the top of the duct. In normal flow, one gate in the straight flow path is open—pivoted upward; the other gate blocks the side opening—pivoted downward. In diverted flow, the gate in the straight flow path is pivoted downward to block forward flow and the gate at the side opening is pivoted upward to permit side flow into the discharge box. Both types of gates may be operated manually or by electric or air-cylinder operators.

Inspection Door and Ports

These are required in the top of each section of conveyor at strategic points, such as on either side of flow gates, after bends or flow diverters, over discharge boxes, and after flow feeders. These permit checking for flow conditions at these points and presence of materials that prevent flow such as lumps, foreign or wet materials. The airflow should be off when these doors are used, even then, care should be taken when opening these to prevent uncontrolled flow of aerated material through the door. The lower end of each plenum chamber should also be fitted with an inspection door to permit inspection of the underside of the porous medium for excessive atmospheric dirt, for material leakage through the porous medium, and for the presence of water.

System Capacity

The greatest advantage of the air-activated gravity conveyor is its high discharge capacity for its relatively small size and the absence of moving parts. Vendors tables make no reference to angle of inclination, airflow, or air pressure, but are intended only as a general guide to its possibilities. Actual capacities, as stated before, must be determined by test. Since most tests are made on 4-inch- or 6-inch-wide conveyors, the results of these tests can be extended to determine the capacity at other conveyor widths by noting that the conveying capacity varies approximately as the square of the conveyor width. Thus, doubling the width increases the capacity about four times for the same airflow rate per square foot of porous surface at a given pressure. The amount of mechanical equipment required to operate the conveyor is small and near-horizontal conveying can be used to clear many interferences within a plant by changing its slope or by changing the direction of flow. See Figure 14–8.

FIGURE 14–8. Installation of air-activated gravity conveyors, showing changes in direction. (Fuller Company/GATX.)

Design Considerations

The disadvantages of the air-activated gravity conveyor are its limitations in regard to the type material it can handle, the possibility of blockage of the pores of the porous membrane, the difficulty in cleaning the porous membrane, and the time and care necessary in replacing the porous membrane.

The following details must be given consideration when applying an air-activated gravity conveyor: (1) When material is supplied from an open hopper or transport, venting should be at the outlet end of the conveyor and the material feed must be throttled at the inlet end to ensure this direction of airflow. Otherwise, dust will pour out of the supply hopper or transport hatch. (2) When a conveyor is selected on the basis of a minimum angle of inclination, make certain that clearance is available to increase this angle in case a change in material characteristics occurs. (3) When material is supplied from full hoppers or bins, provide means for controlling or limiting flow into the conveyor, otherwise material may pack and block the conveyor. (4) When material flow must be cut off completely, arrange to stop airflow to the conveyor in addition to using a gate to stop material flow. (5) Provide an independent air supply unit for each group of conveyors serving a particular process, if possible. (6) When the powder may contain foreign material or tramp iron, provide a material trap (Figure 14–9).

FIGURE 14–9. Material trap. (Kennedy Van Saun Corporation.)

NATURAL-FREQUENCY (VIBRATING) CONVEYORS

A natural-frequency conveyor consists of a spring-supported trough or pan that is reciprocated by a motor-driven eccentric and connecting rod at a speed that is tuned to resonance with the natural frequency of the system comprising the vibrating mechanism and its contents. The springs are uniformly displaced along the trough and set at an angle so that forward motion also gives the conveyed material an upward thrust. Backward motion causes the material to fall on the pan, but downstream from its original position. The impulse of the drive arm is tuned so that it is synchronized with the forward impulse of the springs and becomes simply an exciter to maintain the springs in motion. The energy in the springs is uniformly applied along the trough thus driving the material forward at a uniform rate.

Out-of-phase forces that are created during starting and stopping are absorbed by shock absorbers connected to the drive arm by a system of levers. The shock absorbers compensate for change in vertical elevation of the pan between no load and full load. The drive arm and pan connector bushings are made of rubber to further absorb these out-of-phase forces.

The pan can be totally enclosed (Figure 14–10) and can handle many powdered materials that cannot be handled by other forms of

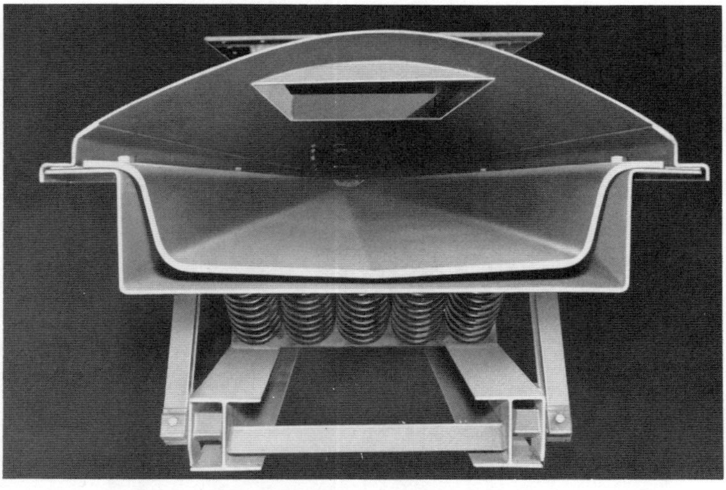

FIGURE 14–10. Vibrating conveyor trough for hot materials. (Carrier Vibrating Equipment Division of Rexnord, Inc.)

conveyors. Its only disadvantage, disregarding cost, is the fact that it requires flexible sock connections at each inlet and discharge spout to accommodate the reciprocating motion. These connections can be made airtight by using hypalon, or neoprene impregnated nylon sheeting to fabricate these socks and fastening them to the conveyor spouts with standard band clamps. A felt-lined cuff at each connection helps to reduce wear caused by the motion of the conveyor. All stitching of seams on these materials should be made with nylon thread.

Capacity of System

The pans are sized to advance the material at a definite rate determined by test for each type of material conveyed. Bulk density, conveying speed and bed depth, as obtained by test, determine the size of conveyor pan required to obtain a given rate of flow. Since the conveyor is a volumetric device, the pan must be sized to accommodate the largest surge load expected in service. Material must be delivered to the conveyor at a uniform rate, or in small increments as from a dust filter hopper, so that it does not become plugged with material. Plugging can cause overstroking and break the conveyor parts. One method of preventing plugging by avalanching material is to insert a flat plate across each inlet connection so that material flow into the trough is restricted and can be discharged off the plate at a limited rate.

Although these conveyors are intended for horizontal installation, they may sometimes be installed pitched downward at a slight angle toward the discharge end in order to increase the capacity of a marginally sized unit or to decrease the amount of material adhering to the pan, or the amount of residue in the pan at shutdown.

The only other way of changing the capacity of these units is to change the amplitude of the eccentric and the spring constant of the entire system of trough supports.

Conveyor Drive

A stroke plate is usually attached to the vibrating pan to guide the setting for natural frequency and to detect system troubles (Figure 14–11). At rest, the stroke plate shows two parallel, vertical lines displaced by a distance equal to the stroke of the eccentric. At natural frequency, three lines are visible, one elongated center line with a shorter line and a diagonal pattern on each side of it. At understroke, the center line is split and the diagonal patterns are separated by a gap so that four lines are visible. At overstroke, the center line is split and the diagonal patterns overlap so that four lines are visible. A variable pitch drive sheave is adjusted to bring the drive to its natural frequency.

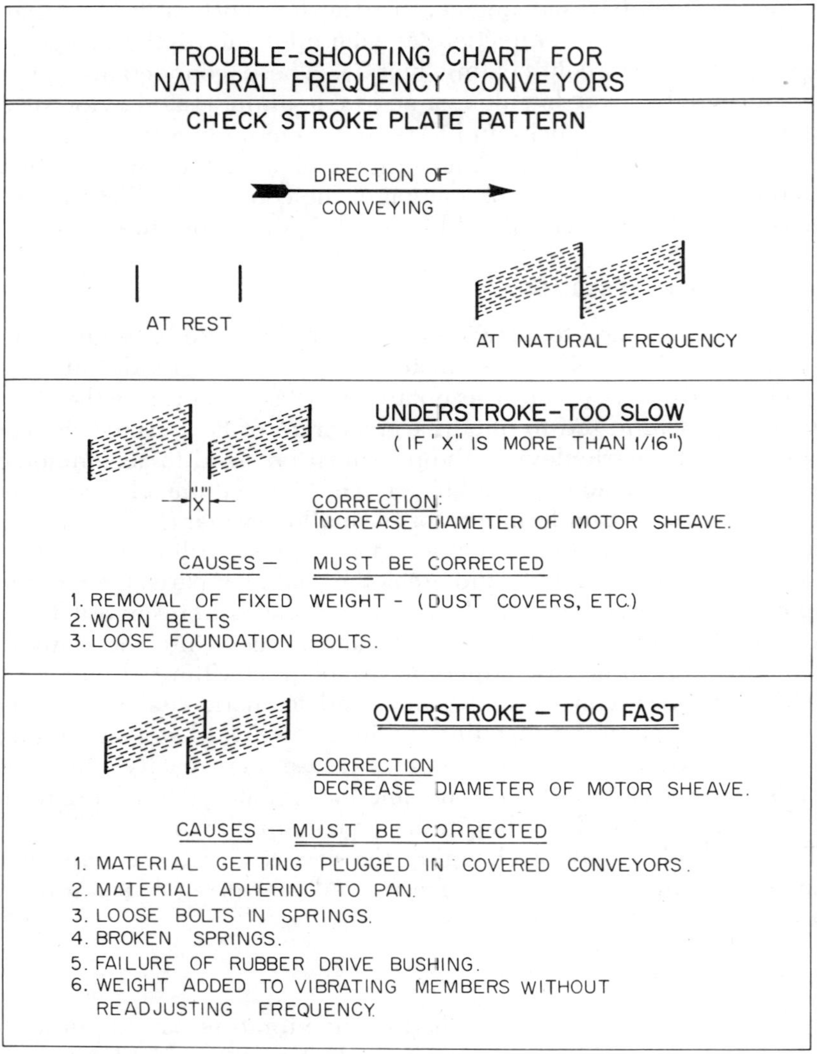

FIGURE 14–11. Trouble-shooting chart for natural-frequency conveyors. (Carrier Vibrating Equipment Division of Rexnord, Inc.)

STRUCTURAL SUPPORTS

The design of supports is dependent upon conveyor location since they must absorb the dynamic reaction of the spring system. These forces are usually absorbed by the earth for conveyors installed on or at grade level. However, when a conveyor is to be installed on a building frame, consideration must be given to isolating or counterbalancing the dynamic reaction. The various types of conveyor supports are shown in Figure 14–12. The frame of the standard conveyor rests directly on the floor or on a base that can absorb the reactive forces.

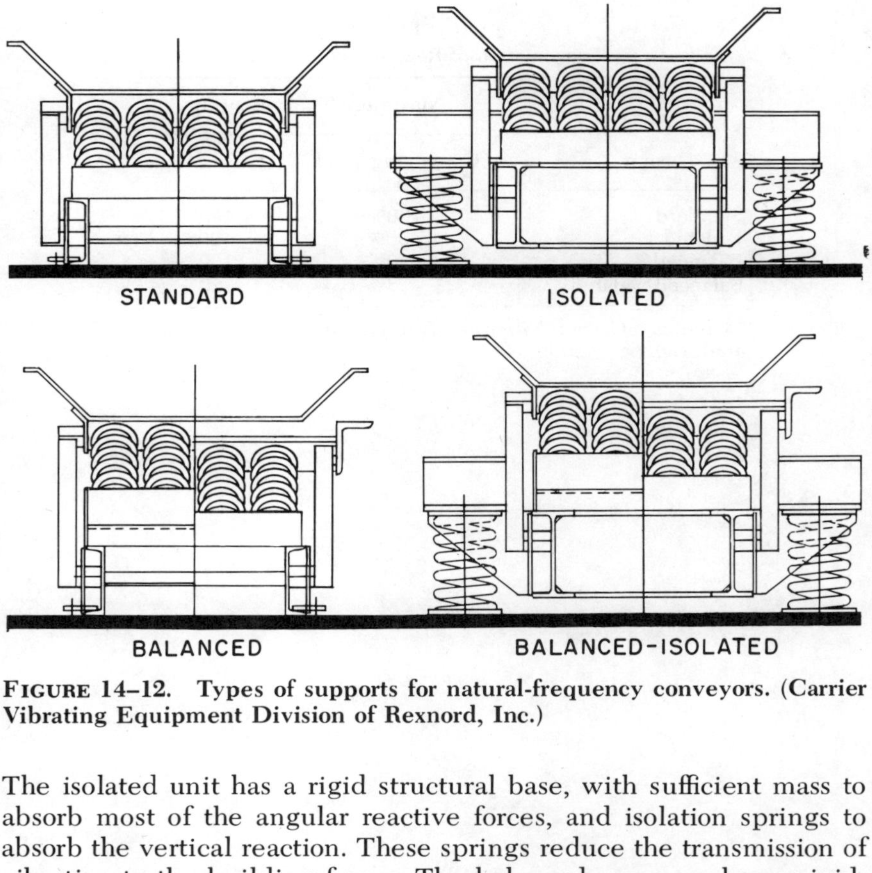

FIGURE 14–12. Types of supports for natural-frequency conveyors. (Carrier Vibrating Equipment Division of Rexnord, Inc.)

The isolated unit has a rigid structural base, with sufficient mass to absorb most of the angular reactive forces, and isolation springs to absorb the vertical reaction. These springs reduce the transmission of vibration to the building frame. The balanced conveyor has a rigid, continuous counterweight frame running the full length of the conveyor, on each side of the pan, and mounted on a duplicate spring system tuned to resonance with the vibrating pan frequency. This frame is positively vibrated 180 degrees out of phase with the conveyor frame by separate connecting arms on the common drive shaft. This arrangement ensures that the thrust of each set of trough support springs is countered by the opposite reaction of each set of balancing frame springs. If maximum reduction in vibration transmission is desired, the balanced type of conveyor may also be isolated. All isolated conveyors, regardless of type, must be designed to suit each specific installation.

The degree of transmission of reactive forces to the supporting structure is given in Table 14–1.

TABLE 14–1
Transmission of Reactive Forces*

	Maximum Transmissibility	
Design	Conveying Deck	Material Load
Standard	100%	100%
Isolated	20%	20%
Balanced	10%	100%
Balanced-isolated	2%	20%

* Courtesy Carrier Vibrating Equipment Division of Rexnord, Inc.

15 Electrical Controls and Control Devices

The requirements for design, installation, materials, apparatus, conductors and wiring methods of electrical distribution systems are completely covered in the National Electrical Code prepared by the National Fire Protection Association. Compliance with the Code requirements is mandatory in most municipalities and deviation from these requirements is limited and requires special permission of a local electrical inspector. The wiring for control systems must also comply with the Code requirements, but the design of control circuitry to meet the functional requirements of a pneumatic conveying system is limited only by the designers knowledge of available control equipment, his imagination and his ingenuity.

The controls available today range from conventional, heavy-duty, industrial type equipment to the most modern modular plug-in type relays and solid state miniaturized electronic switching components. Components that can meet the functional requirements of a pneumatic conveying control system are available in both classes of equipment, but the selection is dependent on such factors as (1) ready availability of replacements and parts; (2) type and training of maintenance personnel available for maintaining the equipment; (3) availability of servicemen for trouble-shooting if plant labor is not available; (4) current-carrying capacity of available components.

ELECTRICAL CONTROL SYSTEMS

The cost of electrical controls and wiring represents almost fifty percent of the capital investment for many pneumatic conveying systems. The simplicity of the piping required to convey materials to remote areas of a plant tends to shadow the complicated and extensive wiring sometimes required to reach equipment in those areas. Also, the electrical controls for a system are so intimately related to the mechanical functions that a knowledge of this subject is required for proper preparation of plans and specifications, analysis of vendors quotations and drawings, operational analysis of the system and trouble-shooting. It is also imperative that the purchaser of the system should know his control requirements and prepare basic plans and specifications to guide the vendor. The rather elementary treatment of electrical control requirements for pneumatic conveying systems contained herein should enable a purchaser to discuss the subject intelligently with prospective vendors.

The best way to select controls for a pneumatic conveying system is to prepare both a list of basic mechanical operations required to start and stop the system and a list of functional requirements desired by the production department to facilitate their operations. A preliminary wiring diagram that translates these requirements into an integrated control system will reveal the complexity of what originally appeared to be a simple conveying system. It is not unusual to find that a major portion of the final cost of installing the system is for the electrical controls.

Types of Control Systems

The basic types of control systems are manual, semiautomatic and fully automatic.

Under manual control, level indicators and alarms signal the operator that a process bin requires material. The operator then positions gates to permit material from the source of supply to enter the system and be directed to the point of discharge; he then pushes buttons to start each piece of conveying equipment. Upon receiving an indication or alarm that the receiver at the point of discharge is full, he proceeds to shut down the conveying equipment in reverse order to the starting sequence, allowing the proper time delay to permit airflow to clean out the conveying line.

Under semiautomatic control, the operator aligns equipment to permit the flow of material from the source of supply to the point of discharge, and then pushes a single button or turns a selector switch that initiates the starting of all other equipment in the proper sequence. When the receiver at the point of discharge is full, the high-level indicator shuts down all equipment in the system in the proper sequence, and with the proper time delay to permit airflow to clean out the line. The operator must then reposition gates to their original setting prior to startup, or to the setting required for a different operation.

In a fully automatic system, the low-level indicator in the receiver or bin at the point of discharge initiates startup of the system. All gates required for flow of material—from source of supply, through the conveying system, and to the receiver at the point of discharge—are positioned automatically. The conveying equipment is then started automatically in proper sequence. Upon signal from a high-level indicator that the receiver at the point of discharge is full, the equipment is shut down in reverse order to the starting sequence, again with the proper time delay to permit airflow to clean all material out of the line.

The sequence of operations for all these control systems is basically the same and is as follows:

1. Set controls to type of function to be performed, such as unloading, reclaiming, or batching.
2. Position diverter gates to direct material to proper destination, or move hose to proper connections.
3. Start air movers to establish flow through the conveyor pipelines, or into a blow-tank, and the dust-filter vent system.
4. Start equipment that feeds material into the conveying system—downstream equipment first; upstream equipment last.
5. When high level is indicated at the destination, stop upstream material feeders first and downstream last. Allow adequate time for complete cleanout of auxiliary conveyors.
6. Allow a preset time for airflow to clear out the conveying line. Then stop the air movers, including dust-filter exhausters. In some blow-tank systems, residual air will clear the line; in others, the air mover may have to be bypassed.
7. Isolate the material from the conveying system by closing bin or feeder gates.

The control system incorporates such devices as time-delay relays, level alarms and indicators, pressure-differential switches, motion switches, solenoid valves, electropneumatic relays, and reed switches. The complexity of even a simple system is revealed by the electrical diagram. To comply with National Electrical Code requirements, the cost of electrical controls and wiring, for many systems, might be almost 50 percent of the total capital investment.

Typical Control Functions

An exact statement of required functions must be specified for manual, semiautomatic and automatic control so as to leave no doubt as to what the vendor is to furnish. Some typical requirements are as follows: (1) pushbutton start-stop of each motor-driven unit; (2) automatic positioning of gates and valves prior to startup; (3) automatic shutdown at high level in a particular bin; (4) automatic startup at low level in a particular bin; (5) conveying to stop when certain specified operating conditions occur; and (6) dual control from certain operating stations, with master control at a particular operating station.

Almost any functional requirement can be incorporated into an electrical control system by judicious use of the vast array of electrical and electronic relays and components available. In most cases, all these devices are inserted into a low voltage control circuit designed to energize a particular solenoid only when certain functional requirements are met. The armature of the solenoid then actuates a contractor to energize a higher voltage circuit to run a motor, to apply a brake, or to open or close a group of contacts and interlock with one or several additional control circuits. In general, each circuit will contain a relay coil or a motor starter coil with pushbutton contacts, relay, and control device contacts arranged across the supply line to open or close the circuit and actuate equipment to fulfill a functional requirement. Some of these contacts may be in a control panel, such as pushbuttons, selector switches or sequencing relays, others may be on external control devices, or on remotely located pushbuttons and relays.

Power Supply

Since three phase alternating current is the predominant power source in this country, discussion of control circuits will be confined to this application. The control principles involved can easily be extended to other sources of electricity. The control circuit for most three-phase a-c motors is generally single-phase a-c taken from across two of the supply wires. This circuit always includes the motor starter coil, the overload trip contacts and a pushbutton station or other control device to energize the circuit. Obviously, all devices in the circuit would have to operate at the supply voltage. The control voltage should be lower than the supply voltage—preferably 110 volts—for safety and to include a greater variety of control devices that are not available for the higher voltage. Two methods are available to obtain the lower voltage: (1) an individual control transformer and a low-voltage holding coil may be furnished for each motor starter; or (2) a single control transformer for energizing the entire control system may be furnished and each motor starter equipped with a low voltage holding coil. The transformers should be sized to supply all current-drawing control devices that are included in a particular circuit. In developing a control circuit that includes a motor-starter solenoid, the overload trip contacts should be inserted between the solenoid and the supply line in such a manner that they can not be by-passed.

COMPONENTS OF CONTROL SYSTEMS

Pushbutton Control Stations

In addition to the standard motor start-stop stations, many other types of pushbutton control units are available with multiple contacts having various combinations of normally open or normally closed contacts or with momentary or maintained contacts. The most common of these are the lock-stop and jog buttons for motor control.

The lock-stop button must be used within sight of all remotely controlled motors and is fitted with a locking device to keep the circuit open. This permits personnel to work on the motor safely. It has one set of normally closed contacts that should be connected to the fused supply line, ahead of all other control devices.

The jog pushbutton is used to by-pass a series of contacts in a circuit (except overload trip contacts) in order to operate a motor independently of all other equipment. It has one set of normally closed (N.C.) contacts and one set of normally open (N.O.) contacts. When pushed, the N.O. contacts are closed and held closed as long as pressure is maintained on the button, the N.C. contacts are opened simultaneously. Releasing the button stops the motor and restores the circuit to its normal condition. The jog button should be connected ahead of all control contacts that are to be by-passed, but after a lock-stop pushbutton, if one is installed.

Jog buttons should be installed on the following types of equipment: (1) On feeders that withdraw material from the source of supply. Interlock them so that the feeders cannot be operated unless conveying air is flowing through the system. (2) On remotely operated and controlled diverter or flow gates, so that limit switches on these units may be properly adjusted in an emergency. Normally, a manual operating device should be provided for this purpose. (3) On shaker motors of dust filters—to clean bags before entering the unit for servicing. (4) On all screw conveyors—to relieve jamming, or to empty them before opening the troughs for servicing or inspection. (5) On belt-driven fans or blowers—to check belt tension without starting up the entire conveying system, or to clean out a conveying line with the material feeder stopped.

Selector Switch Units

Selector switches may be obtained in various forms and types depending on the number and position of the contacts. Their purpose is to maintain a selected contact or group of contacts in the closed or open position to perform a related series of control functions, but permit the rapid switching of the contacts and re-grouping to perform a different series of control functions, when the same mechanical equipment is used to handle several materials. The most common selector switches are the two- and three-position cam-operated types in which the turn of a knob to left, right, or center makes one set of contacts and breaks another set. The center position, when used, is usually "off"—all contacts are open. Another common type, that is functionally similar, is the selector pushbutton unit in which a sleeve around the button is turned in one direction to permit the button to function normally, or turned in the opposite direction to hold the normally closed set of contacts open, so that depressing the button only closes the normally open set of contacts momentarily. The standard uses for these types of switches are for motor control functions as follows: (1) Switching from "off" position to "manual" or "automatic" control. (2) Switching from "off" position to "forward" or "reverse." (3) Switching from "off" position to "open" or "close." (4) Switching from "safe" position to "jog" or "run." The two-position switch equivalent for these functions eliminates the center or "off" position.

When the same pieces of equipment are used for conveying several materials, the controls must be interlocked in such a manner that only equipment required to convey each specific material will be in operation. This is done by using a special drum type selector switch with multiple rows of normally open contacts. These contacts are closed in a pre-determined sequence by merely turning a knob to the proper position as marked on the face of the switch. The closed contacts in each position of the drum can be selected as required to suit the circuitry and permanently recorded on a selector switch development diagram for the type of switch used. This diagram should appear on all schematic wiring diagrams to aid in maintenance of the system.

Use selector switches to perform the following preliminary operations: (1) Setting flow-diverter gates to the proper position. (2) Opening or closing gates from source of material supply or to receiving bins. (3) Setting safety devices to permit proper startup sequence. (4) Setting interlocking relays for a particular operation where the system can convey more than one material, or can convey material to several receivers.

Indicator Lights

Indicator lights are available in two types: the full voltage type and the transformer type. The full voltage type is equipped with a candelabra size screw base and operates at 110 volts to suit the recommended control voltage. The transformer type has a 110 volt to 6 volt transformer built below the base of a 6 volt bayonet type socket and uses a bayonet type pilot light. This type is recommended where long bulb life is desired and equipment is subject to mechanical shock.

Lenses of different colors, sizes and types are available to suit the intent and service of the lights. Pushbutton sized lights are used as pilot lights at motor control pushbutton stations or as trouble indicating or ground-detector lights on control panels. Miniature lights are used on the flow diagrams of graphic panels to indicate levels in bins, equipment in operation, flow path of a diverter gate, or energization of a particular circuit. Some equipment control circuits may have two indicating lights of different color and size; a white light to identify the equipment on the graphic flow diagram and a red pilot light at its pushbutton station.

The proper methods of connecting indicator lights are as follows: (1) Connect operational indicator lights of motor-driven equipment across starter coils and overload trip contacts so that on trip-out of overload relays, the indicator lights will be off, thus showing the operator the nature of the trouble. (2) Connect trouble- and alarm-indicating lights in such a manner that each fault will be indicated and the alarm sounded, if more than one fault should occur simultaneously. (3) Connect level-indicator lights in such a manner that they are "on" when material is at paddle or probe level and "off" when material drops below this level. (4) Connect high-level indicating lights in such a manner that failure of power to the indicator would prevent startup of the conveying system. (See page 324 for connecting level display units.)

Time-Delay Relays

Time-delay relays are used where a definite time interval must elapse before the next successive operational sequence can be performed. There are many types of timing relays that utilize several electrical and mechanical principles for their operation. Regardless of operating principle, delay can be obtained two ways: delay after energization of the relay, or delay after deenergization. The several most commonly used types of relays and the manner in which delay is effected are as follows, in order of increasing range of time delay. (1) The operation of an electronic timer is based on the fact that a condenser requires a definite time to charge or discharge. The circuit operates on rectified a-c current. By connecting suitable values of capacity and variable resistance in the grid circuit of a triode, anode conduction can be delayed for varying periods. At the firing voltage of the tube, current passes through it to energize a relay coil and actuate a set of contacts. (2) The operation of a thermal relay is based on expansion or contraction of a bimetallic element by energizing or deenergizing a heating element. Time is required for the heating element to attain the desired temperature and transfer its heat to the element. The common motor-overload-trip relay operates on this principle. (3) A fluid dashpot relay consists of a solenoid with its armature connected to a dashpot—a piston moving in a cylinder filled with fluid. A variable leak port in the piston restricts the flow of fluid from one side of the piston to the other, thus retarding movement of the armature and producing a time delay. (4) A pneumatic timing relay consists of a solenoid and a flexible bellows for retarding movement of its armature. Adjustment of time delay is provided by an adjustable air vent to regulate the speed of air admission to the bellows. (5) The synchronous-motor-driven reset timer is actually an electric clock that drives a device to trip the contacts closed, or open them after an adjustable time interval. A built-in clutch drives the contact making device. Releasing the clutch, resets the mechanism for the next operating interval. There are several methods of starting and resetting the unit as well as a wide variety of contact operational sequences available to obtain any control function. The time delay period is selected by turning a knob and pointer to a setting on a calibrated dial. An elapsed time pointer can be furnished that moves from the time set-point to zero as time expires.

Each of these timers can be applied to a control circuit depending on the length of time-delay required, the necessity for repeatability of setting, cost considerations, versatility and type of maintenance personnel available. The industrial pneumatic timing relay is the most rugged and versatile of these timers, but requires considerable manipulation to obtain a given setting. It can be equipped with auxiliary contacts that are timed out, or auxiliary contacts that can operate without a time delay. The motor-driven timer is more accurate, can be readily adjusted to the set point, can be face-mounted on a control panel and is low enough in cost to be competitive with the pneumatic timing relay. The electronic timer requires electronic tubes, manipulation of its potentiometer to obtain a setting and may present a maintenance problem to the average industrial electrician. (Static controls are available, but adjustment of a potentiometer is required to obtain a time setting.)

The following conditions usually require the use of time-delay relays: (1) At startup, to establish airflow before admitting material into the feeding system. (2) At shutdown, to maintain airflow for cleanout of the conveying line for a definite period after stopping material feed to the system. (3) On mechanical conveying systems, to start downstream conveyors before starting conveyors that discharge into them. (4) On mechanical conveying systems, to permit emptying downstream feeders and conveyors after stopping material flow from the source of supply. (5) At shutdown, to determine the shaking time of a single-compartment, cloth-tube dust collector. (6) On systems with dual control, to allow sufficient time for startup at the subcontrol station before the control is transferred to the master control station. (7) On alarm circuits, to prevent operation of the alarm unless trouble is of sustained duration.

Control Relays

These relays are simply small, solenoid operated switching units. The solenoid armature, or plunger is equipped with an insulated bar or several tiers of bars carrying movable contacts. A stationary, insulated contact structure is mounted above the laminated magnet frame of the solenoid and carries the stationary contacts and terminal screws. In operation, closing the circuit energizes the coil and causes the solenoid plunger to snap upward and carry the movable contacts with it. As the plunger moves upward, it closes all normally-open contacts and also opens all normally-closed contacts. If the circuit is opened, the plunger instantly drops, restoring all contacts to their original position.

Relays are obtainable in single- and multiple-pole construction, up to a twelve-pole unit. Some have contacts that can be changed from normally-open to normally-closed by simply moving the connecting wire to another terminal screw. Contact operation can be made to overlap—the normally open contact closes before the normally closed contact opens—if required and specified.

Control relays link the different functions of a system and are used whenever control functions must follow a definite sequence. The usual applications in pneumatic conveying are as follows: (1) To start the operation of dust filters and vent systems before starting conveying systems. (2) To start downstream conveying equipment before starting material feeders into a conveying system. (3) To start airflow in a conveying system before starting material feeders. (4) To receive signals from various types of actuators and in turn, actuate contacts in several other circuits. These actuators may be level indicators, limit switches, pressure or vacuum switches, timers, or any other single-contact device that must operate several sets of contacts in various circuits. (5) On an important equipment unit at a remote location, a separate relay should be installed, energized from the load side of the motor circuit breaker so that the conveying-system control circuit cannot be energized unless the circuit breaker for that unit is closed. (6) To interlock controls where more than one source of power supply must be used because of widely separated locations of equipment. The system cannot be started unless all sources of power supply for equipment connected to the system are energized. When a control transformer at each source of power is energized, a relay is energized to close a contact in the main control circuit. The main control circuit is energized only when all sources of power are energized.

Latching Relays

Latching relays are used to lock a set or group of contacts into the open or closed position until a particular set of operations is completed. Upon completion of the operations, the relay is restored to its original position manually or electrically. Two types of latching relays are available: magnetic latching and mechanical latching. In both types, the momentary energization of a solenoid actuates a contact-bearing armature. The armature is held in its operating position, after coil power is removed, by a permanent magnet on the magnetic type, and by a mechanical latch on the mechanical type. Reset is accomplished by energizing a release coil to oppose the magnetism, or trip the latch mechanism, depending on the type of relay used. In some cases, the mechanical latch type may be reset by manually tripping the locking mechanism.

Latching relays may be used to deal with the following circumstances: (1) To permit completion of the shaking cycle of single-compartment dust collectors that are actuated only on shutdown of the system. (2) To prevent interrupting operation where a system is under dual control from control panels where different operations can be performed.

Auxiliary Contacts

Auxiliary contacts are extra contacts that can be field installed on control relays or motor starters without removing the device from its enclosure. They are made for attachment to special mounting brackets or to the contact bar of the switch. The number of auxiliary contacts that can be added to a specific starter or contactor should be checked with the manufacturer. The use of an auxiliary contact on the starter of a motor is the most direct way of interlocking it with another motor. The contact is wired into the control circuit of the other motor so that the starting of one initiates the starting of the other. This method is simpler than using a relay, but it can only be used where no time delay is required before starting a particular piece of equipment.

Limit Switches

The most common type of limit switch in pneumatic conveying work is the lever operated, sensitive, snap-action contact mechanism enclosed in a case. The lever can be obtained in various configurations to suit the mechanical action available to actuate it. The ends of the levers have roller cams to eliminate wear due to sliding action of the actuating mechanism. The majority of limit controls have single-pole, double-throw switches with terminals to a normally-open contact, a normally-closed contact and a common terminal. Both contacts may be used for control functions. If necessary, double-pole double-throw switches can be obtained, but any extra contacts may require use of a control relay. In applying limit switches, care must be taken to install them so that allowance is made for overtravel of the mechanical actuating device. Limit switches are used on the following types of equipment: (1) On diverter gates, to stop the drive motor and to initiate other control functions when proper alignment is obtained. (2) On manually operated, remote shut-off gates or valves, to indicate whether they are open or closed and to align interlocking contacts that will permit a mechanical operation to proceed. (3) On doors, or closures on bins and important equipment to indicate whether they are open or closed. (4) On manually operated gates that have a normal position and an emergency position, to indicate the gate position and align circuit contacts accordingly to effect each type of operation.

POWER UNITS

Electrical power units are low-voltage (115 volts to 230 volts, a-c), oil-submerged control units that are used to regulate dampers or operate butterfly valves, blast gates, or air valves. In operation, an induction motor drives a main shaft through a speed-reducing gear train. Cam-operated limit switches stop the motor at the limits of rotation. Rotation is either 90 degrees or 180 degrees. Auxiliary switches may be mounted within the unit to perform additional control functions depending on the position of a crank arm fastened to the shaft.

Movement of the shaft may be (1) unidirectional, with limit stops 180 degrees apart; (2) reversible, with switches arranged for stopping, restarting in the original direction, or reversing at any point; (3) proportional, with a built-in potentiometer responding to external controller action.

The units are made for applications requiring low torque at high speed or a higher torque at low speed. The speeds may be fixed or adjustable, as specified.

The crank arm may be connected to operate any combination of kinematic linkage within the torque limitations available, but the linkage must permit operation of the crank arm through its full angular range even though the full range may not be employed. This is done by slotting the linkage through the portion of the stroke where no motion occurs. Installation must be made with the power unit upright and the shaft in a horizontal position. The linkage should be designed to move dampers and valves only through the control range—not full stroke.

When specifying a power unit the following information is required: (1) type—unidirectional, reversible or proportional; (2) torque rating; (3) timing; (4) limits of travel; (5) fixed or adjustable speed; (6) voltage and frequency.

CONTROL APPLICATIONS

The following additional details should be given consideration when developing the design of a control installation for pneumatic conveying:

One source of power supply for all equipment and controls should be used whenever possible, so as to insure energization of all equipment required to operate the conveying system.

On exhaust fans that maintain negative pressure on a dust collector at the receiver of a pressure system, install an air-flow switch in the suction duct to interlock the conveying-system starting circuit. This is to insure maintenance of negative pressure in the dust collectors during the entire operating period of the pressure type conveying system, from startup until after the blower is shut down.

Figure 15–1 shows a typical motor connection and control circuit that includes most of the components described above. It also shows the schematic wiring diagram for a two-way diverter gate control circuit that illustrates the application of a wide variety of electrical devices. This control was developed to permit the alignment of a pipe switch for reclaiming material by vacuum from a yard storage bin for delivery to a process, with control from a process-center control panel at all times, except when control is transferred to a yard control panel. A yard operator at the yard control panel may elect to unload hopper cars sending material directly to the process area, or to reclaim material from the storage bin for delivery to process. He can wrest control away from the process control panel by means of a mechanical latching relay.

The mechanical latching relay contains two coils; a latching coil and a reset coil. The latching coil receives a momentary energizing pulse from a pushbutton and, in this case, the armature closes a series of contacts and locks them in that position even after the energizing pulse is terminated. The reset coil can be energized so that its armature will trip the latch mechanism and allow the relay to reset to its original position. The energized reset coil armature also closes a series of contacts to perform other control functions. However, here the reset coil is

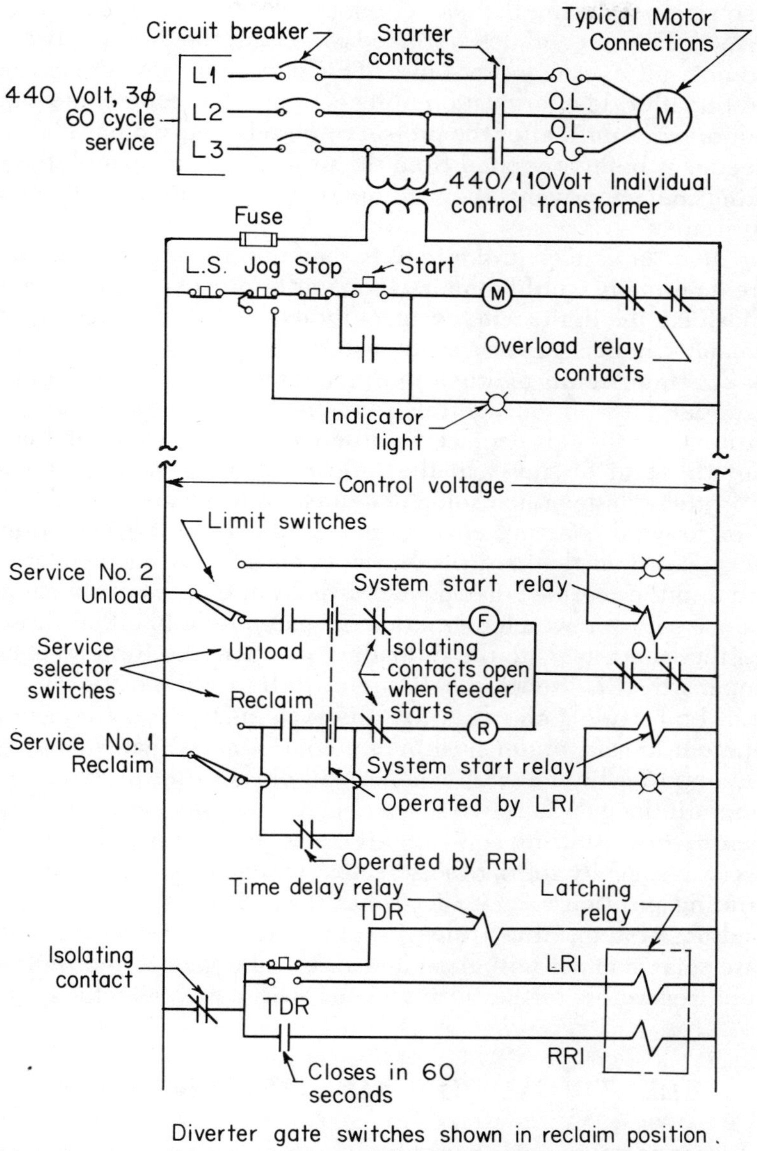

FIGURE 15–1. Circuit diagram showing application of control components.

de-energized for a predetermined period by an open time-delay-relay contact until the rotary feeder under the filter-receiver is started. The feeder starting circuit then energizes a relay that opens isolating contacts in several circuits, on other equipment, to prevent operation of that equipment during the conveying period. The open isolating contacts keep the reset coil de-energized even after the time-delay relay is timed out and its contacts re-close. This particular time-delay relay is timed out after de-energization of its coil, and the contact mechanism cannot re-close until after the pre-set time-delay period has elapsed. If the feeder is being operated from the process control panel, the open isolating contact prevents the yard operator from taking over control of the system.

The diverter gate control circuit is shown in the reclaiming position where it normally would remain when under process center control. In this position, the limit switch contact lights a position indicator light at both panels and energizes a system-start relay coil that closes contacts in the starting circuits of other equipment to permit that equipment to start under pushbutton control from the process center panel. The opposite limit switch contact is closed to set up a circuit that will permit the gate to travel to the unloading position later when the latch-in relay and service selector switch contacts are closed and the "F," or forward, starting coil is energized. When the yard operator obtains control at the yard panel, the latch-in relay contacts are both closed and the release relay contact is held open thus permitting him to use the selector switch to operate the gate motor in either direction. When the gate travels to the unloading position, the limit switch contact opens the "F" starting coil circuit to light the position indicator light at both panels and energize the system-start relay that permits equipment to start under pushbutton control at the yard panel.

The opposite limit switch contact will close to set up a circuit that will permit the gate to travel to the reclaiming position later when the "R," or reverse, starting coil is energized. This coil may be energized in two ways: (1) by setting the selector switch at the yard panel to the reclaiming position to close its contact, or (2) by shutting down the unloading system, thus releasing the latching relay so that the release-relay contact will close and return the gate to the reclaiming position regardless of the setting of the yard panel selector switch.

ELECTRICAL CONTROL SPECIFICATIONS

Electrical control panels should be specified in detail and should conform with the NEMA Standard type, consistent with the amount of protection desired. Panels for outdoor service should have a hinged, external, weathertight door and a hinged control door. All doors to panels should have handles fitted with lock and key to prevent unauthorized access to the interior. A graphic panel should include sufficient interior space for all control relays, with space for a few additional ones; otherwise a separate relay-mounting panel may be required. The control voltage should be lower than the power-supply voltage for safety—preferably 110 volts.

The following items should be specified: (1) a complete, easily readable, graphic flow diagram with a color scheme based on plant preference or standards; (2) miniature indicating lights and selector switches incorporated into the graphic flow diagram so that the operator can visualize the equipment setup for performing all conveying system operations; (3) miniature lights on the graphic flow diagram to indicate the flow positions of all diverter gates or pipeline switches; (4) start-stop pushbuttons and red indicating lights as required for manual control of remote motor-driven equipment; (5) a system of level indicating lights incorporated into the graphic flow diagram to show required levels in each bin, silo, or dust hopper.

For manual operation, the level-indicating lights should be used in conjunction with an alarm horn to warn the operator of high or low bin levels. An initial signal should sound the horn and cause the indicating light to flash "on" and "off." When the signal is acknowledged by the operator pressing a button, the horn should stop sounding and the indicator light should stop flashing, but remain "on" until material level is changed.

Further specifications should call for: (6) Wiring tied in neat bundles or enclosed in wiring racks and connected to numbered terminal blocks. Each wire should be numbered using standard commercial numbering tags at terminal posts of relays, transformers and other panel-mounted equipment. (7) A ground-detector system and terminal blocks to divide the control circuits into two or more separate circuits, to permit isolating them for ground test by removal of leads from binding posts. (8) A test circuit and test pushbutton for checking all indicating lights on the panel except motor-running and ground-test lights. (9) Nameplates of black phenolic-laminated plastic, with white letters to identify the service of each pushbutton station, indicating light, selector switch, and equipment outline shown on the graphic flow diagram. (10) Relays furnished as required to perform all specified control functions, and mounted on a sheet-steel plate attached to the interior of the control panel. Where union labor problems may occur, the panel should be assembled and wired by union labor affiliated as designated by the purchaser.

The arrangement of the control panel should be given consideration. The face of the panel should be arranged essentially in three sections. The upper section should bear the graphic flow diagram of the systems controlled by the panel. The central portion should bear the special service pushbuttons or selector switches for resetting control circuits, silencing alarms, and performing temporary operations to clear troubles. It may also mount trouble lights. The lower portion of the face, above standard table height, should mount all motor-starting pushbutton stations and their pilot lights.

The type of panel may vary depending on plant preference. It may be a simple insert to be mounted as part of the control panel for a non-related process. It may be a wall-mounted panel with front opening doors. A larger panel may be floor mounted with front doors, or freestanding away from a wall, with rear or side doors, depending upon the width of the panel.

CONTROL AND INDICATING DEVICES

Control, indicating, and protective devices are required to integrate the various remotely located components of a pneumatic conveying installation into a conveying system. Material-level indicators at supply and receiving bins serve to guide manual operation of the system or may be used to start and stop the system automatically. A motion switch driven by direct belt or chain drive from a rotary airlock feeder proves that the rotor is turning freely and indicates when the feeder is jammed. A flow switch in the exhaust-fan ductwork proves that the fan is in operation. A differential-pressure switch across a dust filter is used to detect clogging of the filter. Torque-arm overload-release switches and indicator lights indicate when a drive motor for a conveyor or other remote unit is overloaded or is not operating. All these devices require electrical connections from their terminals back to a control center, and increase the cost of electrical installation. A system that at first appears to require only a simple control panel may be found to require an auxiliary relay panel and an extensive electrical-control system after it has been shown that certain control and protective devices are required for proper operation.

Material Level Control

The most important of these control devices is the material-level detector at storage bins and feed hoppers. For pneumatic conveying service, almost all these are for point-level detection: they can indicate only the presence of material at the detecting element. When close control of bin level is necessary, several level detectors may be required, located at different points around the periphery of the bin and at different elevations within the bin in order to obtain a composite picture of material level. Where extra-high or extra-low level in a bin presents a hazard to other equipment or to the maintenance of production, two level indicators may be installed at these points, wired as required, to ensure performance in case one detector fails to operate.

The manner of transferring the signal from the detecting element to the external circuit varies with the principle of operation of the detector. Several types of detectors are commercially available as follows:

1. Diaphragm type. A flexible metal or plastic diaphragm is mounted on a vented housing that contains a spring- or lever-loaded contact assembly or snap switch. Movement of the diaphragm by material lying against it causes actuation of the contacts to operate external control circuits. The unit is always mounted in an opening in the bin wall at the desired level.

2. Rotating-paddle type. A synchronous motor or torque motor drives a paddle that is attached to its rotor shaft. Paddle rotation is stopped when it is immersed in material and restarts when material falls away from it. A built-in centrifugal clutch actuates a set of contacts to stop the motor and operate external control circuits.

3. Capacitance type. The detecting probe is essentially a capacitor that forms part of an inductance-capacitor bridge circuit within a detector unit. The detector unit also contains a transistorized oscillator that is coupled to the bridge and maintains a frequency to keep the bridge balanced. Material in contact with the probe increases the capacitance and unbalances the bridge. Part of the bridge output is rectified and signaled to an electronic trigger circuit which converts the signal to a current that can operate a relay to actuate external control circuits.

4. Piezoelectric type. A tuning fork is driven mechanically at its resonant frequency by a piezo-ceramic crystal, mounted on the fork membrane, that is subjected to an alternating electrical field. Another crystal, similarly mounted, receives the mechanical vibrations and converts them to electrical oscillations that are fed back to an amplifier to operate a switching relay. Material lying against the probe reduces the frequency of the oscillations and, hence, the current to the amplifier. The relay is then de-energized to actuate the external control circuits. See Figure 15–2.

5. Radioactive type. A shielded source of radioactive material (cesium 137, radium 226, or cobalt 60) is located at the desired level, outside the bin, and directed to a gamma-ray detector that is mounted diametrically opposite the source in the same horizontal plane. Any material above this plane, within the bin, will reduce the amount of

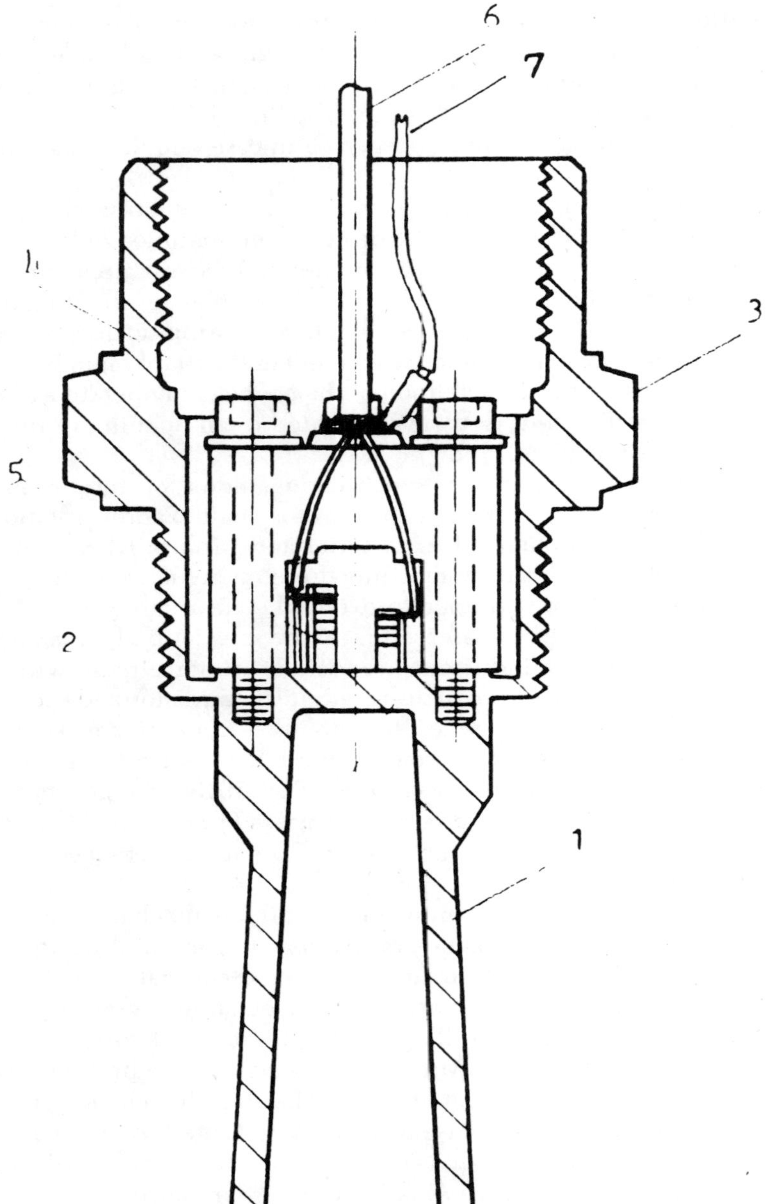

FIGURE 15–2. Probe for piezoelectric type level detector. (Endress + Hauser, Inc.)

radiation received by the detector and cause a relay to actuate the external control circuits.

The principal limitations on the use of these detectors are as follows:

Diaphragm type. The membranes are sensitive to variations in air pressure within a bin and may provide false level indications. They are not recommended for use on bins that receive material from pneumatic conveying lines.

Rotating-paddle type. (1) The paddle may not be stopped if the material is aerated. (2) Shaft length for vertical installation is limited to about 10 ft. (3) Paddles must be protected from the impact of avalanching material. (4) The rotating shaft must be suitably guarded, and its entrance to the actuator or motor housing sealed against powder leakage so as to prevent the shaft from seizing, especially if the bin is under air pressure. (5) Pipe guards around the shaft must have a dust seal at the paddle end, except when the shaft is flexible, so that material cannot pack inside the guard. (6) The housing for the motor or activating device must be mounted at the end of the shaft so that access for servicing may be difficult. (7) The motor life is limited, especially on systems that are in continuous service.

Capacitance type. (1) Its sensitivity must be adjusted for each type of material and may require adjustment for the material characteristics from day to day. (2) The material may build up on the probe and affect its sensitivity. (3) The entire system, from probe to relay housing, must be sealed against moisture. (4) The sensitivity is affected by vibration of the probe, especially on bins equipped with vibrating bin dischargers. However, on some designs, the detection unit may be disconnected from the probe and separately mounted on a fixed surface. (5) Ambient temperature is limited to 160° F for display units and 200° F for detector electronics.

Piezoelectric type. (1) It cannot be used in deep bins due to tine breakage under material pressure. Depth limited to about 20 ft if installed vertically. (2) Ambient temperature limited to 140° F for electronics and 200° F for piezo crystals. Electronics can be designed for 200° F, if specified.

Radioactive type. (1) Subject to drift depending on the half-life of the radioactive source. (2) Openings into the bin must be secured and locked while the source is in use. (3) Source must be removed when the bin must be entered for servicing. (4) Warning signs must be posted in the area. (5) Large quantities of radioactive source material within a plant requires AEC licensing and control.

The location of a level indicator must be determined with regard to its own limitations and the limitations presented by the characteristics of the material that will actuate it. The following precautions should be observed when locating bin-level indicators of the paddle or probe type: (1) Detecting elements should be mounted vertically where possible so that unanticipated changes in level due to material characteristics can be compensated for by changing the length of probe or paddle shaft. (2) Detecting elements should be located where they will be most sensitive to changes in material level, but out of the path of flowing materials that would result in false level indications. (3) Detecting elements should be located away from bin walls and baffles so that material cannot bridge between them. (4) Detecting elements mounted horizontally in sides of bins, and low-level vertically mounted elements, should be protected from avalanching material and bridging by a protective baffle designed to shed powder. (5) The housings for the actuating mechanism or relay should be completely sealed from moisture due to condensation.

Modern capacitance probe units are made with the detector amplifier circuits on a miniature circuit board attached to the head of the probe thus eliminating the troublesome wiring connections that formerly had to be made between probes and remotely located detector units. In addition, level display units are available with two lights, one red and one green for each probe, to indicate safe or unsafe conditions. The usual modes of operation are high-level fail-safe and low-level fail-safe. Fail-safe operation means that the control relay in the detector unit is de-energized upon power failure, upon loss of contact of material with the low-level probe and upon contact of material with the high-level probe. The display units are reversible so that the red signal lamp is at the top for high-level fail-safe and at the bottom for low-level fail-safe indications.

The piezoelectric type of probe has supplanted the capacitance type on systems handling sticky materials or materials that can build up on the probe. The vibrating tines tend to shed material and prevent buildup.

Motion Switches

Motion switches are generally used to shut down a conveying system upon failure or jamming of a rotary airlock feeder or screw feeder. They are intended to prove that the rotor or screw shaft is in motion, since it is still possible for the drive motor to continue in operation upon failure of a shear pin, chain or belt while the rotor or screw is stalled. Several types of motion switches are available, all of which require connection to a chain or belt drive.

The operating principles of a few of these switches are as follows: (1) A rotating shaft drives a pump impeller to develop oil pressure against a spring-loaded plunger actuating a sensitive switch. A bypass valve is adjusted to regulate the set pressure that is proportional to the speed of the impeller. (2) A rotating shaft and a set of bevel gears drive a horizontal-revolving impeller that creates a vortex in a body of oil. The vortex swings a vertically hanging vane so that a cam on the vane shaft actuates a sensitive switch (Figure 15–3). (3) A shaft drives a very small generator that sends an electrical signal to a transistorized amplifier circuit and operates a relay connected to the external electrical circuit. (4) A shaft drives a miniature centrifugal-flyball governor linked to a spring-loaded collar. The collar actuates a sensitive switch when motion stops. (5) A shaft drives an aluminum disc past pivoted electromagnets in such a position that the contacts remain open as long as motion continues. This type of unit is actually a standard plugging switch. (6) The field of a shaft-mounted permanent magnet that rotates within a copper cup induces a voltage which causes current to flow in the periphery of the cup. The field caused by the current interacts with the field of the permanent magnet to rotate the cup so that a bracket mounted on it is forced against a spring-loaded contact arm. When the torque is great enough to overcome the spring force, the contact arm is forced against a fixed contact that remains closed as long as the shaft is turning faster than a given speed. This type of unit is called a speed switch. See Figure 15–4. (7) A shaft-driven impeller rotates within a chamber filled with hydraulic fluid and creates a drag on a mating impeller that is separated from it by a small gap. The mating impeller starts to turn as the shaft speed and the fluid drag increases and actuates a set of spring-loaded contacts.

The selection of a motion switch requires consideration of the following limitations: (1) Oil-actuated units on outdoor service require oil changes to suit the season and may prove sluggish even if a year-round oil is used. (2) The spring-loaded fly-ball governor type will not function if the spring breaks. (3) A speed-increasing chain or belt drive is normally required for operation of any motion switch, since rotary and screw feeders operate at relatively low speeds. (4) Motion or speed switches that depend on the drag of an hydraulic fluid to effect motion of an impeller are subject to leakage and require factory replacement of the fluid since the properties of the fluid are usually proprietary.

Other Control Devices

Flow switches are vane-operated switches similar to those used to indicate airflow on air-conditioning units. They are useful to prove actual airflow through a duct at a remote location, especially when using belt-driven fans where belt slippage can occur. For very small

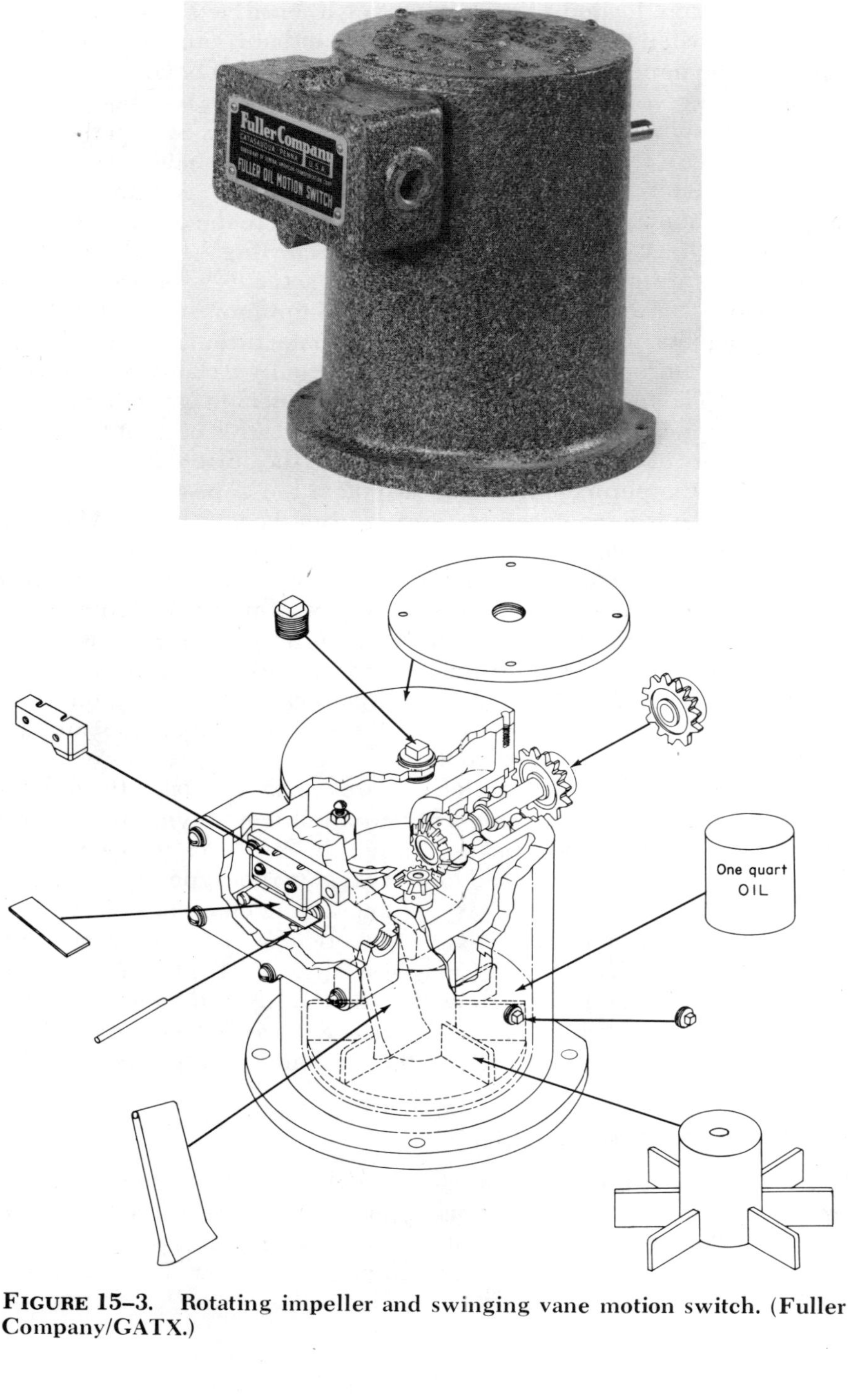

FIGURE 15–3. Rotating impeller and swinging vane motion switch. (Fuller Company/GATX.)

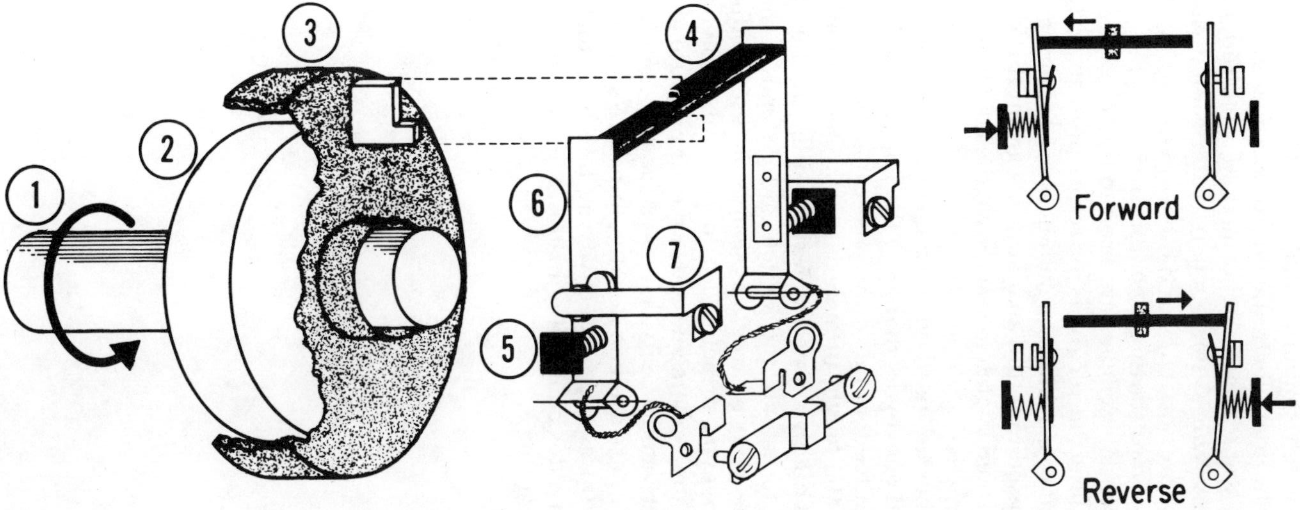

FIGURE 15–4. Simplified view showing principle of operation of an electromagnetic speed switch. (Allen-Bradley Company.)

airflow, a vane switch similar to that used for vacuum-tube-cooling service is available from electronic equipment manufacturers.

In a few cases, water-flow switches may be required, especially on high-pressure conveying systems where water-cooled sliding-vane or reciprocating compressors supply the conveying air, and water flow must be proved before starting a compressor.

Pressure-differential switches are used to detect excessive pressure drop across dust filters at receivers, or across protective dust filters on the suction side of rotary blowers on negative-pressure systems. High pressure drop is used to sound an alarm to warn of this condition or to actuate an interlocked switch to shut down the conveying system until the trouble is corrected.

Pressure switches are used on mixing systems to control the accessory material-feed conveyors that keep the hoppers to the pneumatic-conveying system feeders full. Since the capacity of these systems is very sensitive to the height of material in the feed hopper, and since level detection of highly aerated material is difficult, the conveying-air pressure is used to actuate the switch and do one of the following, depending on the type of accessory conveyor used: either start and stop a screw conveyor, or actuate a low-pressure solenoid valve to release air to the atmosphere and lower the air pressure on the plenum chamber of air-activated gravity conveyors, so that material flow stops.

Vacuum switches are used on negative-pressure systems to maintain a constant suction pressure in a conveying line by actuating solenoid valves to position gates controlling material outflow from storage bins into the conveying line.

III SYSTEM OPERATION

16 Inspection, Testing, and Cleanout

Inspection of a pneumatic conveying system should be made before construction starts. Standard, conventional equipment may be inspected when received at the jobsite; equipment built in accordance with the purchaser's special detailed requirements should be examined for conformity with specifications at the fabricator's plant. When this inspection is waived due to the expense involved, the equipment should be accepted conditionally so that the responsibility for correcting mistakes remains with the fabricator.

MOTOR-DRIVEN EQUIPMENT

To check the operating speed of belt- or chain-driven equipment, multiply the motor speed by the gear reducer ratio, then multiply the product by the ratio of the diameters of the sheaves or sprockets. Fit sheaves and sprocket hubs on shafts or measure bores and key-ways to insure proper sizing. When checking chain and belt lengths against sheave and sprocket diameters and center distances, allow for the takeup that will occur when the belts and chains will be in tension.

Make sure the base slots of motors and driven equipment permit enough adjustment. This is especially important when variable-pitch sheaves are spring-loaded, because of the greater adjustments that may be necessary to get sheave diameters corresponding to the full speed range. If the variable-pitch sheaves are set manually (not spring-loaded) and the speed must be set in the field, have three belt lengths on hand (the calculated size and the next smaller and larger standard sizes), so that the right length can be found by trial and error. For chain-driven equipment, have sprockets of three sizes that would fit a common quick-detachable hub.

Rotary Feeders

To check the capacity of rotary feeders fill one pocket level with material and pour it into a calibrated container (or fill the pocket from a calibrated container). Avoid packing the material.

The apparent capacity of the feeder is the volume of one pocket multiplied by the number of pockets on the rotor and by the operating speed. This capacity should be greater than the pneumatic conveying rate by factors ranging from 1.1 for positive-pressure to 2.0 for negative-pressures systems. The true density of the material will be reduced, of course, when the feeder is operating, because the material will be aerated. High rates of speed will foreshorten filling and further reduce capacity.

Blowers and Compressors

Find the inlet capacities of blowers and compressors at the anticipated discharge pressure and operating speeds from the characteristic curves or by calculation. Besides verifying capacities, this information permits checks of the size of dust collectors, air-to-material ratios, conveying line velocities, and air losses.

Compare the horsepowers of drive motors with horsepower requirements at design conditions. There should be enough excess horsepower to unplug conveying lines.

Screw Conveyors and Feeders

Find the volumetric capacity of screw feeders by multiplying the volume delivered per revolution from Table 16–1 by the design speed, by the pitch factor, and by a slippage factor of 0.90. The volumetric capacity multiplied by the apparent weight of the material should equal, or be slightly greater than the capacity of the conveying system. To find the actual volume delivered per revolution of a screw conveyor, multiply the value from the table by the percentage of trough

TABLE 16–1
Standard-Pitch Screw-Feeder Capacity for 100% Trough Loading
(ft^3/hr/1 rpm)

Screw Dia. (in.)	Standard Pipe Size Shaft (in.)									
	1	1¼	1½	2	2½	3	3½	4	5	6
4	1.56	1.44								
6			5.30	4.97						
9			18.99	18.49	17.85					
10			26.28	25.73	25.02					
12				45.28	44.42	42.99	41.89			
14					71.67	70.15	68.73	67.10		
16						106.35	104.72	102.86		
18						153.06	151.20	149.10	146.5	
20							209.50	204.00	201.0	197.5
24							366.40	363.60	356.4	348.6

loading when the level is just below the shaft bearings. The volumetric capacity of the conveyor is this volume multiplied by the design speed and the pitch factor.

The pitch of a screw feeder or conveyor is standard (equal to 1) when the distance between flights is equal to the screw diameter. The pitch factor is the ratio of this distance to the screw diameter on short pitch screws. For multiple-flight screws, the pitch factor is 1 if the pitch of the main screw is standard.

CONTROL PANELS

The control panel should be tested before it is installed, by energizing all circuits with the rated voltage of the control system. Following the schematic diagram, check every circuit.

Generally each circuit will contain a relay or a motor starter coil with pushbutton contacts and relay or control device contacts arranged across the supply lines to open or close the circuit according to functional requirements. Some of these contacts may be in the panel, such as pushbutton or sequencing relays, while others may be on external control devices or on remotely located pushbuttons and relays. The external contacts will be connected to a terminal wiring strip by numbered binding posts during installation.

When a circuit is tested, the objective is to get continuity from the control source through the motor starter coil and the motor overload relays to the common return wire of the source. These tests can be made before installation by jumping all the open contacts at the terminal posts, by energizing or hand-operating the plungers of the relays or motor starters of the circuit, or by pushing the circuit's start or jog button. Continuity is proved when all the contacts of the circuit close and the relay or starter coil is energized.

If the relay or motor starter initiates the sequencing of other circuits, check out each of the interlocked circuits for continuity in the proper operational sequence. Next, check the de-energizing devices by (a) opening circuits at safety-stop pushbuttons, at lock-stop or stop buttons, and at the contacts or jumper wires that represent external control devices, or by (b) de-energizing the first relay or motor starter coil in the sequence of interlocked circuits.

Set all time-delay relays at the approximate timing for their function.

In order to complete the circuits shown in Figure 16–1 when testing only the control panel, several jumper wires must be installed across the panel terminals to simulate closed circuits on external equipment. To avoid a short circuit, a test lamp, not a jumper wire, must be placed across the terminals to starter coil M to simulate the missing coil.

When testing the control panel together with its associated external

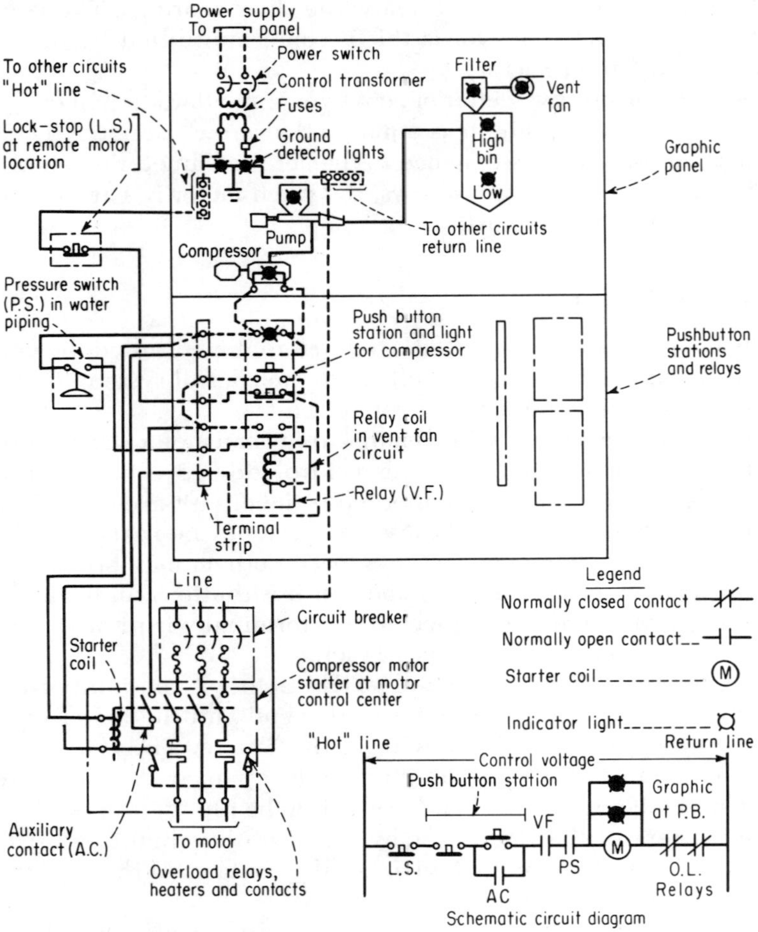

FIGURE 16–1. Control circuit for a sliding-vane air compressor.

equipment for continuity, pressure switch *PS* may be closed by water pressure or by lowering its contact setting. Relay coil *VF* may be closed by pushing its plunger, by activating the vent fan circuit, or by applying current from an external source directly to the coil terminals.

When all contacts are closed and the start button is pushed, the starter coil will energize and the indicator lights will show "On." The circuit may then be opened by one of several possible ways: pushing the lockstop or stop button; stopping the vent fan, or opening the circuit to relay coil *VF*; shutting off the cooling water to the compressor, or setting the pressure switch *PS* high enough to open its contacts.

Dust Collectors

Large dust collectors are usually shipped in sections and erected in the field.

The bolt spacing of mating parts (such as dust hopper and bag housing flanges, or blower connections to plenum chambers) that were not made in the same shop should be checked for alignment. See if the slope of dust-hopper walls, especially at corners, is enough to ensure a free flow of material to the discharge spout. Look at the size, fit, gasketing and fastening devices of access doors. Filter-tube ferrules should be clean and undamaged. Check center punch marks often used as guides when bending sheet metal—occasionally these punch holes in the plate.

Check the sequencing, timing and timing change apparatus of cleaning cycle programmers. See that dampers operate freely and seat tightly, and check the linkage to the drive mechanism and the operating time of the drive motor.

The material, diameter and length, reinforcement, stitching, and ground wire (if static charges must be dissipated) of filter tubes or bags should also be examined. The bags should be installed just before startup, and never shipped assembled in the collector, because they may be damaged by welding sparks and are likely to absorb moisture.

Diverter Gates

Make a quick check of the operating speed of diverter gates so that control relays may be timed to prevent the premature operation of parts of the system before a gate completes its movement. Turn the unit by hand, releasing the brake manually if necessary, and check the actuation of the limit switches with a test flashlight or a low-voltage continuity tester to detect the actual making point of the limit contacts. Compare the actual position of the gate with respect to the inlet and outlet ports and the contact point of the limit switch by looking through the inspection ports on the gate housing. Reset the limit switches if the gate is not aligned.

INSPECTION AFTER INSTALLATION

After all the equipment is installed, prepare for the startup, following detailed plans. A sound- or battery-powered communication system that connects remote areas with the control center will facilitate the preparation and startup.

General Inspection

Check the following items before accepting the system as being ready for startup:

1. The tightness of foundation bolts.
2. Dust filter bags for proper tension and tight clamping.
3. The insides of such equipment as control enclosures, bins, dust hoppers, dust filters, conveyors and storage silos for lost tools, loose bolts, dunnage, rags and other debris.
4. Equipment access doors, inspection ports and flanged joints on bins, dust collectors and ducts for missing bolts and gaskets and for the tightness of bolts and clamps.
5. The tension of belt and chain drives, the room for takeup, and the presence of safety guards.
6. The accessibility of shaft ends of major equipment and the presence of center holes for tachometer readings.
7. Hangers and anchors at pipe bends for tightness and rigidity.
8. The packing, bolting and fit of packing glands and external shaft seals.
9. All weight and spring-loaded relief valves for free movement. Remove excess weights and store them close by for possible future use.
10. Magnetic devices in control enclosures for free movement. Remove restraining devices installed for shipment.
11. Electric brakes on motors for free movement. Release shipping screws.
12. Fan wheels for free rotation without binding or striking the housing. Access doors should be tight and sealed.

Lubrication

All gear and oil cases should contain the prescribed lubricant and be filled to the proper level. Vent plugs must be unobstructed. Check the position of the lubricant fittings of all ball and sleeve bearings and make sure the fittings are of the right type (in accordance with plant standards); wrong fittings should be replaced. If they cannot be, an adapter should be provided so that one grease gun can be used for all types of fittings.

If fittings are hidden, or partially hidden, paint some markers at strategic places so that the oiler will be able to find them easily. Clear the paint from all fittings. See that oil-lubricated bearings have sight glasses and that they are properly located for visibility. Make sure bearing-oil rings turn freely.

If lubrication fittings are not accessible, make them so by extending them with tubing. All oil cases should be flushed out with a light oil before being refilled with the regular lubricant. But make sure all the light oil is drained so that it will not dilute the regular oil.

Electrical Controls

Tabulate all the motors and list their current ratings. Before installing overload relay heaters at the starters, size them for the current rating, making allowance for the ambient temperature at the starter.

Remove current transformer shunts installed before shipment.

Prevent motors from operating by pushing the lock-stop button at each motor, or by disconnecting all the wires at the load terminals of each motor so that it cannot operate even if its starter is energized.

Install fuses to energize the control system. Close all access covers and enclosure doors and apply control voltage to the entire system. Because all the bins and hoppers will be empty, the low-level lights should indicate "on." The high-level lights, of course, will be "off."

To check rotating-paddle types of high- and low-level indicators, remove their covers and see if the paddle shaft is rotating. If it is not, check the circuits to its drive motor.

If possible, set off probe-type level-indicator elements by hand, and stop the rotation of paddle-type elements by grasping the shaft to see if the control panel lights indicate the right level. If the level-indicating elements cannot be reached, remove the covers at the actuating contacts and simulate high and low levels by jumpering or removing wires to check the level indication at the control panel.

Next, check each relay or motor starter circuit for continuity, sequencing, and safety and control device operation. To make these tests, follow the procedure outlined for the preliminary checking of the control panel, with one exception: this time, use jumper wires or disconnect the lead wires to simulate the required action at the contacts in each remote control or safety device instead of at the control-panel terminal strip.

Limit switches may be manually actuated. Pressure switches may be set to obtain switch action; after the continuity test, these switches should be reset to their normal position.

Power Supply

Install the proper size and type of fuse at each motor disconnect switch. To check motor rotation, reconnect the load terminal wires at each motor starter or release the lock-stop buttons and energize each motor momentarily.

To start a motor of a semiautomatic or a fully automatic system, push in the starter contact block or the plunger manually; stop the motor by pressing the remote lock-stop push button. For manual systems, start and stop a motor by pushing the start and stop buttons in quick succession. Check the rotation of reversible motors in both directions. Check each source of power supply to a common conveying system to make certain that the power-supply interlocking control relays are operative.

After energizing the control system and applying power to all the motor-driven equipment, operate all the diverter gates in both directions to check their brake action and the stop positions of the limit switches. If a diverter gate does not open or close properly, its limit switches or the brake tension may need adjusting. If a gate does not have an external position indicator, open the inspection ports in the gate housing and check the gate position.

Blowers, compressors and other rotating equipment should not be started until after the conveying system has been cleaned out.

CLEANING OUT THE SYSTEM

The general inspection of the conveying system should have resulted in the removal of the debris that remained at the end of construction. Nevertheless, further cleaning of the inside surfaces of bins, hoppers, pipes, feeders and conveyors is required.

When the system is being cleaned out, every precaution should be taken to prevent the entry of dirt into close-clearance rotating equipment. The bottom discharge connection of bins and dust hoppers should be temporarily blinded by clamping a sheet metal plate between the flanges. A fine-mesh screen should be temporarily installed at the suction flange of close-clearance blowers and compressors, even if an air filter is in front of it, because the dirt in the suction line must be caught. No blower or compressor should be started without a filter in its suction line.

Temporary screens should have long handles that stick out past the flanges so that they can be spotted easily. These handles also make installing and removing the screens easier. Open suction lines should be covered with heavy expanded metal screens to keep refuse out.

The insides of silos, bins, large ducts, dust-collector tube sheets, and compartments should be swept by plant personnel or by a tank cleaning contractor. Where cleanliness is critical, the inside surfaces of all the large equipment should be washed with a detergent or an antiseptic cleaning solution (such as citric acid).

The ferrules of dust-collector tube sheets should be cleaned before the filter tubes are attached. Filter tubes should be installed just before the startup and only after the insides of the dust collectors have been cleaned.

When a startup occurs during winter, all the ducts and spouts that connect outdoor equipment with equipment inside buildings should be blanked off on the inside end, even if the duct or spout can be closed off by a diverter gate. The blinds will keep moisture from condensing and collecting in the outdoor equipment and prevent problems of wet filter bags and ice or water pools in the bottom of dust hoppers and bins, or in the star-wheel pockets of rotary feeders.

The covers of screw conveyors, or other types of mechanical conveyors, can be removed so that the insides can be cleaned. When covers cannot be totally removed, the inside of the conveyor can be cleaned out with an air hose.

After the preliminary cleanout, blow out all the air ducts and conveying lines by starting up all the air movers. Relief valves, flow switches, pressure switches, dust collector timers, and the cleaning-cycle control system's operation and setting may also be checked at this time. All duct and conveying lines under positive pressure should also be tested for leaks by applying soapy water to joints and seals where leaks may occur.

Leaks in negative-pressure systems may be detected by listening for hissing noises at joints and seals, or by feeling for airflow at places that are likely to leak. The inlet end of a suction line may be blanked off temporarily to obtain maximum negative pressure during this test.

In the case of an outdoor dust collector under negative pressure, leaks may be found by hosing the collector with water, starting at its roof and aiming specifically at joints and access openings, and then opening the collector to look for wet spots.

The final cleaning out of the conveying line, or the first cleaning out of a closed, inaccessible conveyor, should be done by scouring it with several hundred or thousand pounds of the material that the system will normally convey.

This material will be sacrificed for the cleaning. It should be run through the system until it arrives clean at the discharge point. All the material that is collected should be cleaned out of the bins and dust hoppers, and either transported to a dump or classified as a degraded, nonsanitary product and sold for applicable uses. Any of the material that is trapped in conveyors can be blown out with compressed air.

17 Starting Up Pneumatic Conveyors

DETERMINATION OF CONVEYING RATE

The most important test on a tight and properly functioning conveying system is the test for conveying rate. The purchaser must give some thought to this matter at the time of purchase so that he may have a basis for invoking the terms of the vendor's guarantee if the conveying rate is not obtained. The principal methods of testing are as follows: (1) The time to unload a transport that has been weighed in and out on a registered scale is recorded. This is done for several transports over a period of time covering various weather conditions. A chart showing the summation of the total material received versus the summation of the total unloading time should show a line of constant slope. A complete record of downtime during unloading should be maintained, including periods of reduced load due to failure of material to flow from the transport, or from some other cause. If necessary, flow inducers or manual labor should be used to keep the transport unloading hose full at all times while discharging cargo. (2) All material entering the system is weighed in calibrated weigh hoppers or bins making due allowance for material trapped in equipment such as dust filters, rotary feeders, dead legs of chutes, and screw conveyors. (3) A downtime recorder is used to obtain actual operating time of calibrated volumetric feeders. The calculated volume delivered is checked against the calibrated volume of the receiving bins or hoppers. Conversion of volume to weight should be made by actually weighing a known volume of the aerated material. This method is approximate, since aerated densities are variable during operation. Allowance must be made for material accumulated within equipment. The calculated volume delivered by the conveying system feeder may be compared to the amount of material supplied by a batching process where each ingredient is weighed. Here again, a chart of total weight batched versus calculated weight delivered by the conveying system feeder between low and high levels in the receiving bin should have a constant slope. (4) The time to fill a bin or silo of known volume may be determined and the volume converted to weight using the aerated density of a material sample from the bin.

When several transport unloading stations supply material to a single conveying line, the conveying rate should be determined at each station, since the rate will vary from station to station, because of the differences in conveying line resistance.

When material has not previously been fed to the system for cleaning it out, the feed rate should start slowly and be increased gradually. This can be done on systems where the feeders are equipped with a varispeed drive, or where bins have cut-off gates on the inlet side of the feeder. On transports, the outlet gates should be opened slowly; pressurized transports should be unloaded using the lowest air pressure that can deliver material; units with air-activated gravity conveyors should have air supplied to the plenum at the lowest pressure that can deliver material. As soon as conveying can proceed without trouble, the conveying rate should be increased gradually until rated pressure or vacuum is reached on the conveying line. Continue all operations at the design pressure until the conveying rate has definitely been established to the satisfaction of all concerned with the project.

BASIC OPERATING PROCEDURE

The following startup procedure is common to all types of pneumatic conveying systems. The low level lights or alarms at the receiving bin or weigh hoppers indicate that material can be received. If the hoppers or bins are empty, the diverter gates are set to direct the flow of material to the proper hopper and bin. Next, start the blower or air mover to establish airflow through the conveying line—on a blow-tank system, start the air compressor. Finally, start equipment that feeds the material into the conveying system.

When the high-level light or alarm indicates that the receiver is full, stop the flow of material into the system. After allowing time for the airflow to clear out the conveying line, stop the airflow. In blow-tank systems, the residual air will empty the blow tank. Finally, isolate the material from the conveying system by closing bin or feeder gates.

The principle involved in these operations is to start airflow first and stop it last so that no material is trapped in any part of the conveying line. A conveying line filled with dead material may be very difficult to clean out without taking apart pipelines.

UNLOADING TRANSPORTS

Pneumatic conveyors are commonly used to unload transports. Unloading procedures vary with the type of transport because some may be unloaded by air pressure, some by vacuum, while others may be emptied mechanically or by gravity. The following procedure is general enough to apply to the unloading of any of these types.

Inspection

After the transport has been weighed, record the weight, noting also the source of material and the transport number from the bill of lading. If the car seals are intact, open the hatch cover and see how the material is piled and packed. Take a sample for tests.

Connecting the Transport

Be sure the outlet valves are closed before opening the valve cover plates. After opening the cover plates, connect the hose adapters or pickup nozzles to the transport, and the discharge hose between the transport and the plant conveying system.

Open the vent hatches and install venting dust filters if the transport is under vacuum or when air pressure must be relieved. Connect the air supply hose that will aerate the transport, and attach the flow inducers where they will be most effective. Finally, connect the flow inducers to the power supply system.

Preliminary Aeration

Aerate the contents of the transport at the highest pressure possible for five minutes or more. This should unpack the material and make it easy to open the discharge gate.

Unloading

Start the plant and the transport's built-in conveying system. Slowly open the transport's outlet valve. If the valve cannot be opened, aerate the transport here with the highest pressure possible, and work the valve manually. After the valve opens, lower the air pressure to the minimum that will control the flow of material into the conveying system. Set the discharge rate with the outlet valve, then lock or clamp it. If the conveying system becomes overloaded, shut off the flow until the system becomes cleared, then reset the outlet valve for a lower flow rate. Overloading will blow the vacuum relief valve of negative-pressure systems, and the pressure relief valve at the blower discharge of pressure systems. It is good practice to operate the system at low rates until the dust filter has built up a proper filter cake. Then, the discharge rate may be increased to the design rate.

If the flow out of the transport stops or slows when the outlet valve is open, raise the aeration air pressure or start the flow inducers. The sides of transports that are internally lined or coated should never be struck. Vibratory flow inducers should not be operated when the outlet valve is closed or the line is plugged.

Setting Maximum Flow

Relief valve settings should permit conveying at the highest operating suction or pressure that does not overload the air-movers or plug the conveying lines. Unloading hoses throb because of the pulsing of flowing material. When gages stop pulsing, the entrance to the unloading hose may be blocked or the conveying lines plugged.

Clearing Plugged Lines

Lumpy or compacted material may plug lines and hoses. To unplug either, shut off the flow of material by closing the outlet valve, and let the pressure or suction clear it. If plugging persists, shut down the system, disconnect the unloading hose, and probe for lumps at the outlet valve. If lumps are found, more delays can be expected and unloading will be prolonged. Any future load may have to be screened when the transport is filled.

In general, a plugged line may be cleaned by applying the highest available pressure or vacuum to the system with all material flow cut off. Relief valves may have to be set higher than normal to do this, but the pressure or vacuum must not exceed the design limits of the blower or compressor. The horsepower drawn by the blower or compressor will increase during clean-out and may overload the drive motor, tripping out the starter. By allowing the overload heaters and the motor to cool off, the air movers may be started and restarted several times until the conveying line is cleared.

If plugging persists, the approximate location of the compacted powder should be found by rapping the line at suspected blockage points. Several methods may be used to clear this type of blockage. The air movers may be restarted and the piping hammered along the blocked section until the line is clear. A rectangular slot may be cut in the blocked section of pipe and a stiff wire snake inserted into the line to loosen the material. The slot may be closed with a cover plate and U-bolts and the air movers restarted until the line is cleared. A commercial power-driven snake, such as is used for cleaning sewer lines, may be used to clear the line from one end, preferably the lowest end of the conveying line and the air movers restarted to clear out the line. The pipeline at the blocked sections may be dismantled and cleaned mechanically or manually.

When restarting, operate at a reduced flow rate and observe the pressure gages for signs of plugging as the conveying rate is slowly increased. If plugging recurs at a particular spot, the layout of the piping at that location may have to be redesigned. Plugging commonly occurs at bends having too short a radius, at kinks in the lines, wherever a flange gasket protrudes, or where weld metals or other objects may be wedged.

CHECKING THE CONVEYING AND VENTING SYSTEMS

Some of the equipment and operating conditions that should be checked during the startup of a material-into-air conveying system are shown in Figure 17–1.

Dust Leakage

After the conveying rate has been set and the material appears to be flowing steadily, check the entire system for dust leaks at joints, access doors, cover plates, chutes, receiving bin vents, and dust-collector exhaust stacks.

A light plume of dust leaving the air exhaust of a dust collector having new, woven-fabric bags is normal and will persist until a filter cake forms in the interstices of the fabric. If the flow does not lessen, look for a torn or an improperly attached bag. If leaks are not troublesome, conveying may continue and the leaks marked for later investigation and repair. On dust filters with felt tubes, check for poor tube installation, loose bag clamps or torn bags.

Pressure or Material Buildups in Filters and Receivers

Receiving bins and dust filters should be checked periodically for pressure buildups. This can be done with a manometer, or by opening

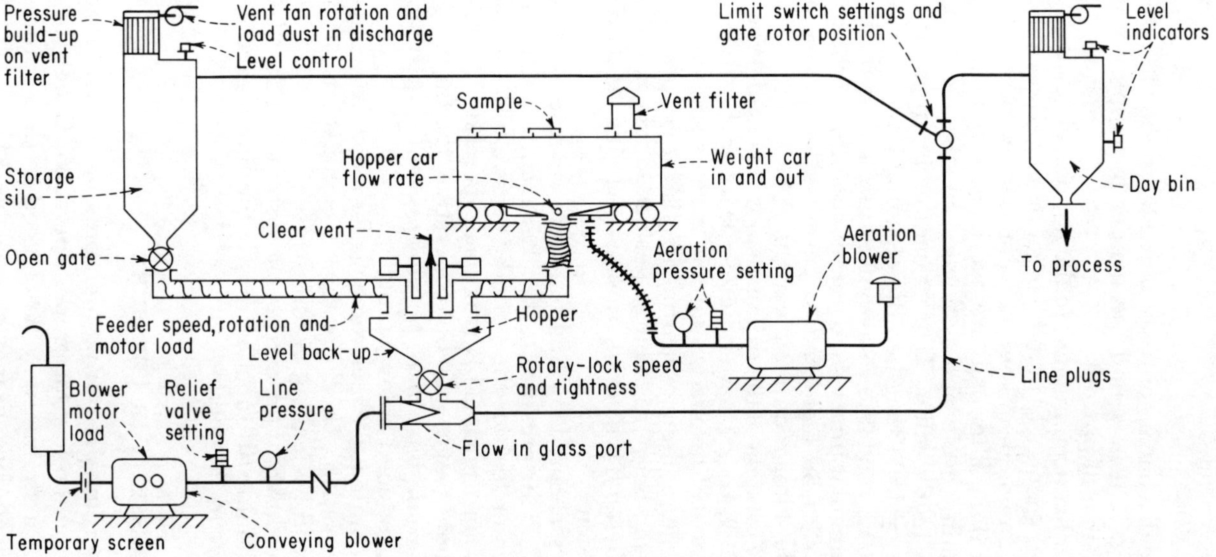

FIGURE 17–1. Checking and setting critical points of a material-into-air system.

inspection ports and noticing whether air is sucked in or dust is blown out. If a pressure buildup is apparent, make the following investigations:

1. See if the discharge conveyor or feeder of the filter dust hopper is operating. If it is, sound the hopper by rapping it to detect an accumulation of powder.
2. If powder has accumulated in the hopper, check the discharge for a plugged spout, a blanked flange, or an air leak into the hopper that is big enough to keep powder from getting out.
3. Find out if the filter bags are overloaded by manually advancing the cleaning cycle and noticing how the cleaning affects the pressure in the filter. If overloading is apparent, shorten the cleaning cycle or lengthen the cleaning time, or do both. On an intermittent unit, the conveying system may have to be shut down at regular intervals to permit cleaning, or the dust filter size may have to be increased to permit a longer operating period before the filter requires cleaning.
4. See if the rotation of the exhaust fan is reversed. When its rotation is reversed, a fan sometimes appears to be rotating correctly because it still discharges air and creates suction.
5. If two or more filter compartments discharge into a common hopper and the hopper is being overloaded, check the cleaning sequence to see if the compartments are being cleaned in too quick a succession.
6. On filters cleaned by high-pressure reverse-air jets, check the compressed air pressure at the filter manifold. Operate the timer and listen or feel for the periodic escape of air at each solenoid valve exhaust port. (This is normal operation.) If no air escapes, the solenoid valves may be inoperative, the timer fuse may have blown, an air filter may be clogged with dirt and the supply pressure reduced below normal.
7. Open the unit and check the bags for moisture accumulation due to condensation. Check for dust accumulations on the clean air side caused by leaking or torn bags.
8. After shutdown of the conveying system, the dust filter cleaning cycles should still function during the line cleanout period, or a little beyond this period.

Low Compressed Air Pressure

On dust filter units requiring high-pressure air for cleaning, the air pressure may be too low for proper operation. Check for leaks in the supply piping, a strainer may be clogged or the compressor may be faulty. The solenoid valves may be stuck open because of dirt or frozen water in the line, or they may be held open because of improper wiring. A ruptured diaphragm on a solenoid valve may be the cause of excessive air leakage.

Rotating Equipment

Exhaust fan drive motors should be checked for heating or other evidence of overloading. This may be caused by excess airflow through the low resistance of the new, clean filter media on the dust filter, or because the blower on a positive-pressure system acts as a booster, or because the fan is oversized. In the first case, the exhaust fan discharge should be temporarily throttled until a filter cake builds up and the dust filter pressure drop becomes constant. A permanent damper may have to be installed on the discharge of an oversized fan. A larger-size motor and speed-up of the fan may be required where positive pressure in a bin or receiver must be made negative.

If the conveying pressure or vacuum appears to fluctuate periodically and material flow to the system is known to be steady, check the belt drives on the blower or vacuum pump for slippage or loose belts. Feel the belt guards for overheating or sniff for the odor of burning rubber and, if necessary, stop the conveying system to feel the belts.

ACCESSORY EQUIPMENT

The startup troubles that sometimes beset accessory equipment, such as screw feeders and screw, belt, and air-activated gravity conveyors, are overfeeding, underfeeding, poor venting and mechanical defects.

Defects at the inlet end of conveyors are detected by observing the effects of the rate of flow at the discharge end. Normally, the input feed rate of properly designed conveyors are self-regulating. The discharge valve of a transport or storage bin is opened wide so that the inlet end of the conveyor is filled. Feed flow is usually regulated to screw feeders by closely spaced or tapering screw flights, to belt conveyors by inlet spouts that are sealed on three sides to discharge through an adjustable gate in the direction of flow, and to air-activated gravity or vibrating conveyors by inlet weirs or gates.

Overfeeding may be evidenced by the blowing of relief valves in material-into-air systems; by the backing up of material into the rotary feeder inlet hopper, which causes dust to spew from joints after vents become blocked; by the backing up of material into screw or belt conveyor housings, which trips out the motor or distorts the housing; by the sounding of the high-level alarm of the feed hopper; or by the empty conveying line of mixing systems, which occurs when the feed hopper vent line becomes blocked.

If overfeeding is caused by excessive conveyor speed, lower the speed by changing the sheaves or sprockets. For an air-activated gravity conveyor, reduce the air pressure to the plenum chamber or to the feeder weir.

If aeration air is flushing material through the conveyor, eliminate or reduce the air pressure in the bin or transport that is feeding the conveyor, slow the screw conveyor, install a shroud at the discharge of the screw feeder (if a shroud is already present, reduce clearances or extend it to cover more screw flights), or decrease the size of the opening at the regulating end of air-activated gravity conveyors.

If a large spout opening to a belt conveyor or a high gate setting to an air-activated gravity conveyor is causing overfeeding, make the opening smaller or reset the flow gate. Holes or spaces in the porous membrane of an air-activated gravity conveyor may cause overfeeding by passing too much air into the material causing it to flow uncontrollably.

Underfeeding may be signaled by the failure of the pneumatic conveying system to come up to the design vacuum or pressure, by the low level in the feed hopper, by the hollow sound when the conveyor casing is rapped, by the slow filling of the blow tank, or by the lack of material in the sight port at the injector fitting of a material-into-air system.

First, check the supply bin for material—it may be empty! If underfeeding is being caused by a stoppage at the conveyor inlet, check the discharge gate of the bin or transport, probe for lumps in the metering section of the conveyor, or try the flow inducers. Increase the speed or tighten the drive belts, if the conveyor is too slow. An air-activated gravity conveyor may need the air pressure in the plenum chamber raised. If underfeeding is caused by slippage of highly aerated material in screw conveyor flights, reduce the air pressure or cut off the air supply to bin aeration pads ahead of the conveyor.

Inadequate venting at the discharge of a conveyor may be suspected when: dust spews from the joints and openings of the conveyor housing; the impeller pumps of an air-mixing system lose capacity; there is no flow in the conveying line of an air-mixing system having double-entry rotary feeders; the blow tank of an air-into-material system is being partially filled; or when the rotary feeders of a positive-pressure material-into-air system lose capacity.

If the vent filter is too small, lengthen the cleaning time, reduce the dust load by moving the filter away from the spot where the air blows back in a positive-pressure system, install a larger capacity filter, or convert the filter to a continuous type. If the vent pipe is plugging, put a level controller in the hopper that receives the feed conveyor discharge, and eliminate horizontal runs in the pipe; reduce the speed of the feed conveyor if it is overfilling the vented hopper. Increase the pitch of sloping ducts to make them self-cleaning. When practicable, make the vent lines vertical for their entire length. An exhauster will improve the venting; when installing it, interlock the controls so that the exhauster and the cleaning cycle controls start before the accessory conveyor. Do not depend on air pressure build-up to relieve itself through the filter.

The most frequent mechanical defects of accessory equipment, and the ways that they can be detected and corrected, are listed in Table 17–1.

TABLE 17–1
Accessory Conveying Equipment—Detection and Correction of Mechanical Defects

Mechanical Defect	Detection	Correction
	Screw Feeders and Conveyors	
Fixed type ball or roller bearings on each end of a shaft that is deflected due to load or high temperature.	Motor load is excessive or irregular; motor cannot start; shaft is broken.	Install fixed bearing at drive or thrust end and a floating bearing at the free end of the shaft.
Internal hanger bearings made of soft material, such as Micarta, Nylon, or Arguto wood.	Screw flights rub on trough.	Use hard cast-iron hanger bearings and hardened steel couplings on shaft.
Cover bolts or clamps spaced too far apart.	Dust spewing from covers.	Insert additional cover bolts or clamps.
Cover support angles intermittently welded to U-trough housing.	Dust spewing from spaces between welds.	Continuously weld all joints on housings.
Omission of deflector shield between packing seal on trough end and outboard ball or roller bearing.	Drive motor overloaded; motor cannot be started; bearing runs hot.	Install a deflector shield whenever the trough may be under air pressure.
Omission of pressure-break baffle over screw flights in bins.	Drive motor overloaded; motor cannot be started.	Install inverted V-shaped baffle over flights so that material enters the screw from the sides, below the baffle.
Ball or roller thrust type bearing clamped to the shaft with a split adapter.	Shaft pulls through adapter; end flight jams against tail end of trough.	Use shouldered shaft with cylindrical bore mounted type of bearing or a flanged thrust block.

TABLE 17-1 (*Cont.*)

Mechanical Defect	Detection	Correction
	Screw Feeders and Conveyors	
Shroud at discharge end is too short to seal a feeder that handles aerated material.	Material flushes through feeder even when it is not rotating; feeder capacity varies.	Shroud should enclose at least 6 pitches of flighting to prevent or retard flushing.
Inspection doors fastened with piano-type hinges.	Dust spews from joints of hinge.	Locate pintle type hinges beyond door joint and install bolts or clamps on hinge side of door for tightness.
	Air-Activated Gravity Conveyors	
Flow-control gate at entrance to conveyor omitted.	Overfeeding of downstream equipment; plugging of conveyor casing.	Install adjustable flow gate at inlet end of conveyor.
Uncoated fabric aeration pad used where moisture is present.	Material leaks into air plenum because of holes caused by mildew; overfeeding caused by excessive aeration.	Use ceramic, or silicone- or neoprene-impregnated fabric.
Clean-out and inspection port at lower end of air plenum and atop material chute omitted.	Leakage blocks air plenum and causes loss of conveying.	Install ports with quickly detachable covers.

BLOW-TANK SYSTEMS

Since a blow-tank conveying system usually operates at high air pressure, the air receiver and the blow tank should be insured, unfired pressure vessels. Before material is conveyed, their structural and operating safety must be assured.

These vessels must be equipped with A.S.M.E. standard safety valves, which must not be set higher than the maximum allowable working pressure stamped on the vessel. The settings must be checked during the preliminary test of the system. Bypass the safety and normal control devices by manually opening or closing switch contacts, or temporarily jumping open contacts to make air pressure tests and to set the controls without admitting material into the system. The steps of the test can be followed in Figure 17–2.

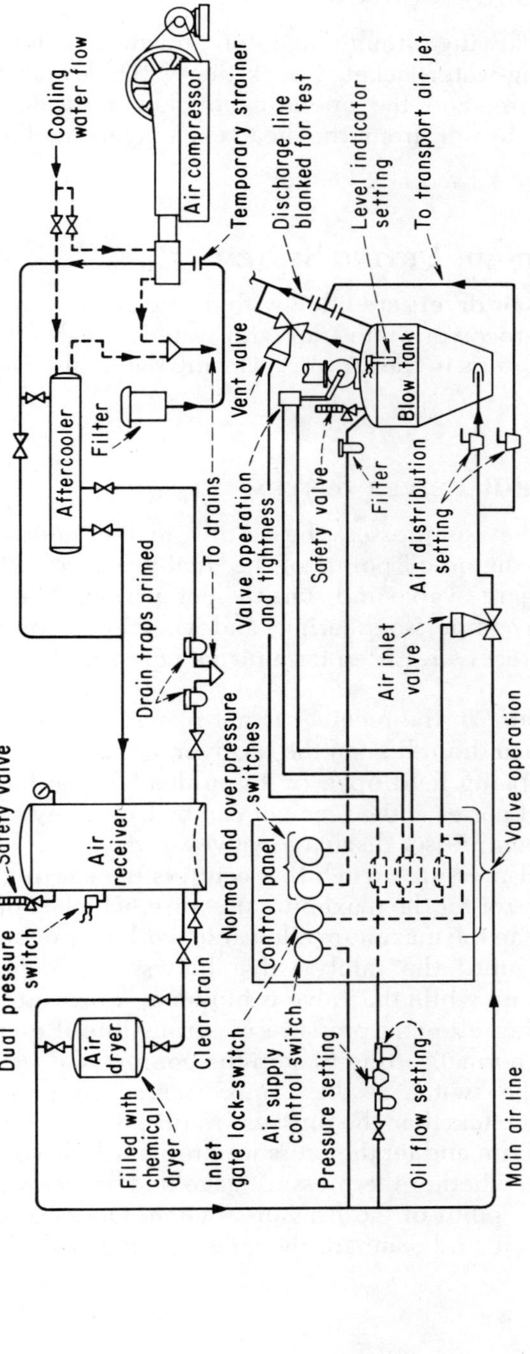

FIGURE 17–2. Checking and setting important features of a blow-tank system.

Compressed-Air Cooling System

Start water flowing through the inter- and after-coolers and the compressor cooling-water jacket. Check the water flow to the drain, the supply pressure, and the operating of flow indicators and control valves. Vent the air from the heat exchangers and cooling water jackets.

Compressed-Air Drying System

Be sure the air dryers are filled with dessicant and drained of liquid. Prime the air-receiver drain trap with water to seal it. Crack open all the low-spot drains in the air lines, closing them only after air is blowing out.

Compressor-Receiver System

To check the compressor, the setting of the safety valve on the receiver, and the on-off points of the dual-pressure switch, close the receiver discharge valve and start the compressor. See whether oil is flowing to the compressor bearings and whether the pressure is building up in the receiver. Listen for unusual noises at the compressor and the receiver.

If the pressure in the receiver is not rising, look for a missing pipe plug, or an open drain line on the receiver, or for a compressor suction valve that is being held open by the unloader or a faulty valve diaphragm. Loud noises at the receiver can be caused by excessive clearances at the compressor discharge valves.

Set the dual-pressure switch that controls the starting and stopping of the compressor for the maximum pressure, which should be at least 10% higher than the maximum allowable working pressure. Let the air pressure rise until the safety valve blows, and note whether the pressure goes up while the valve is blowing. A pressure buildup 10% higher than the relieving pressure is permissible. Reset the pressure switch to the normal setting to stop the compressor.

Next, set the switch at the compressor starting pressure, which should never be less than the pressure required at the blow tank. Open the receiver drain and let the pressure drop until the compressor starts again, and see whether the pressure gage on the receiver corresponds with the "On" point of the pressure switch. Operate the compressor until it shuts off, and compare the gage reading with the set point.

Blow-Tank Operating Valve Settings

Check blow-tank operating valves and controls as follows:

Fill lubricators in the auxiliary air line to the valve actuating air cylinders. Admit air to the auxiliary air line; then bleed off air to create a flow that will permit setting the lubricating oil feed rate. Close the bleed valve. Set the air pressure regulator, if fitted, for proper supply pressure to the air cylinders.

Operate each air cylinder for its full stroke several times and observe the corresponding movement and seating of the valve which it operates. Valves located inside the blow tank should be observed through the open tank manhole and the manhole cover replaced when certain that the valves seat properly. The air cylinders may be operated temporarily by manually making proper tubing connections to an air supply line, by setting the control panel for manual operations, if so fitted, or by opening the air supply solenoid valves using a 110-volt extension cord to energize the solenoid, thus by-passing all sequencing controls.

Energize the control panel and set it for automatic operation of the tank. Simulate actual operation, without admitting material and with the air receiver discharge valve closed, by placing a jumper wire across the high-level indicator contacts to simulate high material level. The vent and material-inlet valves will close. Then, disconnect a wire, on the low-air pressure switch or weight-operated empty-tank switch, to de-activate the circuits that close the air supply valve and open the vent and material-inlet valves. Place a jumper wire across the contacts of the air-pressure switch to simulate the availability of sufficient air-supply pressure. The air-supply solenoid valve should then open to the discharging position.

Restore the tank valves to the filling position by removing the jumper wire from the high-level indicator contacts and replacing the wires on the low-air-pressure switch or the weight-operated empty-tank switch. The air-supply solenoid valve should close and the vent and material-inlet valves should open. De-energize the control panel and remove the jumper wire from the air-pressure switch so that the proper receiver pressure could start the blow cycle.

Blow-Tank and Conveying-Line Tightness

Test the blow tank and conveying line for air leaks by blanking off the end of the conveying line and placing a jumper wire across the overpressure switch contacts. Simulate a full tank as described above.

However, no air will enter the tank because the receiver discharge valve is closed. Open the receiver discharge valve slowly and check for leaks in the conveying line; in the interconnecting air piping between the receiver and blow tank; and at the material-inlet and vent valve seats. Allow pressure to build up slowly until the blow-tank safety valve opens; then open wide the receiver discharge valve. The blow-tank pressure should not rise more than 10% above the safety valve set pressure.

Set the pressure switches while the safety valve is blowing, remove the jumper wire from across the overpressure switch contacts. The air-supply solenoid valve should close immediately and the blow-tank safety valve should stop blowing shortly thereafter. The overpressure switch should be set tentatively at least 5 psi below the safety valve setting and air released from the blow tank to check the switch setting. Remove the jumper wire from across the high-level indicator contacts and simulate an empty tank by replacing the terminal wires on the low-pressure switch or weight-operated empty-tank relay. Set the low-pressure switch or the weight-operated empty-tank relay tentatively at the vendor's recommended setting and note the blow-tank pressures at which the air supply solenoid valve is cut off and the vent and material-inlet valves are opened.

When all air is released from the blow tank, remove the closure at the end of the conveying line. The system is now ready to receive material.

OPERATING THE BLOW-TANK SYSTEM

The air distribution in a blow tank must suit the characteristics of the material and make the discharging time as short as possible. How the air is distributed depends on the design of the blow tank. The air may enter directly above the material level, near the entrance of the conveying line (by way of aeration nozzles, pads, or perforated sill plates), or through a transporting nozzle that is aimed into the conveying line.

For some materials, the air should enter at all three points, for others only at one point. To find the best distribution, try all the entry points, taking simultaneous pressure readings at the following places: the air receiver, the blow tank, and the entrance to the transporting nozzle. The readings should be plotted and the curves examined for ways to improve the discharge cycle by changing the distribution or reducing the airflow (Figure 17–3). The dashed curve is for a system with a high-pressure compressor-receiver air supply handling a highly aerated material. The material is discharged so rapidly that the pressure

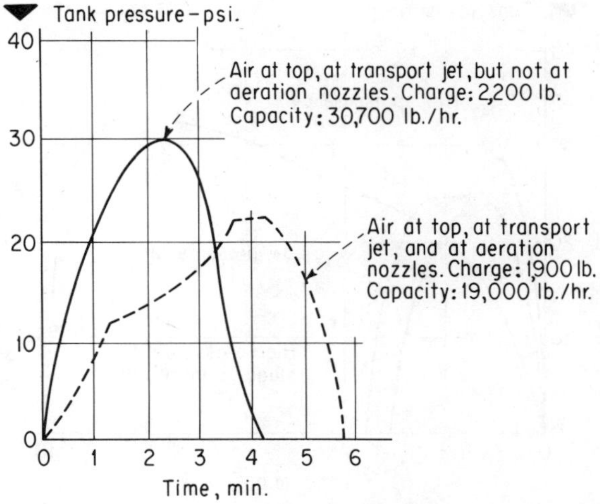

FIGURE 17–3. Performance of a blow tank improved by changing the air-distribution pattern.

cannot build up to its maximum. The overall effect is to prolong the discharge cycle. The removal of side-entering agitator nozzles resulted in obtaining increased capacity as shown by the solid curve. The procedure for obtaining proper air distribution and maximum discharge capacity is called "tuning the system." High air pressure on top of a material that compacts easily could plug the conveying line. If a material cannot be moved unless mixed with air, it may become necessary to supply air at all the distribution points.

FILLING THE BLOW TANK

Start the air compressor and build up the pressure in the air receiver system. When the compressor stops, set the controls for manual operation and fill the tank with material until the inlet valve closes. Open any accessible plug or port in the top of the tank to check the level. Reset the level controller, if necessary.

STARTING THE BLOW CYCLE

Switch to manual control so that only one operating cycle is completed. After starting the cycle, take pressure readings every 10 or 15 seconds at the air receiver, the blow tank, and the aeration nozzles. Plot the readings for a clue to the type of troubles that may be encountered.

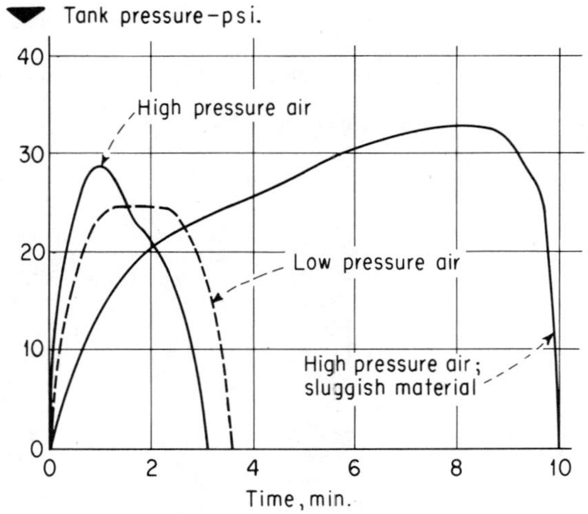

FIGURE 17–4. Typical blow-tank discharge pressure-time curves.

Some typical blow-tank discharge-pressure versus time curves are shown in Figure 17–4. The first curve is typical for a high-pressure compressor and receiver air-supply system. It rises sharply to a peak, then drops slowly as the tank is being emptied. The second curve, which is typical for a blower air-supply system, rises sharply to the maximum blower pressure and levels off before dropping as the tank is emptied. The third curve is of a high-pressure compressor and receiver air-supply system handling a problem material. The discharging time is unusually long because the material is difficult to fluidize, which makes it unsuitable for handling in a blow-tank system.

MIXING SYSTEMS—THE IMPELLER SCREW PUMP

The two most common types of mixing feeders—the impeller screw pump and the double-entry rotary feeder—are different enough to require separate discussions.

The impeller screw pump is a highly refined, high-speed screw feeder that compresses material into a seal space, then forces the seal through a check valve into a mixing chamber (Figure 17–5). High-pressure air nozzles at the back of the mixing chamber blow the material into the entrance of the conveying line.

The drive end of the impeller screw is supported in an overhung position at the bottom of an inlet hopper by two sets of outboard bear-

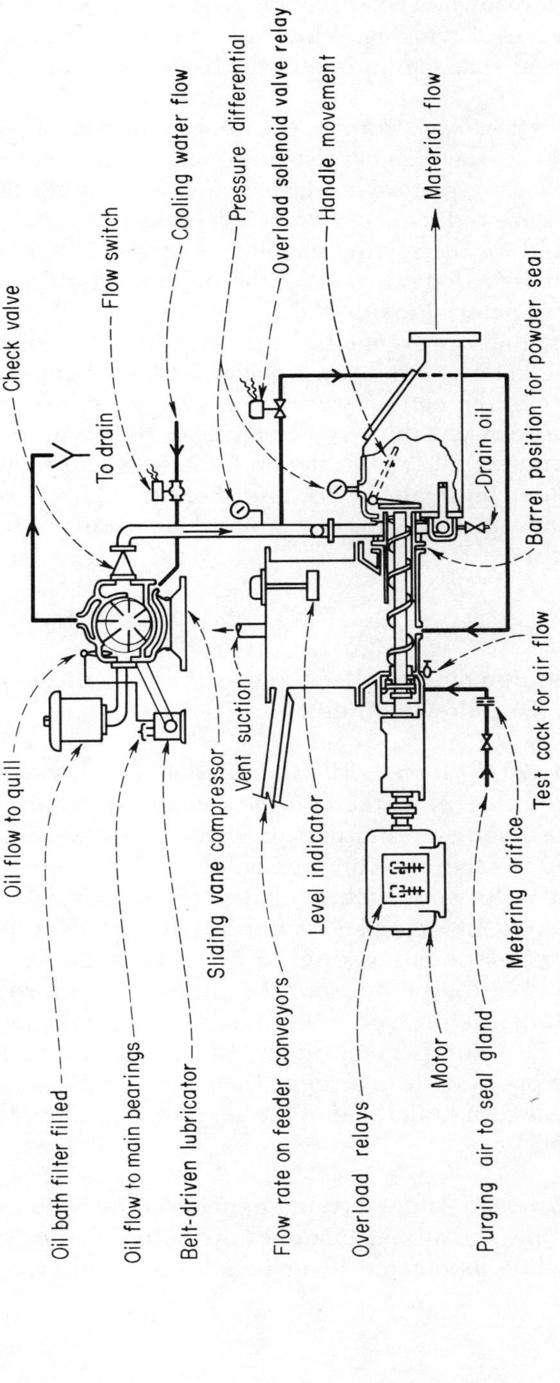

FIGURE 17–5. Checking and setting critical points on an impeller-screw pump of an air-mixing system.

ings. The pitch of the screw flights decrease gradually in the direction that the powder flows. The discharge end of the screw rotates in a closely fitting barrel housing, which is adjustable axially to vary the length of the seal space and prevent the backflow of air from the mixing chamber.

The drive-shaft passage through the inlet hopper is sealed by a rotating mechanical seal, which consists of a carbon ring bearing against a lapped steel wearing-plate. The air that cools and purges the seal flows through a pressure reducing orifice in a valveless branch line from the supply manifold for the mixing chamber nozzles. Other branch lines supply air to points along the axis of the impeller to relieve overloads by aerating the material inside the screw housing.

Air to the main supply manifold comes directly from the water-cooled, sliding-vane compressor. Special oil-bath type suction filters are supplied with the compressor to keep the casing free of dirt. Mechanical lubricators are directly driven from the compressor shaft to ensure a continuous oil feed to the end bearings and to an air-intake quill for cylinder lubrication. Because there is no air receiver, pressure fluctuations in the supply manifold are transmitted into the mixing chamber.

Operating Procedures

After the equipment has been inspected and cleaned, start the equipment in the following order:

Compressed Air System. Fill the suction filter with the recommended oil to the level marked on the casing. After filling the force-feed lubricators, set a preliminary feed rate. Set the oil flow at the minimum after starting the compressor.

Start cooling water flowing through the compressor jacket. Prime the air traps at the base of the oil and water drip legs in the pipes between the compressor and the pump. Slightly open the drain valves at the air supply manifold of the pump; start the compressor; close the valves.

The conveying line pressure will be low because the air in the line flows freely. The pump manifold pressure will be twice as high because of the pressure drop through the nozzles. Check the airflow to the rotating seal at the inlet end of the screw by opening the test cock on the seal housing.

Pumping System. After starting the material into the inlet hopper, check its flow into the mixing chamber by feeling the resistance at the check valve when its outside lever is held down with a slight hand force and by observing the pressure rise on the gages connected to the supply manifold and the mixing chamber (Figure 17–5).

Since the pump screw is connected directly to the motor shaft, its speed is fixed. Its capacity, therefore, is a function of the size of the screw and the density of the material. Maximum capacity occurs when the rotary-compressor discharge pressure remains constant and below the relieving pressure of the safety valves. Pump capacity is lowered when the powder in the inlet hopper is highly aerated. A low or erratic conveying line pressure, air blowing back into the inlet hopper, and low power drawing by the pump motor are signs of undercapacity.

If underfeeding is caused by the excessive aeration of the material before it enters the inlet hopper, lower both the aeration pressure in the bin or transport from which material is being taken and the air pressure to the air-activated gravity conveyors that deliver to the pump.

If the aeration is excessive after the material enters the inlet hopper, check the setting of the overload relays (the solenoid valves may be letting air into the screw too soon); look for open bypass valves around the solenoid valves; and inspect the suction on the inlet-hopper vent duct. Also adjust the pump barrel, increasing the length of the material seal at the discharge end of the screw.

If the inlet hopper is empty, check the conveyors for stoppages, the bin or transport for material hangup, the level controller of the hopper, and the vent system.

Operational Problems

Problems that may occur on this type of system over a long period of time are as follows: When two pumps are located alongside each other and both receive material from a common inlet hopper equipped with one high level control, the control may give a false indication of high level when only one pump is operating because a head of material builds up over the dead material at one side of the common hopper and actuates the level detector.

When the discharge lines from two pumps are cross-connected so that they can be served by either pump, material may be blown back into an inoperative pump through leaky isolating valves and enter the air nozzles and piping. Upon re-starting the inoperative pump, powder in the air supply manifold may be blown out through the nozzle orifices thus enlarging them.

When the pump is conveying abrasive materials, the capacity and conveying line pressure will decrease almost imperceptibly over a long period of time, due to wearing of the screw flights and of the interior of the barrel housing. Excessive wear may even expose the interior of the balancing chamber and cause loss of the balancing weights. One method of compensating for wear is to bleed air from the air supply manifold until the conveying line pressure increases, then suddenly starts to decrease. This is the point of optimum air-to-material ratio; operation may now continue, but at reduced capacity, until such time as the impeller screw can be replaced.

The air supply manifold, piping, and drip legs may become loaded with oil because of inattention to the feed setting on mechanical lubricators, or because of leakage of oil from the oil bath suction filters directly into the compressor. Leakage in the filter casing may be caused by corrosion if the oil is not changed regularly and water accumulates in it.

The bottom of the radial compressor sliding-vane slots will accumulate dirt and force the vanes outward against the cylinder causing excessive wear of the vanes. These slots should be checked periodically for cleanliness and the vanes discarded when wear exceeds 3/8 of an inch.

MIXING SYSTEMS—DOUBLE-ENTRY ROTARY FEEDER

The double-entry feeder is a rotary feeder with open-ended pockets. Air is admitted into each pocket from both ends of the rotor just before it rotates into the discharge position. The air thus mixes thoroughly with the material trapped in the pocket. The air sweeps the pocket clean when it reaches the transition that discharges the mixture into the conveying line. A relief vent outlet, located between the material inlet and the material outlet connections, discharges the high-pressure air that is trapped in the empty pocket so that material can enter and fill it.

Because the double-entry feeder uses high-pressure air, (1) it has more pockets than other types of feeders (the additional vanes act as a labyrinth seal between the material inlet and the air entry points), and (2) its bearings are housed in the inlet-air plenum chambers (the high-pressure air keeps out the dust that would otherwise enter the bearing seals).

This type of feeder must be supplied from a shallow feed hopper, and the air relief line must always be clear so that it can vent freely. If the vent line becomes blocked or the head of material at the inlet becomes excessive, the feeder pockets cannot be filled because the air cannot escape, and conveying stops completely. The sensitivity of this mixing feeder to the head of material at the inlet is so great that a discharge-pressure switch that starts and stops the feed can be used to control the material level in the feed hopper. To get the same discharge pressure (flow rate), the level in the hopper must be higher when the material is aerated than when it is not, because the lower density must be offset by a higher head.

Index